W9-AFH-929

McGraw-Hill Circuit Encyclopedia and Troubleshooting Guide

McGraw-Hill Circuit Encyclopedia and Troubleshooting Guide

Volume 4

John D. Lenk

McGraw-Hill

New York San Francisco Washington, D.C. Auckland Bogotá
Caracas Lisbon London Madrid Mexico City Milan
Montreal New Delhi San Juan Singapore
Sydney Tokyo Toronto

McGraw-Hill

A Division of The **McGraw·Hill** *Companies*

Copyright © 1998 by The McGraw-Hill Companies, Inc. All rights reserved. Printed in the United States of America. Except as permitted under the United States Copyright Act of 1976, no part of this publication may be reproduced or distributed in any form or by any means, or stored in a data base or retrieval system, without the prior written permission of the publisher.

1 2 3 4 5 6 7 8 9 0   FGR/FGR   9 0 2 1 0 9 8 7

ISBN 0-07-038117-8 (PBK)
ISBN 0-07-038116-X (HC)

The sponsoring editor for this book was Steve Chapman, the editing supervisor was Andrew Yoder, and the production supervisor was Sherri Souffrance. It was set in Palatino Roman by Jana Fisher through the services of Barry E. Brown (Broker—Editing, Design and Production).

Printed and bound by Quebecor/Fairfield.

McGraw-Hill books are available at special quantity discounts to use as premiums and sales promotions, or for use in corporate training programs. For more information, please write to the Director of Special Sales, McGraw-Hill, 11 West 19th Street, New York, NY 10011. Or contact your local bookstore.

Dedication

Greetings from the Villa Buttercup.
To my wonderful wife, Irene.
Thank you for being by my side all these years!
To my lovely family, Karen, Tom, Branden, Justin, Michael, and Robin.
And to our Lambie and Suzzie, be happy wherever you are!
To my special readers: May good fortune find your doorway,
bringing you good health and happy things.
Thank you for buying my books!
And special thanks to Steve Chapman, Stephen Fitzgerald, Leslie Wenger,
Ted Nardin, Mike Hays, Lisa Schrager, Patrick Hansard, Peter Mellis,
Mary Murray, Carol Wilson, Florence Trimble, Fran Minerva, Jane Stark, and
Robert McGraw of McGraw-Hill for making me an international bestseller again!

This is book number 89.
Abundance!

Contents

Introduction

When you have finished reading this encyclopedia, you should be able to recognize more than 500 circuits that are commonly used in all phases of electronics. You also will understand how the circuits operate and where they fit into electronic equipment and systems. This information alone makes the book an excellent one-stop source or reference for anyone (student, experimenter, technician, or designer) who is involved with electronic circuits.

However, this book is much more than a collection of circuits and descriptions. First, all circuits are grouped by function. With each functional group, the book has a practical guide for testing and troubleshooting that type of circuit. Thus, if you are building a circuit, and the circuit fails to perform as outlined, you are given a specific troubleshooting approach to locate problems.

Second, you can put the circuit to work as it is, because actual circuits with proven component values are given in full detail. On those circuits where performance (such as frequency range, power outputs, etc.) depend on circuit values, you are given information as to how circuit values can be selected to meet a certain performance goal. This information is particularly useful to the student or experimenter, but it also can be an effective time-saver for the designer.

Acknowledgments

Many professionals have contributed to this book. I gratefully acknowledge the tremendous effort needed to produce this book. Such a comprehensive work is impossible for one person, and I thank all who contributed, both directly and indirectly.

I give special thanks to the following: Alan Haun of Analog Devices, Syd Coppersmith of Dallas Semiconductor, Rosie Hinojosa of EXAR Corporation, Jeff Salter of GEC Plessey, John Allen, Helen Cox, and Linda da Costa of Harris Semiconductors, Ron Denchfield and Bob Scott of Linear Technology Corporation, David Fullagar and William Levin of Maxim Integrated Products, Fred Swymer of Microsemi Corporation, Linda Capcara of Motorola, Inc., Andrew Jenkins and Shantha Natarajan of National Semiconductor, Antonio Ortiz of Optical Electronics Incorporated, Lawrence Fogel of Philips Semiconductors, John Marlow of Raytheon Electronics Semiconductor Division, Anthony Armstrong of Semtech Corporation, Ed Oxner and Robert Decker of Siliconix Incorporated, Amy Sullivan of Texas Instruments, Alan Campbell of Unitrode Corporation, Sally and Barry E. Brown (Broker), and Andrew Yoder (best-selling author).

I also wish to thank Joseph A. Labok of Los Angeles Valley College for help and encouragement throughout the years.

And a very special thanks to Steve Chapman, Stephen Fitzgerald, Leslie Wenger, Patrick Hansard, Peter Mellis, Ted Nardin, Mike Hays, Lisa Schrager, Mary Murray, Carol Wilson, Judy Kessler, Monika Macezinskas, Florence Trimble, Fran Minerva, Jane Stark, Fred Perkins, Robert McGraw, Judith Reiss, Charles Love, Betty Crawford, Jeanne Myers, Peggy Lamb, Thomas Kowalczyk, Suzanne Barbeuf, Jaclyn Boone, Kathy Green, Donna Namorato, Julie Lucas, Sherri Soufrance, Allison Arias, and Midge Haramis of the McGraw-Hill Professional Publishing organization for having that much confidence in me.

And to Irene, my wife and Super Agent, I extend my thanks. Without her help, this book could not have been written.

Circuit sources and addresses

The source for each circuit is given at the end of the circuit description. The source information includes the publication title and date, as well as the page number or numbers. This information makes it possible for you to contact the original source for further information on the circuit or circuit component. To this end, the complete mailing address and telephone numbers for each source are included in this section. When writing, give complete information, including publication date and page number of the original source. Notice that all circuit diagrams have been reproduced directly from the original source, without redrawing, by permission of the original publisher in each case.

AIE Magnetics
701 Murfreeboro Road
Nashville, TN 37210
(615) 244-9024

Analog Devices
One Technology Way
PO Box 9106
Norwood, MA 02062-9106
(617) 329-4700
Fax (617) 326-8703

Dallas Semiconductor
4401 S. Beltwood Parkway
Dallas, TX 75244-3292
(214) 450-0400

EXAR Corporation
2222 Qume Drive
PO Box 49007
San Jose, CA 95161-9007
(408) 434-6400
Fax (408) 943-8245

GEC Plessey Semiconductors
Cheney Manor
Swindon, Wiltshire
United Kingdom SN2 2QW
0793 51800
Fax 0793 518411

Harris Semiconductor
PO Box 883
Melbourne, FL 32902-0883
(407) 724-7000
Fax (407) 724-3937
1-800-442-7747

Linear Technology Corporation
1630 McCarthy Boulevard
Milpitas, CA 95036-7487
(408) 432-1900
Fax (408) 434-0507
1-800-637-5545

Magnetics Division of Spang and
    Company
900 East Butler
PO Box 391
Butler, PA 16003
(412) 282-8282

Maxim Integrated Products
120 San Gabriel Drive
Sunnyvale, CA 94086
(408) 737-7600
Fax (408) 737-7194
1-800-998-8800

Motorola, Inc.
Semiconductor Products Sector
Public Relations Department
5102 N. 56th Street
Phoenix, AZ 85018
(602) 952-3000

National Semiconductor Corporation
2900 Semiconductor Drive
PO Box 58090
Santa Clara, CA 95052-8090
(408) 721-5000
1-800-272-9959

Optical Electronics, Inc.
PO Box 11140
Tucson, AZ 85734
(602) 889-8811

Philips Semiconductors
811 E. Arques Avenue
PO Box 3409
Sunnyvale, CA 94088-3409
(408) 991-2000

Raytheon Company Semiconductor
    Division
350 Ellis Street
PO Box 7106
Mountain View, CA 94039-7016
(415) 968-9211
Fax (415) 966-7742
1-800-722-7074

Semtech Corporation
652 Mitchell Road
Newbury Park, CA 91320
(805) 498-2111

Siliconix Incorporated
2201 Laurelwood Road
Santa Clara, CA 95054
(408) 988-8000

Unitrode Corporation
8 Suburban Park Drive
Billerica, MA 01821
(508) 670-9086

Substitutions and cross-reference tables

Substitutions can often be made for the semiconductor and IC types that are specified on the circuit diagrams. Newer components, not available when the original source was published, might actually improve the performance of the circuit. Electrical characteristics, terminal connections, and such critical ratings as voltage, current, frequency, and duty cycle must, of course, be taken into account if experimenting without referring to substitution guides.

Semiconductor and IC substitution guides can usually be purchased at electronic parts-supply stores. In the absence of any substitution guides, the following cross-reference tables will help in locating possible substitute ICs.

General Cross References

INDUSTRY TYPE	RAYTHEON DIRECT REPLACEMENT	RAYTHEON FUNCTIONAL REPLACEMENT	INDUSTRY TYPE	RAYTHEON DIRECT REPLACEMENT	RAYTHEON FUNCTIONAL REPLACEMENT
ADVFC32		RC4153	ICL7660		RC4391
ADOP07	OP-07		ICL7680		RC4190
ADOP27	OP-27		ICL8013		RC4200
ADOP37	OP-37		LF155	LF155	
ADREF01	REF-01		LF156	LF156	
ADREF02	REF-02		LF157	LF157	
AD101	LM101		LH2101	LH2101	
AD558		DAC-4888	LH2108	LH2108	
AD565	DAC-8565		LH2111	LH2111	
AD581		REF-01	LM101	LM101	
AD586		REF-02	LM111	LM111	
AD647		RC4207	LM108	LM108	
AD654		RC4152	LM124	LM124	
AD707		RC4077	LM148	LM148	
AD708		RC4277	LM324	LM324	
AD741	RC741		LM331		RC4152
AD767		DAC-4881	LM348	LM348	
AM686		RC4805	LM368-5.0		REF-02
AM6012	DAC-6012		LM368-10		REF-01
CA124	LM124		LM369		REF-01
CA324	LM324		LM607		RC4077
CA139	LM139		LM741	RC741	
CA339	LM339		LM833	RC5532	
CA741	RC741		LM1458		RC4558
CS3842		RC4190	LM1851	LM1851	
CMP-04		LM139	LM1851		RC4145
CMP-05		RC4805	LM2900	LM2900	
DAC-08	DAC-08		LM2901		LM339
DAC-10	DAC-10		LM2902		LM324
DAC-80		DAC-4881	LM3900	LM3900	
DAC-100		DAC-10	LP165	LP165	
DAC-312	DAC-6012		LP365	LP365	
DAC0800	DAC-08		LT-1001	LT-1001	
DAC0801	DAC-08		LT-1012	LT-1012	
DAC0830		DAC-4888	LT-1012		RC4097
DAC-888		DAC-4888	LT-1019		REF-01
DAC1208		DAC-4881	LT-1019		REF-02
DAC1218		DAC-6012	LT-1024		RC4207
DAC1219		DAC-6012	LT-1028		OP-37
DAC1230		DAC-4881	LT-1054		RC4391
DAC8222		DAC-4881	LT-1070		RC4190
HA-OP27	OP-07		LT-1084		RC4292
HA-OP27	OP-27		MAX400		RC4077
HA-OP37	OP-37		MAX630	RC4193	
HA-3182	RC3182		MAX630		RC4190
HA-4741	RC4741		MAX634	RC4391	
HA-5147		OP-47	MC1741	RC741	
HSOP07	OP-07		MC1747	RC747	
HSOP27	OP-27		MC3403	RC3403	
HSOP37	OP-37		MC4558	RC4558	

RAYTHEON

General Cross References (Continued)

INDUSTRY TYPE	RAYTHEON DIRECT REPLACEMENT	RAYTHEON FUNCTIONAL REPLACEMENT	INDUSTRY TYPE	RAYTHEON DIRECT REPLACEMENT	RAYTHEON FUNCTIONAL REPLACEMENT
MC4741	RC4741		SG741	RC741	
MPREF01	REF-01		SI-9100		RC4292
MPREF02	REF-02		SSM-2134		RC5534
MPOP07	OP-07		TA7504	RC741	
MPOP27	OP-27		TA75339	LM339	
MPOP37	OP-37		TL494		RC4190
MP108	LM108		TL496		RC4190
MP155	LM155		TL497		RC4190
MP156	LM156		TL510		RC4805
MP157	LM157		TSC9400		RC4151
NE5532	RC5532		TSC9401		RC4151
NE5534	RC5534		TSC9402		RC4151
OPA156		LM156	UC1842		RC4292
OPA27		OP-27	VFC-32		RC4153
OPA37		OP-37	XR-2207	XR-2207	
OP-02		RC741	XR-2208		RC4200
OP-04		RC747	XR-2211	XR-2211	
OP-07	OP-07		XR-3403	RC3403	
OP-14		RC4558	XR-4136	RC4136	
OP-16		LF156	XR-4194	RC4194	
OP-27	OP-27		XR-4195	RC4195	
OP-37	OP-37		XR-5532	RC5532	
OP-77	OP-77		XR-5534	RC5534	
OP-97		RC4097	µA101	LM101	
OP-200		RC4207, RC4277	µA108	LM108	
			µA111	LM111	
OP-207		RC4207			
OP-227		RC4227	µA124	LM124	
OP-270		RC4227	µA139	LM139	
PM-108	LM108		µA148	LM148	
PM-139	LM139		µA324	LM324	
			µA339	LM339	
PM-148	LM148				
PM-155	LM155		µA348	LM348	
PM-156	LM156		µA741	RC741	
PM-157	LM157		µA747	RC747	
PM-339	LM339				
PM-348	LM348				
PM-741	RC741				
PM-747	RC747				
RC4136	RC4136				
RC4151	RC4151				
RC4152	RC4152				
RC4558	RC4558				
RC4559	RC4559				
REF-01	REF-01				
REF-02	REF-02				
REF-05		REF-02			
REF-10		REF-01			
SE5534		RC5534			
SG101	LM101				
SG124	LM124				

RAYTHEON

Precision Operational Amplifier Cross Reference

ANALOG DEV.	RAYTHEON	PACKAGE	ANALOG DEV.	RAYTHEON	PACKAGE
AD OP-07AH	*OP-07AT	TO-99	AD OP-37AH/883	OP-37AT/883B	TO-99
AD OP-07AH/883	*OP-07AT/883B	TO-99	AD OP-37AQ	OP-37AD	CERAMIC
AD OP-07CN	*OP-07CN	PLASTIC	AD OP-37AQ/883	OP-37AD/883B	CERAMIC
AD OP-07CR	*OP-07CM	SO-8	AD OP-37BH	OP-37BT	TO-99
AD OP-07Q/883	*OP-07D/883B	CERAMIC	AD OP-37BH/883	OP-37BT/883B	TO-99
AD OP-07DN	*OP-07DN	PLASTIC	AD OP-37BQ	OP-37BD	CERAMIC
AD OP-07EN	*OP-07EN	PLASTIC	AD OP-37BQ/883	OP-37BD/883B	CERAMIC
AD OP-07H	*OP-07T	TO-99	AD OP-37CH	OP-37CT	TO-99
AD OP-07H/883	*OP-07T/883B	TO-99	AD OP-37CH/883	OP-37CT/883B	TO-99
AD OP-07Q	*OP-07D	CERAMIC	AD OP-37CQ	OP-37CD	CERAMIC
AD OP-07AQ	*OP-07AD	CERAMIC	AD OP-37CQ/883	OP-37CD/883B	CERAMIC
AD OP-07AQ/883B	*OP-07AD/883B	CERAMIC	AD OP-37EN	OP-37EN	PLASTIC
			AD OP-37FN	OP-37FN	PLASTIC
AD OP-27AH	OP-27AT	TO-99	AD OP-37GN	OP-37GN	PLASTIC
AD OP-27AH/883	OP-27AT/883B	TO-99			
AD OP-27AQ	OP-27AD	CERAMIC	AD707AQ	*RC4077FD	CERAMIC
AD OP-27AQ/883	OP-27AD/883B	CERAMIC	AD707CH	*RM4077AT	TO-99
AD OP-27BH	OP-27BT	TO-99	AD707CH/883	*RM4077AT/883B	TO-99
AD OP-27BH/883	OP-27BT/883B	TO-99	AD707CQ	*RM4077AD	CERAMIC
AD OP-27BQ	OP-27BD	CERAMIC	AD707CQ/883	*RM4077AD/883B	CERAMIC
AD OP-27BQ/883	OP-27BD/883B	CERAMIC	AD707JN	*RC4077FN	PLASTIC
AD OP-27CH	OP-27CT	TO-99	AD707JR	*RC4077FM	SO-8
AD OP-27CH/883	OP-27CT/883B	TO-99	AD707KN	*RC4077EN	PLASTIC
AD OP-27CQ	OP-27CD	CERAMIC	AD707KR	*RC4077EM	SO-8
AD OP-27CQ/883	OP-27CD/883B	CERAMIC	AD707SH	*RC4077AT	TO-99
AD OP-27EN	OP-27EN	PLASTIC	AD707SH/883B	*RC4077AT/883B	TO-99
AD OP-27FN	OP-27FN	PLASTIC	AD707SQ	*RC4077AD	CERAMIC
AD OP-27GN	OP-27GN	PLASTIC	AD707SQ/883	*RC4077AD/883B	CERAMIC
			AD707TH	*RC4077AT	TO-99
AD OP-37AE	OP-37AL	LCC	AD707TH/883B	*RC4077AT/883B	TO-99
AD OP-37AE/883	OP-37AL/883B	LCC	AD707TQ	*RC4077AD	CERAMIC
AD OP-37AH	OP-37AT	TO-99	AD707TQ/883	*RC4077AD/883B	CERAMIC

BURR BROWN	RAYTHEON	PACKAGE	BURR BROWN	RAYTHEON	PACKAGE
OPA27AJ/883	*OP-27AT/883B	TO-99	OPA37AJ	*OP-37AT	TO-99
OPA27BJ/883	*OP-27BT/883B	TO-99	OPA37AJ/883	*OP-37AT/883B	TO-99
OPA27CJ	*OP-27CT/883B	TO-99	OPA37AZ	*OP-37AD	CERAMIC
OPA27AJ	*OP-27AT	TO-99	OPA37AZ/883	*OP-37AD/883B	CERAMIC
OPA27AZ	*OP-27AD	CERAMIC	OPA37BJ	*OP-37BT	TO-99
OPA27BJ	*OP-27BT	TO-99	OPA37BJ/883	*OP-37BT/883B	TO-99
OPA27BZ	*OP-27BD	CERAMIC	OPA37BZ	*OP-37BD	CERAMIC
OPA27CJ	*OP-27CT	TO-99	OPA37BZ/883	*OP37-BD/883B	CERAMIC
OPA27CZ	*OP-27CD	CERAMIC	OPA37CJ	*OP-37CT	TO-99
OPA27EP	*OP-27EN	PLASTIC	OPA37CJ/883	*OP-37CT/883B	TO-99
OPA27FP	*OP-27FN	PLASTIC	OPA37CJ/883	*OP-37CD/883B	CERAMIC
OPA27GP	*OP-27GN	PLASTIC	OPA37CZ	*OP-37CD	CERAMIC
OPA27GU	*OP-27GM	SO-8	OPA37EP	*OP-37EN	PLASTIC
OPA27GZ	*OP-27GD	CERAMIC	OPA37FP	*OP-37FN	PLASTIC
OPA27AZ/883	*OP-27AD/883B	CERAMIC	OPA37GP	*OP-37GN	PLASTIC
OPA27BZ/883	*OP-27BD/883B	CERAMIC	OPA37GU	*OP-27GM	SO-8
OPA27CZ/883	*OP-27CD/883B	CERAMIC			

* Denotes functionally equivalent types.

RAYTHEON

Precision Operational Amplifier Cross Reference (Continued)

LTC	RAYTHEON	PACKAGE	LTC	RAYTHEON	PACKAGE
OP-07AH	OP-07AT	TO-99	LM108AH	LM108AT	TO-99
OP-07AH/883B	OP-07AT/883B	TO-99	LM108AH/883B	LM108AT/883B	TO-99
OP-07AJ8	OP-07AD	CERAMIC	LM108AJ8/883B	LM108AD/883B	CERAMIC
OP-07AJ8/883B	OP-07AD/883B	CERAMIC	LM108H	LM108T	TO-99
OP-07CN8	OP-07CN	PLASTIC	LM108H/883B	LM108T/883B	TO-99
OP-07CS8	OP-07CM	SO-8	LM108J8/883B	LM108D/883B	CERAMIC
OP-07EN8	OP-07EN	PLASTIC			
OP-07H	OP-07T	TO-99	LT1001ACH	LT-1001ACT	TO-99
OP-07H/883B	OP-07T/883B	TO-99	LT1001ACN8	LT-1001ACN	PLASTIC
OP-07J8	OP-07D	CERAMIC	LT1001AMH/883B	LT-1001AMT/883B	TO-99
OP-07J8/883B	OP-07D/883B	CERAMIC	LT1001AMJ8	LT-1001AMD	CERAMIC
			LT1001AMJ8/883	LT-1001AMD/883B	CERAMIC
OP-27AH	OP-27AT	TO-99	LT1001CH	LT-1001CT	TO-99
OP-27AH/883B	OP-27AT/883B	TO-99	LT1001CN8	LT-1001CN	PLASTIC
OP-27AJ8	OP-27AD	CERAMIC	LT1001CS8	LT-1001CM	SO-8
OP-27AJ8/883B	OP-27AD/883B	CERAMIC	LT1001MH	LT-1001MT	TO-99
OP-27CH	OP-27CT	TO-99	LT1001MH/883B	LT-1001MT/883B	TO-99
OP-27CH/883B	OP-27CT/883B	TO-99	LT1001MJ8	LT-1001MD	CERAMIC
OP-27CJ8	OP-27CD	CERAMIC	LT1001MJ8/883B	LT-1001MD/883B	CERAMIC
OP-27CJ8/883B	OP-27CD/883B	CERAMIC			
OP-27EN8	OP-27EN	PLASTIC	OP-227EN	*RC4227FN	PLASTIC
OP-27GN8	OP-27GN	PLASTIC	OP-227GN	*RC4227GN	PLASTIC
			OP-227AJ	*RM4227BD	CERAMIC
OP-37AH	OP-37AT	TO-99	OP-227AJ/883B	*RM4227BD/883B	CERAMIC
OP-37AH/883B	OP-37AT/883B	TO-99			
OP-37AJ8	OP-37AD	CERAMIC			
OP-37AJ8/883B	OP-37AD/883B	CERAMIC			
OP-37CH	OP-37CT	TO-99			
OP-37CH/883B	OP-37CT/883B	TO-99			
OP-37CJ8	OP-37CD	CERAMIC			
OP-37CJ8/883B	OP-37CD/883B	CERAMIC			
OP-37EN8	OP-37EN	PLASTIC			
OP-37GN8	OP-37GN	PLASTIC			

*Denotes functionally equivalent types.
NOTE: LTC OP-227 contains two die in a 14-pin package.
Raytheon's 4227 is a monolithic IC in an 8-pin package.

RAYTHEON

Precision Operational Amplifier Cross Reference (Continued)

PMI	RAYTHEON	PACKAGE	PMI	RAYTHEON	PACKAGE
OP07AJ	OP-07AT	TO-99	OP77AJ	OP-77AT	TO-99
OP07AJ/883	OP-07AT/883B	TO-99	OP77AJ/883	OP-77AT/883B	TO-99
OP07AZ	OP-07AD	CERAMIC	OP77AZ	OP-77AD	CERAMIC
OP07AZ/883	OP-07AD/883B	CERAMIC	OP77AZ/883	OP-77AD/883B	CERAMIC
OP07CP	OP-07CN	PLASTIC	OP77BJ	OP-77BT	TO-99
OP07CS	OP-07CM	SO-8	OP77BJ/883	OP-77BT/883B	TO-99
OP07DP	OP-07DN	PLASTIC	OP77BRC/883	OP-77BL/883B	LCC
OP07DS	OP-07DM	SO-8	OP77BZ	OP-77BD	CERAMIC
OP07EP	OP-07EN	PLASTIC	OP77BZ/883	OP-77BD/883B	CERAMIC
OP07J	OP-07T	TO-99	OP77EP	OP-77EN	PLASTIC
OP07J/883	OP-07T/883B	TO-99	OP77FP	OP-77FN	PLASTIC
OP07RC/883	OP-07L/883B	LCC	OP77FS	OP-77FM	SO-8
OP07Z	OP-07D	CERAMIC	OP77GP	OP-77GN	PLASTIC
OP07Z/883	OP-07D/883B	CERAMIC	OP77GS	OP-77GM	SO-8
OP27AJ	OP-27AT	TO-99	PM108AZ	LM108AD	CERAMIC
OP27AJ/883	OP-27AT/883B	TO-99	PM108AZ/883	LM108AD/883B	CERAMIC
OP27AZ	OP-27AD	CERAMIC	PM108AJ	LM108AT	TO-99
OP27AZ/883	OP-27AD/883B	CERAMIC	PM108AJ/883	LM108AT/883B	TO-99
OP27BJ	OP-27BT	TO-99	PM108ARC	LM108AL	LCC
OP27BJ/883	OP-27BT/883B	TO-99	PM108ARC/883	LM108AL/883B	LCC
OP27BRC/883	OP-27BL/883B	LCC	PM108DZ	LM108D	CERAMIC
OP27BZ	OP-27BD	CERAMIC	PM108DZ/883	LM108D/883B	CERAMIC
OP27BZ/883	OP-27BD/883B	CERAMIC	PM108J	LM108T	TO-99
OP27CJ	OP-27CT	TO-99	PM108J/883	LM108T/883B	TO-99
OP27CJ/883	OP-27CT/883B	TO-99			
OP27CZ	OP-27CD	CERAMIC	PM2108AQ	LH2108AD	CERAMIC
OP27CZ/883	OP-27CD/883B	CERAMIC	PM2108AQ/883	LH2108AD/883B	CERAMIC
OP27EP	OP-27EN	PLASTIC	PM2108Q	LH2108D	CERAMIC
OP27FP	OP-27FN	PLASTIC	PM2108Q/883	LH2108D/883B	CERAMIC
OP27FS	OP-27FM	SO-8			
OP27GS	OP-27GM	SO-8	OP207AY/883	*RM4207BD/883B	CERAMIC
OP27GP	OP-27GN	PLASTIC	OP207AY	*RM4207BD	CERAMIC
OP37AJ	OP-37AT	TO-99	OP227AY	*RM4227BD	CERAMIC
OP37AJ/883	OP-37AT/883B	TO-99	OP227AY/883	*RM4227BD/883B	CERAMIC
OP37AZ	OP-37AD	CERAMIC	OP227BY/883	*RM4227BD/883B	CERAMIC
OP37AZ/883	OP-37AD/883B	CERAMIC	OP227GY	*RC4227GN	PLASTIC
OP37BJ	OP-37BT	TO-99			
OP37BJ/883	OP-37BT/883B	TO-99			
OP37BRC/883	OP-37BL/883B	LCC			
OP37BZ	OP-37BD	CERAMIC			
OP37BZ/883	OP-37BD/883B	CERAMIC			
OP37CJ	OP-37CT	TO-99			
OP37CJ/883	OP-37CT/883B	TO-99			
OP37CZ	OP-37CD	CERAMIC			
OP37CZ/883	OP-37CD/883B	CERAMIC			
OP37EP	OP-37EN	PLASTIC			
OP37FP	OP-37FN	PLASTIC			

* Denotes functionally equivalent types.
NOTE: PMI's OP207/227 contains two die in a 14-pin package.
Raytheon's 4207/4227 is a monolithic IC in an 8-pin package.

RAYTHEON

General Purpose Operational Amplifier Cross Reference

Raytheon	PMI	FSC	AMD	Motorola	National	RCA	Signetics	T.I.
LH2101A LH2111 LM101A LM111 LM124		µA101A µA111 µA124	LH2101A LH2111 LM101A LM111 LM124	LM101A LM111 LM124	LH2101A LH2111 LM101A LM111 LM124	CA101A CA111 CA124	LH2101A LM101A LM111 LM124	LM124
LM139 LM148 LM301A LM324 LM339	PM139 PM148 PM339	µA139 µA148 µA301A µA324 µA339	LM139 LM148 LM301A LM324 LM339	LM139 LM301A LM324 LM339	LM139 LM148 LM301A LM324 LM339	CA139 CA301A CA324 CA339	LM139 LM148 LM301A LM324 LM339	LM139 LM301A LM324 LM339
LM348 LM2900 LM3900 RC3403A RC4136	 OP-09	µA348 µA2900 µA3900 µA3403 µA4136	LM348	 MC3403	LM348 LM2900 LM3900		LM348	LM348 LM3900 MC3403 RC4136
RC4156 RC4157 RC4558 RC4559		µA148* µA148/ 348* µA4558 µA4558*		MC4741 MC4741* MC4558 MC4558*	LM348* LM348*			LM348* LM348* RC4558 RC4559
RC4741N RM4741D RC5532 RC5532A RC5534				MC3-4741-5 MC1-4741-2			NE5532 NE5532A NE5534	NE5532 NE5532A NE5534
RC5534A RC741 RC747 RC747S	 OP-02 OP-04 OP-04	 µA741 µA747 µA747		 MC1741 MC1747	 LM741 LM747 LM747	 CA741 CA747	NE5534A CA741 CA747	NE5534A

*Functional Equivalent

RAYTHEON

Data Conversion Cross Reference

Raythen	PMI	AMD	Motorola	NSC	Devices	Analog Power	Micro-Datel
DAC-08AD	DAC-08AQ	AMDAC-08AQ	MC1408L8	DAC-08AQ	AD-1508-9D	MP-7523*	DAC-IC8BC*
DAC-08D	DAC-08Q	AMDAC-08Q		DAC-08Q	AD-1508-9D	MP-7523*	DAC-IC8BC*
DAC-08ED	DAC-08EQ	AMDAC-08EQ		DAC-08EQ	AD-1408-8D	MP-7523*	DAC-IC8UP*
DAC-08EN	DAC-08EP	AMDAC-08EN		DAC-08EP			DAC-IC8UP*
DAC-08CN	DAC-08CP	AMDAC-08CN	MC1408P6	DAC-08CP			DAC-IC8UP*
DAC-10BD	DAC-10BX			DAC-1020 LD*	AD7520/30/33*	MP-7520/30/33*	DAC- HF10BMM*
DAC-10CD	DAC-10CX			DAC-1021/22LD8*	AD7520/30/33*	MP-7520/30/33*	DAC- HF10BMM*
DAC-10FD	DAC-10FX			DAC-1020 LCN*	AD7520/30/33*	MP-7520 30/33*	DAC- HF10BMC*
DAC-10GD	DAC-10GX			DAC-1021/22LCN*	AD7520/30/33*	MP-7520/30/33*	DAC-HF10BMC*
DAC- 6012AMD		AM6012ADM		DAC-1220 LD*	AD6012ADM	MP-7531/41*	DAC-HF12BMM*
DAC- 6012MD	DAC-312 BR*	AM6012DM		DAC-1221/22LD*	AD6012DM	MP-7531/41*	DAC- HF12BMM*
DAC- 6012ACN		AM6012ADC		DAC-1220 LCN*	AD6012ADC	MP-7531/41*	DAC- HF12BMC*
DAC- 6012CN	DAC-312FR*	AM6012DC		DAC-1221/22LCN*	AD6012DC	MP-7531/41*	DAC- HF12BMC*
DAC-8565DS*				MC3412L	DAC-1208 AD-I*	AD565JD/BIN	
DAC-8565JS*				MC3412L	DAC-1280 HCD-I*	AD565JD/BIN	
DAC-8565SS*					DAC-1280 HCD-I*	AD565SD/BIN	

*Functional Equivalent

RAYTHEON

Special Functions Cross Reference

Raytheon	Teledyne	Analog Devices	EXAR	Motorola	Datel	Burr Brown
RC4151	4780*	AD451*	XR4151		VFQ-1C*	VFC-32KF*
RC4152	4781*	AD452*	XR4151*		VFQ-2C*	VFC-42BP*
RC4153	4782*	AD537*			VFQ-3C*	VFC-52BP*
RC4200/A		AD539*		MC1494*		4202K* &
						4205K*
XR2207			XR2207			
XR2211			XR2211			
RC4444				MC3416		

*Functional Equivalent

Voltage Regulator and Voltage Reference Cross Reference

Raytheon	EXAR	Maxim	T.I.	Analog Devices	Motorola	NSC
REF-01	REF-01		MP-5501	AD581*	MC1504AU10*	LH0070-0*
REF-01A	REF-01A		MP-5501A	AD581*		LH0070-1*
REF-01C	REF-01C		MP-5501C	AD581*	MC1404U10*	LH0070-2*
REF-01D	REF-01D		MP-5501D	AD581*	MC1404U10*	
REF-01E	REF-01E		MP-5501E	AD581*		
REF-01H	REF-01H		MP-5501H	AD581*	MC1404AU10*	
REF-02	REF-02		MP-5502		MC1504AU5*	LM136-5.0*
REF-02A	REF-02A		MP-5502A			LM136A-5.0*
REF-02C	REF-02C		MP-5502C		MC1404U5*	LM336-5.0*
REF-02D	REF-02D		MP-5502D		MC1404U5*	LM336-5.0*
REF-02E	REF-02E		MP-5502E			LM336A-5.0*
REF-02H	REF-02H		MP-5502H		MC1404AU5*	
RC4190		MAX630*				
RC4193		MAX630*				
RC4391		MAX634*				
RC4194	XR4194CN					
RC4195	XR4195CP				MC1468/	LM325/326*
					MC1568*	

*Functional Equivalent

RAYTHEON

Typical IC packages and pin connections

Not all circuits give power connections and pin locations for ICs (integrated circuits), but you can get this information from manufacturers' data sheets. Also, looking through other circuits might turn up another diagram on which the desired connections are shown for the same IC.

The diagrams show a few typical pin connections. Notice that the functions shown in the following diagrams apply only to that specific IC, and are included to show the normal pin-numbering sequence only. As shown, numbering normally starts with 1 (beginning at the top) for the first pin counterclockwise from the notched (or otherwise marked) end and continues in sequence. The highest number is next to the notch (or mark) on the other side of the IC.

Notice that these guides show only the most common pin-connection configurations, including metal can, DIP (dual-in-line package), SO DIP (small-outline DIP) LCC (leadless chip carrier), multipin DIP, and surface mount.

Connection Information

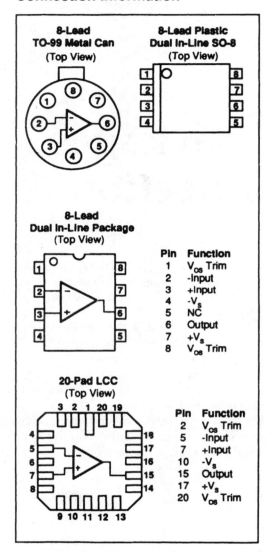

8-Lead TO-99 Metal Can
(Top View)

8-Lead Plastic Dual In-Line SO-8
(Top View)

8-Lead Dual In-Line Package
(Top View)

Pin	Function
1	V_{os} Trim
2	-Input
3	+Input
4	$-V_s$
5	NC
6	Output
7	$+V_s$
8	V_{os} Trim

20-Pad LCC
(Top View)

Pin	Function
2	V_{os} Trim
5	-Input
7	+Input
10	$-V_s$
15	Output
17	$+V_s$
20	V_{os} Trim

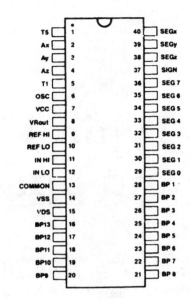

Pin	Name	Pin	Name
1	T5	40	SEGx
2	Ax	39	SEGy
3	Ay	38	SEGz
4	Az	37	SIGN
5	T1	36	SEG 7
6	OSC	35	SEG 6
7	VCC	34	SEG 5
8	VRout	33	SEG 4
9	REF HI	32	SEG 3
10	REF LO	31	SEG 2
11	IN HI	30	SEG 1
12	IN LO	29	SEG 0
13	COMMON	28	BP 1
14	VSS	27	BP 2
15	VDS	26	BP 3
16	BP13	25	BP 4
17	BP12	24	BP 5
18	BP11	23	BP 6
19	BP10	22	BP 7
20	BP9	21	BP 8

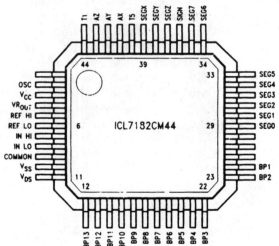

ICL7182CM44

Abbreviations and reference symbols

Most electronics manufacturers and publications outside the United States use some different symbols for electronics components, as well as a different abbreviation system for values or units of measure.

For example, in the illustration, notice the resistor symbol (a rectangular box) at pin 2 of the SL6442. Also notice the capacitor at the same pin. One half of the symbol is solid and the other half is open. The symbol is used wherever polarity must be observed when connecting the capacitor into the circuit. When polarity is of no particular concern, both halves of the symbol are solid, as shown for the capacitor at pin 5 of the SL6442.

Also, the abbreviations for component values are simplified. Thus, μ (Greek letter mu) after a capacitor value represents μF (microfarad), n is nF (nanofarad), and p is pF (picofarad). (The Japanese, and others, often go one step further and use a lowercase u instead of the micro symbol.)

With resistor values, k is thousands of ohms (Ω—Greek letter omega), M is megohms, and the absence of a unit of measure means that ohm is the unit. If μ, n and p are used with an inductance, they represent μH (microhenry), nH (nanohenry), and pH (picohenry). Examples are the 18-nH and 82-nH coils at pin 6 of the SL6442.

For a decimal value, the letter for the unit of measure is sometimes placed at the location of the decimal point. Thus, 3k3 is 3.3 kilohms (kΩ), or 3300 ohms. 2M2 is 2.2 megohms (MΩ), 7μ7 is 7.7 μF, 0μ1 is 0.1 μF, and 3n7 is 3.7 nF.

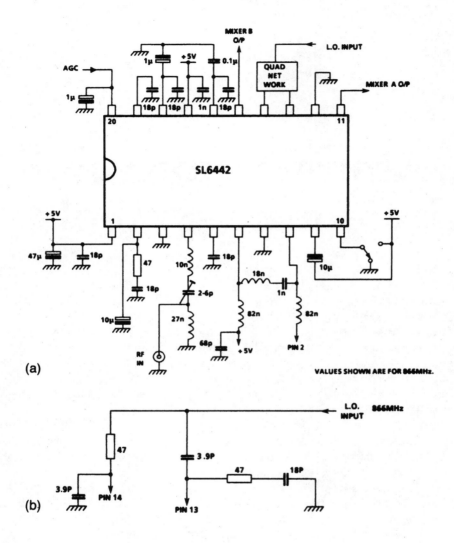

(a)

VALUES SHOWN ARE FOR 866MHz.

(b)

Finding Circuits

The circuits are arranged by type or function, with each group assigned to a separate chapter. For example, Chapter 1 contains digital and microprocessor-supervisory circuits. Chapter 2 contains multiplexer/switch circuits, and so on. The Contents lists each chapter in order. The List of Figures shows the title of each circuit in order.

To find a particular circuit, first note the chapter in which it is likely to appear. Then look for a title that best describes the circuit you want. For example, if you want amplifiers, look in Chapter 6. If you want power supplies, look in Chapter 7. If you want a power supply that is specifically designed for use in micropower and/or battery-operated equipment, look in Chapter 8.

If you want to test or troubleshoot a circuit, begin with the Contents and find the chapter for the appropriate circuit group. For example, if you want to test op amps or comparators, such tests are described in the first sections of Chapter 6, including coverage of test equipment and procedures. If the circuit fails to perform properly (fails to meet the tests), the second sections of Chapter 6 describe amplifier troubleshooting.

Notice that many circuits could appear in more than one chapter. For example, Chapter 2 contains multiplexer circuits. However, chapter 9 contains data-converter circuits with multiplexer functions. So, if you do not locate the circuit you want—even after a careful study of the Contents and List of Figures—use the Index at the back of the book. Here, the circuits are indexed under the different name by which they are known or could be possibly classified. Hundreds of cross-references in the Index will aid you in this search. The Index also lists all of the circuits included in Volumes 1, 2, and 3.

List of Figures

=1=

Digital/microprocessor supervisory circuits

The discussions in this chapter assume that you are already familiar with digital and microprocessor basics (number systems, logic elements, gates, microprocessors, digital test equipment, etc.). If not, read *Lenk's Digital Handbook,* McGraw-Hill, 1992. The following paragraphs summarize both testing and troubleshooting for digital circuits. This information is included so that those readers who are not familiar with electronic test/troubleshooting procedures can both test the circuits in this chapter and locate problems if the circuits fail to perform. Before going into specific testing/troubleshooting, this book starts with a summary of microprocessor-supervisory circuits.

Integrated-circuit (IC) microprocessor supervision

IC microprocessor-supervisory circuits contain analog and digital functions that save considerable design effort and development time. The primary functions in such ICs are concerned with housekeeping required by microprocessors. The functions are vital to microprocessor-based systems because the functions include safeguards against electrical failure. Some microprocessors include the supervisory functions, but such microprocessors cannot always diagnose their own failures. To be most effective and reliable, the monitor circuitry must be in an external supervisory IC.

The most common microprocessor-supervisory functions include: power-on reset, low-voltage reset for glitch and brownout, memory-write protection, power-fail warning, battery backup switchover, and watchdog timer. These functions are not difficult to implement individually, but can be a problem when they all must be included (especially if space is critical). The following review of these supervisory functions shows their relation to the overall microprocessor system.

Power-on reset

When power is applied to a microprocessor, the internal registers "come up" in arbitrary states that contain random data. Applying a RESET command to the microprocessor overcomes this chaos by resetting all internal circuits at a predictable starting point. To ensure a proper startup, the RESET input must be held low for a typical 20 to 120 ms. Although simple, the external reset-timer circuit must keep the RESET signal low while the supply voltage (V_{CC}) is below the minimum level allowed for microprocessor operation.

For example, during startup, the circuit initiates a reset-delay inverval only when V_{CC} reaches that minimum level. If the timed interval begins early (at too low a voltage) or ends too soon, the reset might be overridden by arbitrary states in the digital circuitry while the supply voltage is rising toward the minimum operating level. Power-on reset-timer circuits, therefore, require a timer, a comparator, an accurate voltage reference, and a suitable means for driving the microprocessor RESET input.

Low-voltage (glitch/brownout) reset

Once operating, a microprocessor should continue as long as V_{CC} remains within the specification. To ensure reliable operation, the supervisor IC must also monitor V_{CC} for low voltage or undervoltage (both short-term glitches and longer-term brownouts). These undervoltage conditions are seldom destructive, but they can cause unpredictable operation that leads to a system crash. During a crash, the system or computer fails. The microprocessor cannot be trusted to control its own reset, so the most reliable remedy is to reset the microprocessor with a signal that originates automatically in an external device.

The low-voltage and power-on sections of a typical supervisory IC (such as those produced by Maxim) share the same precision voltage-sensing circuit, and both respond when V_{CC} goes low. When V_{CC} returns to normal, the RESET output to the microprocessor remains low for a timed-delay interval, as is the case when power is first applied (during initial power on).

Memory-write protection

A typical system crash can lead to a potentially worse problem. During the crash, the microprocessor might write "garbage" into the permanent (non-volatile) memory, causing data and program losses that cannot be restored by a simple reset. To prevent such losses, the system must intercept and disable the memory chip-enable (CE) signal during a supply-voltage glitch or brownout. This can be done by applying the CE signal and the output of a low-voltage comparator to a gate, and use the gate output to drive the memory CE input. However, the gate must operate reliably with a V_{CC} as low as 2 V (for a typical 5-V digital system).

Power-fail warning

Full protection for a digital system might require more than a simple low-voltage detection and reset. It might be necessary to provide other actions prior to reset. For example, the microprocessor might need to store the register contents in a non-volatile memory (such as a battery-backed CMOS RAM).

Digital/microprocessor supervisory circuits

Power-fail warning circuits take advantage of the fact that most power-supply regulators have large input-filter capacitors. In a typical 5-V supply, these capacitors charge to between 8 and 10 V. This charge enables the regulator to continue operating for 50 to 100 ms after the primary power is lost (until the capacitor discharges to about 6.5 V, or less for low-dropout regulators). The supervisory IC can be connected to monitor the filter-capacitor voltage. (On Maxim ICs, the power is monitored at a PFI or power-fail-in input.) When the unregulated voltage from the filter capacitor falls below a given value (typically about 7.5 V), an internal comparator issues a PFO (power-fail-out) signal. This allows sufficient time for microprocessor house-cleaning chores before the system initiates a rest.

The PFO is normally connected to the non-maskable-interrupt (NMI) of the microprocessor. This ensures top-priority response. In the supervisory IC, PFI drives one side of an internal CMOS comparator. The other side is driven by a fixed reference (typically 1.3 V). Two external resistors, which divide the capacitor voltage down to the 1.3-V reference, permit the trip point to be set for any desired filter-capacitor voltage.

Battery-backup switchover

CMOS RAM is normally powered by the 5-V supply of the microprocessor. When connected to a 3-V battery in the shutdown or backup mode, the RAM retains its contents but consumes very little power. Backup batteries can be small because the memory current drain (several mA in normal operation) drops to a few microamperes in the backup mode.

The circuit that switches RAM from the main supply to the battery must "stay awake" to switch the RAM back when power is restored. As in the case of a RAM, this switchover circuit depends on the battery; therefore, it must operate on microamperes. Besides low power consumption, the switchover circuit must operate reliably at low supply voltage when the backup battery discharges. The same is true for reset and write-protect circuits, which remain active in the backup mode.

Watchdog timer

Software is usually written as a series of modules interconnected in a continuous loop. During execution, an unforeseen sequence of events can sometimes cause the program to stall within one module, endlessly performing some useless (or possibly harmful) function. A watchdog-timer circuit monitors the program execution and issues a reset command when the stall condition appears.

To use the watchdog, you connect a port of the microprocessor to the timer-rest input the watchdog circuit (which is normally free-running), and configure the software to write data to the port several times per second. Because the circuit interprets missing instructions as a software problem, the watchdog issues a system reset whenever the free-running watchdog times out before receiving the next write instruction.

The optimum timeout depends on the system hardware as well as the software. For example, a longer period at power-up gives the microprocessor extra time to initialize the system before starting the main software loop. In some systems, the watchdog is activated only for certain applications.

Digital/microprocessor circuit testing and troubleshooting

Both testing and troubleshooting for the circuits of this chapter can be performed with conventional test equipment (meters, generators, scopes, etc.), covered in other chapters. However, a logic or digital probe and a digital pulser can make life much easier if you must regularly test and troubleshoot digital devices. So, this section begins with a brief description of the probe and pulser. The description is followed by testing and troubleshooting for the various types of digital circuits.

Logic or digital probe

Logic probes are used to monitor in-circuit pulse or logic activity. By means of a simple lamp indicator, a logic probe shows you the logic state of the digital signal and allows brief pulses to be detected (that you might miss with a scope). Logic probes detect and indicate high and low (1 or 0) logic levels, as well as intermediate or "bad" logic levels (indicating an open circuit) at the terminals of a logic element (the inputs and outputs of a gate, digital-to-analog converter, or a microprocessor, for example).

Not all logic probes have the same functions, and you must learn the operating characteristics of your probe. For example, on the more sophisticated probes, the indicator lamp can give any of four indications: off, dim (about half brilliance), bright (full brilliance), or flashing on and off.

The lamp is normally in the dim state and must be driven to one of the other three states by voltage levels at the probe tip. The lamp is bright for inputs at or above the 1 state and off for inputs at or below 0. The lamp is dim for voltages between the 1 and 0 states, and for open circuits. Pulsating inputs cause the lamp to flash at about 10 Hz (regardless of the input pulse rate). The probe is particularly effective when it is used with the logic pulser.

Logic pulser

The hand-held logic pulser (similar in appearance to the logic probe) is an in-circuit stimulus device that automatically outputs pulses of the required logic polarity, amplitude, current, and width to drive lines and other test points high and low. A typical logic pulser also has several pulse-burst and stream modes available.

Logic pulsers are compatible with most digital devices. Pulse amplitude depends on the equipment supply voltage, which also is the supply voltage for the pulser. Pulse current and width depend on the load being pulsed. The frequency and number of pulses that are generated by the pulser are controlled by operation of a switch. A flashing LED indicator on the pulser tip indicates the output mode.

The logic pulser forces overriding pulses into lines or test points and can be programmed to output single pulses, continuous pulse streams, or bursts. The pulser can be used to force ICs to be enabled or clocked. Also, you can pulse the circuit inputs while observing the effects on the circuit outputs with a logic probe.

Typical testing/troubleshooting application of probe and pulser

A circuit (such as Fig. 1-A) can be tested by applying a pulse at the input and monitoring the output. This can be done with a pulser (or signal generator) at the input and a probe (or scope) at the output. For example, pulses with ECL levels can be applied at the ECL-gate input and TTL pulses can be monitored at the output. If TTL pulses are absent at pin 7 of the 4805 comparator, check for pulses at pins 2 and 3. This procedure will isolate the problem to the gate or to the comparator.

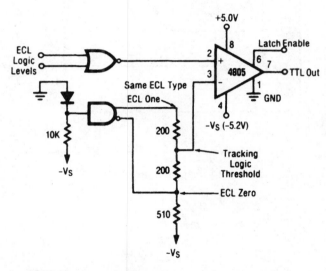

Fig. 1-A   ECL-to-TTL translator (tracking).

General digital-IC troubleshooting tips

The following troubleshooting tips apply to digital circuits where the majority of components are contained in ICs.

Power and ground connections

The first step in tracing problems in a digital circuit with ICs is to check all power and ground connections to the ICs. Many ICs have more than one power and one ground connection. For example, the LTC1043 in Fig. 1-B requires +5 V at pins 4 and 5, and −5 V at pin 17. Also, the LTC1090 in Fig. 1-C has both a dig-

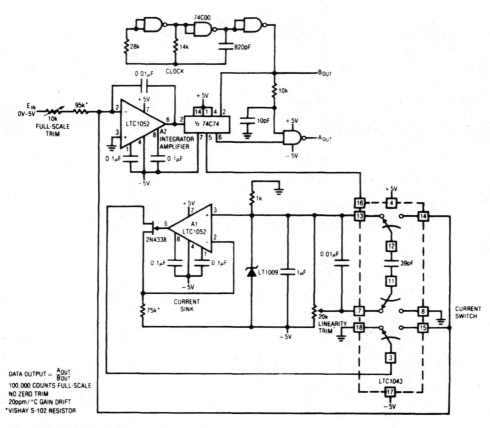

Fig. 1-B 16-bit A/D converter (chopper stabilized).

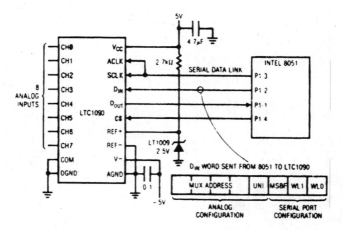

Fig. 1-C Data-acquisition IC with four-wire microprocessor interface.

Digital/microprocessor supervisory circuits

Fig. 1-D   D/A converter with 0- to +10-V unipolar voltage output.

ital ground (DGND) and an analog ground (AGND). Likewise, the DAC-8565 in Fig. 1-D has both an analog common (pin 5) and a digital common (pin 12).

Reset, chip-select, read, write, and start signals

With all power and ground connections confirmed, check that all the ICs receive reset, chip-select, start, and any other function signals, as required. For example, the DAC-4881 in Fig. 1-E requires a chip-select signal at pin 1, as well as address-decode signals at pins 2 and 28. Likewise, the ADC0808/080 in Fig. 1-F requires start, ALE, EOC, and output-enable signals from the microprocessor or control logic. If any of these signals are absent or abnormal (incorrect amplitude, improper timing, etc.) circuit operation comes to an immediate halt.

In some cases, control signals to digital ICs are pulses (usually timed in a certain sequence) and other control signals are steady (high or low). If any of the lines carrying the signals to the ICs are open, shorted to ground, or to power (typically +5 V or +12 V, and 3 V or 3.3 V for newer digital circuits), the IC will not function. So, if you find an IC control pin that is always high, always low, or apparently is connected to nothing (floating), check the PC traces or other wiring to that pin carefully.

As an example, when the DAC4881 is connected as a 12-bit straight-binary digital-to-analog converter (Chapter 6) as shown in Fig. 1-G, the chip-select (pin 1) and address-decode (pins 2, 28) are connected to ground. If the DAC4881 is connected as an 8-bit with complementary input DAC (Fig. 1-E), the chip-select must receive a write ($\overline{WR}$) signal, and the address-decode pins must receive address bits from the microprocessor.

Clock signals

Many digital ICs require clocks. For example, there is a clock at the CP pin of the SAR2504 in Fig. 1-H and a clock at the CP pin of the 74C905 SAR in Fig. 1-I. Figure

General digital-IC troubleshooting tips

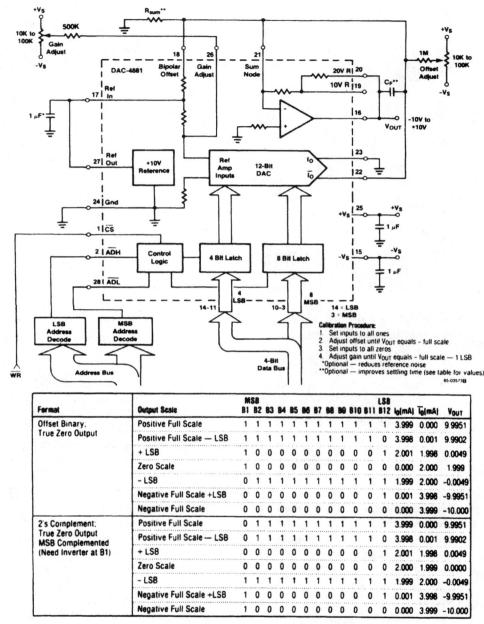

Format	Output Scale	B1	B2	B3	B4	B5	B6	B7	B8	B9	B10	B11	B12	I_O(mA)	$\overline{I_O}$(mA)	V_{OUT}
Offset Binary, True Zero Output	Positive Full Scale	1	1	1	1	1	1	1	1	1	1	1	1	3.999	0.000	9.9951
	Positive Full Scale — LSB	1	1	1	1	1	1	1	1	1	1	1	0	3.998	0.001	9.9902
	+ LSB	1	0	0	0	0	0	0	0	0	0	0	1	2.001	1.998	0.0049
	Zero Scale	1	0	0	0	0	0	0	0	0	0	0	0	0.000	2.000	1.999
	– LSB	0	1	1	1	1	1	1	1	1	1	1	1	1.999	2.000	-0.0049
	Negative Full Scale +LSB	0	0	0	0	0	0	0	0	0	0	0	1	0.001	3.998	-9.9951
	Negative Full Scale	0	0	0	0	0	0	0	0	0	0	0	0	0.000	3.999	-10.000
2's Complement: True Zero Output MSB Complemented (Need Inverter at B1)	Positive Full Scale	0	1	1	1	1	1	1	1	1	1	1	1	3.999	0.000	9.9951
	Positive Full Scale — LSB	0	1	1	1	1	1	1	1	1	1	1	0	3.998	0.001	9.9902
	+ LSB	0	0	0	0	0	0	0	0	0	0	0	1	2.001	1.998	0.0049
	Zero Scale	0	0	0	0	0	0	0	0	0	0	0	0	2.000	1.999	0.0000
	– LSB	1	1	1	1	1	1	1	1	1	1	1	1	1.999	2.000	-0.0049
	Negative Full Scale +LSB	1	0	0	0	0	0	0	0	0	0	0	1	0.001	3.998	-9.9951
	Negative Full Scale	1	0	0	0	0	0	0	0	0	0	0	0	0.000	3.999	-10.000

Fig. 1-E 8-bit D/A converter with microprocessor interface.

1-F(b) shows the clock periods for the ADC in Fig. 1-F. In some cases, the clock comes from an external source (Figs. 1-F and 1-H). In other cases, the clock is part of the circuit.

Digital/microprocessor supervisory circuits

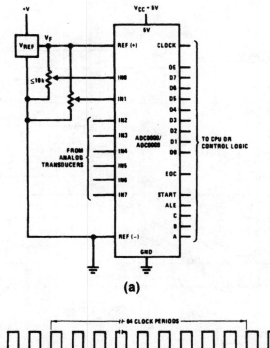

(a)

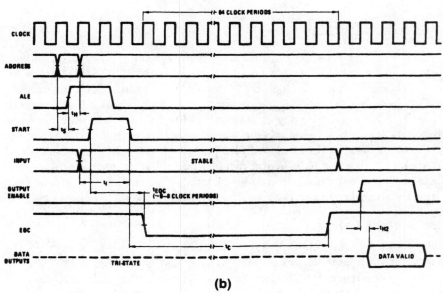

(b)

Fig. 1-F   Ratiometric A/D converter with separate interface.

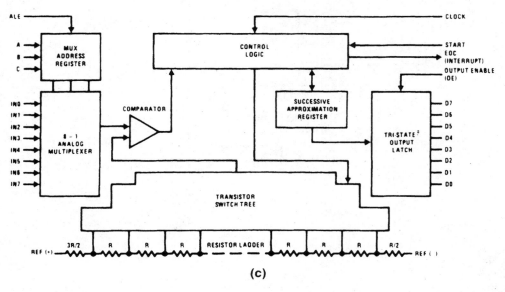

Fig. 1-F Continued.

Generally, the presence of pulse activity on any pin of a digital IC indicates the presence of a clock, but do not count on it. Check directly at the clock pins (typically, all ICs that require a clock are connected to the same clock source).

It is possible to measure the presence of a clock signal with a scope or logic probe. However, a frequency counter provides the most accurate measurement. Obviously, if any ICs do not receive required clock signals, the IC cannot function. On the other hand, if the clock is off frequency, all of the ICs might appear to have a clock signal, but the IC function can be impaired. Notice that crystal-controlled clocks do not usually drift off frequency, but can go into some overtone frequency (typically, a third overtone) beyond the capacity of the IC.

Input-output signals

When you are certain that all ICs are good and have proper power and ground connections, and that all control signals (reset, chip-select, write, start, enable, etc.), and clock signals are available, the next step is to monitor all input and output signals at each IC. This test can be done with either a scope or probe.

Microprocessor-supervisor IC testing/ troubleshooting

The most practical in-circuit test of an IC supervisor is to check the function. (It is thus quite helpful if you understand the function!) The following are some examples.

Digital/microprocessor supervisory circuits

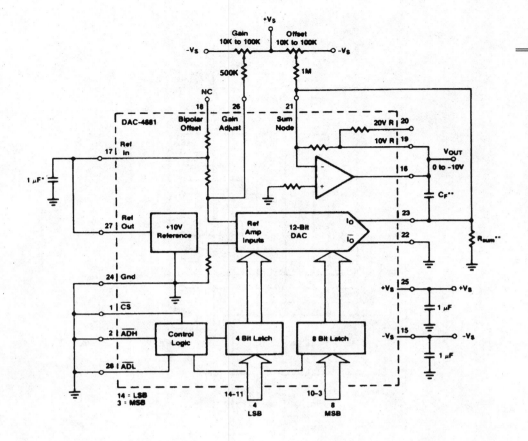

Calibration Procedure:

1. Set inputs to all zeros
2. Adjust offset until V_{OUT} equals 0V
3. Set inputs to all ones
4. Adjust gain until V_{OUT} equals correct full scale value

*Optional — reduces reference noise

**Optional — improves settling time (see table for values)

Format	Output Scale	MSB B1	B2	B3	B4	B5	B6	B7	B8	B9	B10	B11	LSB B12	I_O(mA)	$\overline{I_O}$(mA)	V_{OUT}
Straight Binary. Unipolar with True Input Code. True Zero Output	Positive Full Scale	1	1	1	1	1	1	1	1	1	1	1	1	3.999	0.000	9.9976
	Positive Full Scale — LSB	1	1	1	1	1	1	1	1	1	1	1	0	3.998	0.001	9.9951
	LSB	0	0	0	0	0	0	0	0	0	0	0	1	0.0001	3.998	0.0024
	Zero Scale	0	0	0	0	0	0	0	0	0	0	0	0	0.000	3.999	0.0000
Complementary Binary. Unipolar with Complementary Input Code. True Zero Output	Positive Full Scale	0	0	0	0	0	0	0	0	0	0	0	0	0.000	3.999	9.9976
	Positive Full Scale — LSB	0	0	0	0	0	0	0	0	0	0	0	1	0.001	3.998	9.9951
	LSB	1	1	1	1	1	1	1	1	1	1	1	0	3.998	0.001	0.0024
	Zero Scale	1	1	1	1	1	1	1	1	1	1	1	1	3.999	0.000	0.0000

Fig. 1-G Twelve-bit straight-binary D/A converter.

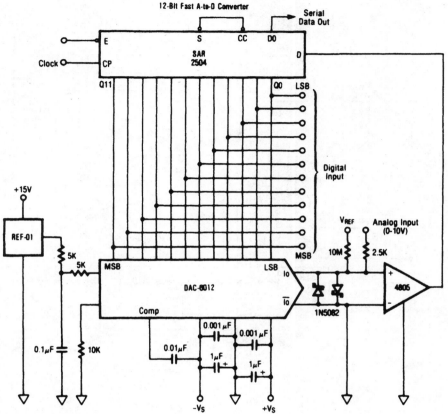

Note:
Device(s) connected to analog input must be capable of sourcing 4.0mA
a buffer may be required

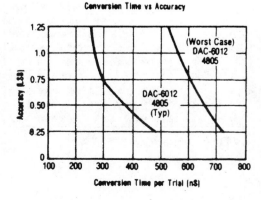

Conversion Time (nS)	Typ	Worst Case
SAR	33nS	55nS
4805	92nS	125nS
Total	375nS	680nS
x 13	4.9µS	8.8µS

Fig. 1-H   Fast 12-bit A/D converter.

Digital/microprocessor supervisory circuits

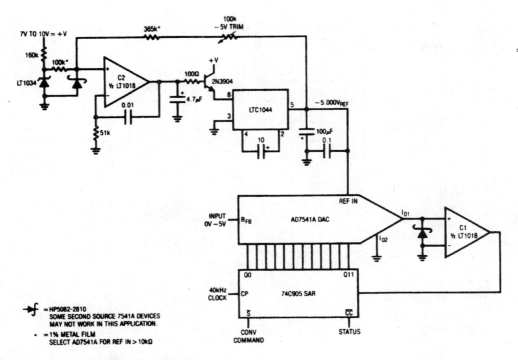

Fig. 1-I   Micropower 12-bit A/D converter.

The circuit of Fig. 1-J shows a DS1231 power-monitor chip that is used with a digital-system power supply. The DS1231 uses a temperature-compensated reference circuit to provide both an orderly shutdown and automatic reset of processor-based systems. The DS1231 operates by monitoring the high-voltage inputs to the power-supply regulators (at a voltage sense point, usually the filter capacitors), and applying RST or $\overline{\text{RST}}$ and $\overline{\text{NMI}}$ control signals to the processor.

The time for the processor shutdown is directly proportional to the available hold-up time of the power supply. Just before the hold-up time is exhausted, the DS1231 unconditionally halts the processor (to prevent spurious cycles) by enabling the rest when V_{CC} falls below a selectable 5 or 10% threshold. (Note that the TOL pin is grounded for a 5% threshold, on this IC.) When power returns, the processor is held inactive until well after power conditions have stabilized, safeguarding any nonvolatile memory in the system from inadvertent data changes.

The DS1231 should produce a RST or $\overline{\text{RST}}$ (both might be required for some systems) and $\overline{\text{NMI}}$ to the microprocessor when voltage is removed from the input. To test this function, apply 10 V to the voltage sense point, and monitor the RST, $\overline{\text{RST}}$, and $\overline{\text{NMI}}$ pins of the DS1231 (be certain that +5 V is applied to the V_{CC} and MODE pins, as shown). Now remove the voltage from the voltage sense point and check that RST, $\overline{\text{RST}}$, and $\overline{\text{NMI}}$ signals appear at the corresponding pins of the DS1231.

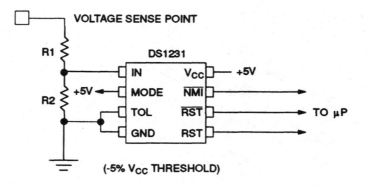

VOLTAGE SENSE POINT

$$V\ SENSE = \frac{R1 + R2}{R2} \times 2.3 \qquad V\ MAX = \frac{V\ SENSE}{VTP\ -} \times 5.0$$

EXAMPLE: V SENSE = 8 VOLTS AT TRIP POINT AND A
MAXIMUM VOLTAGE OF 17.5V WITH R2 = 10K

$$THEN\ 8 = \frac{R1 + 10K}{10K} \times 2.3 \qquad R1 = 25K$$

Fig. 1-J Digital power monitor.

The circuit of Fig. 1-K shows a DS123LP/LPS used to monitor and control the power supply and software execution of a processor-based system, and to provide a pushbutton reset. When V_{CC} falls below a preset level (as defined by the TOL pin connection) the DS1232 outputs RST and $\overline{RST}$ reset signals. With TOL connected to V_{CC}, the reset becomes active when V_{CC} falls below 4.5 V. When TOL is connected to ground, the reset become active when V_{CC} falls below 4.75 V.

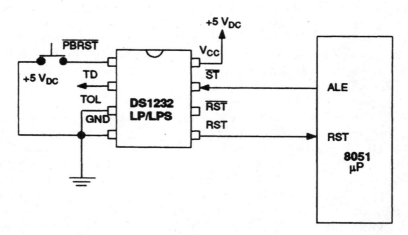

Fig. 1-K Low-power micromonitor with pushbutton.

Digital/microprocessor supervisory circuits

On power up, RST and $\overline{\text{RST}}$ are kept active for a minimum of 250 ms to allow the power supply and processor to stabilize. The pushbutton reset input $\overline{\text{PBRST}}$ requires an active low. Internally, this input is debounced and timed so that reset signals of at least 250 ms (minimum) are generated. The 250-ms delay starts as the pushbutton reset input is released from low level.

To test the low-power reset function, reduce V_{CC} from 5 V to 4.5 V (or 4.75 V, if TOL is connected to ground, as shown), and check that a RST signal appears at the corresponding pin. (A $\overline{\text{RST}}$ should also be available at the corresponding pin, although $\overline{\text{RST}}$ is not used in this circuit.)

To test the manual reset function, return V_{CC} to +5 V. Then press and release the pushbutton. Check that a RST signal appears at the corresponding pin. If required, use a dual-channel scope and check the timing relationship between releasing the pushbutton and the presence of a RST signal. There should be a 250-ms delay starting from the time that the pushbutton is released. The RST signal (and the $\overline{\text{RST}}$ signal) should remain active for a minimum of 250 ms. As a final test, remove and reapply the 5-V V_{CC}, and check that RST and $\overline{\text{RST}}$ signals are generated for a minimum of 250 ms.

The circuit of Fig. 1-L is similar to that of Fig. 1-K, but with the addition of a watchdog-timer function to monitor software execution. The timer function forces RST and $\overline{\text{RST}}$ signals to the active state (shutting down the processor) when the $\overline{\text{ST}}$ input is not stimulated for a predetermined time period (because of some failure in software execution).

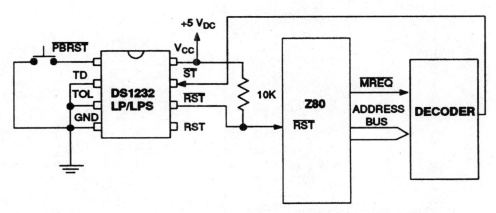

Fig. 1-L Low-power micromonitor with pushbutton and watchdog timer.

The time period is set (by the TD input) to be about 150 ms with TD connected to ground, 600 ms with TD left unconnected, and 1.2 s with TD connected to V_{CC}. The timer starts timing out from the set-time period as soon as the reset signals are inactive. The $\overline{\text{ST}}$ input can be taken from address, data, and/or control signals. When the processor is executing software, these signals are present, which causes the watchdog to be reset prior to time out.

Microprocessor-supervisor IC testing/troubleshooting

Use the same procedures as described for the circuit of Fig. 1-K to test the reset function of the Fig. 1-L circuit. To test the watchdog function, remove the $\overline{\text{ST}}$ signal and check that RST and $\overline{\text{RST}}$ are activated after the appropriate time period. (With TD at ground as shown, the time period should be about 150 ms.) $\overline{\text{ST}}$ can be shorted to ground, or disconnected, whichever is more practical. Use a dual-channel scope to monitor the time delay.

The circuit of Fig. 1-M shows a DS1236 that is used to control a digital-system power supply. The DS1236 generates an $\overline{\text{NMI}}$ for an early warning of power failure to a microprocessor. A precision comparator monitors the voltage level at the In pin, relative to an internal reference. The In pin is a high-impedance input that allows for a user-defined sense point. An external voltage divider interfaces with the voltage signals. The sense point can be taken from the 5-V supply or from a higher voltage level (typically from the power-supply filter capacitors, for an early warning) that is closer to the main system power input. Because the In trip point (V_{TP}) is 2.54 V, the proper values for R1 and R2 can be determined by the equations.

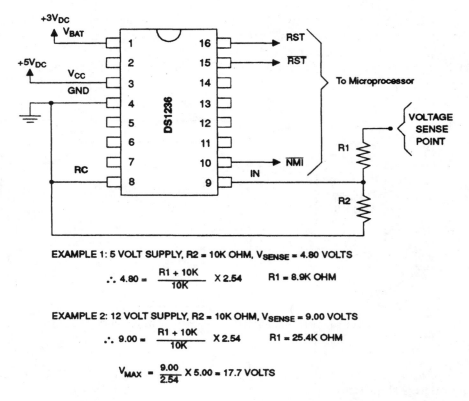

EXAMPLE 1: 5 VOLT SUPPLY, R2 = 10K OHM, V$_{\text{SENSE}}$ = 4.80 VOLTS

$$\therefore\ 4.80 = \frac{R1 + 10K}{10K} \times 2.54 \qquad R1 = 8.9K\ \text{OHM}$$

EXAMPLE 2: 12 VOLT SUPPLY, R2 = 10K OHM, V$_{\text{SENSE}}$ = 9.00 VOLTS

$$\therefore\ 9.00 = \frac{R1 + 10K}{10K} \times 2.54 \qquad R1 = 25.4K\ \text{OHM}$$

$$V_{\text{MAX}} = \frac{9.00}{2.54} \times 5.00 = 17.7\ \text{VOLTS}$$

Fig. 1-M Digital power monitor with early warning.

To test the early-warning function, reduce the voltage at the sense point to the correct value, and check that the RST, $\overline{\text{RST}}$ and $\overline{\text{NMI}}$ functions are activated. Monitor the voltage at the sense point, not at V_{CC}, when making this test. If the reset and interrupt functions are actuated, but not at the correct sense-point voltage, recheck the values of R1 and R2. Keep in mind that the functions should be activated when the voltage at the In pin is 2.54 V. If the functions are not activated at any sense-point voltage (or In voltage), suspect the IC.

The circuit of Fig. 1-N shows a DS1236 that is used to control battery-backup operation for a static RAM (SRAM). First, the DS1236 contains a switch to direct SRAM power from the 5-V supply (V_{CC}) or from an external battery (V_{BAT}), whichever is greater. The switched supply (V_{CCO}) can also be used to battery-back a CMOS processor. Second, RST and $\overline{\text{RST}}$ are activated when there is a power failure (when V_{CC} falls below 4.5 V for 10% operation or below 4.75 V for 5% operation). The same power-fail detection function is used to hold the chip-enable output ($\overline{\text{CEO}}$) to within 0.3 V of V_{CC} or to within 0.7 V of V_{BAT}. This write-protection mechanism occurs when V_{CC} falls below the specified trip point (V_{CCTP}) of 4.5 or 4.75 V.

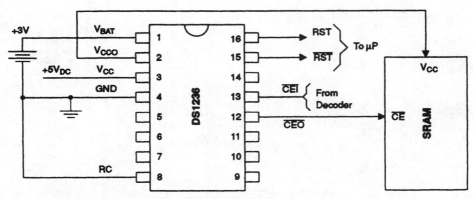

Fig. 1-N Digital memory backup for SRAMs.

To test the reset function, reduce V_{CC} to 4.5 V (or 4.75 V) and check that RST and $\overline{\text{RST}}$ are activated. Continue reducing V_{CC} until the voltage is less than 3 V. RST and $\overline{\text{RST}}$ should remain active, and the voltage at $\overline{\text{CEO}}$ should drop to 3V (within 0.7 V). Finally, increase V_{CC} until the voltage is 5 V, and note that the $\overline{\text{CEO}}$ voltage is also restored to 5 V (within 0.3 V). The changeover at $\overline{\text{CEO}}$ should occur when V_{CC} is increased to about 4 V. RST and $\overline{\text{RST}}$ should be deactivated when V_{CC} reaches 4.5 V (or 4.75 V). If there is no voltage at $\overline{\text{CEO}}$ when V_{CC} is reduced, or the $\overline{\text{CEO}}$ voltage simply follows V_{CC}, suspect the battery. If the battery is good, but the remaining functions are absent or abnormal, suspect the IC.

Microprocessor-supervisory circuit titles and descriptions

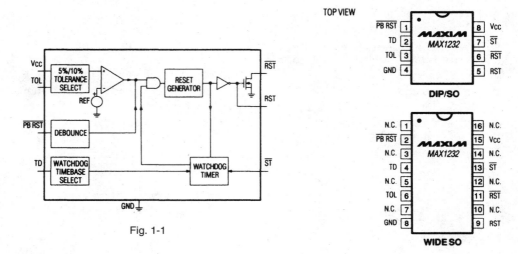

Fig. 1-1

TOP VIEW

Fig. 1-2

NAME	FUNCTION
$\overline{\text{PB RST}}$	Pushbutton Reset Input. A debounced active-low input that ignores pulses less than 1ms in duration and is guaranteed to recognize inputs of 20ms or greater.
TD	Time Delay Set. The watchdog timebase select input (t_{TD} = 150ms for TD = 0V, t_{TD} = 600ms for TD = open, t_{TD} = 1.2sec for TD = V_{CC}).
TOL	Tolerance Input. Connect to GND for 5% tolerance or to V_{CC} for 10% tolerance.
GND	Ground
RST	Reset Output (Active High) - goes active: 1. If V_{CC} falls below the selected reset voltage threshold 2. If $\overline{\text{PB RST}}$ is forced low 3. If $\overline{\text{ST}}$ is not strobed within the minimum timeout period 4. During power-up
$\overline{\text{RST}}$	Reset Output (Active Low, Open Drain) - see RST.
$\overline{\text{ST}}$	Strobe Input. Input for watchdog timer.
V_{CC}	The +5V Power-Supply Input
N.C.	No Connect

Fig. 1-3

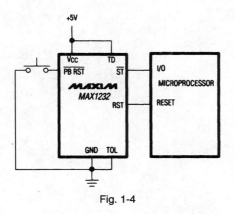

Fig. 1-4

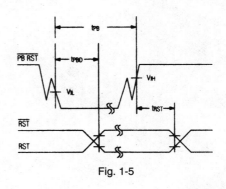

Fig. 1-5

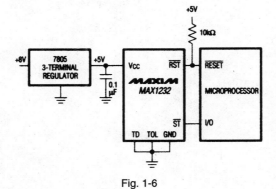

Fig. 1-6

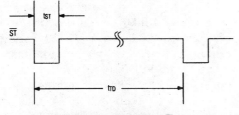

NOTE: t$_{TD}$ IS THE MAXIMUM ELAPSED TIME BETWEEN $\overline{ST}$ HIGH-TO-LOW
TRANSITIONS ($\overline{ST}$ IS ACTIVATED BY FALLING EDGES ONLY) WHICH WILL
KEEP THE WATCHDOG TIMER FROM FORCING THE RESET OUTPUTS
ACTIVE FOR A TIME OF t$_{RST}$. t$_{TD}$ IS A FUNCTION OF THE VOLTAGE AT
THE TD PIN, AS TABULATED BELOW.

CONDITON	t$_{TD}$		
	MIN	TYP	MAX
TD pin = 0V	62.5ms	150ms	250ms
TD pin = open	250ms	600ms	1000ms
TD pin = V$_{CC}$	500ms	1200ms	2000ms

Fig. 1-7

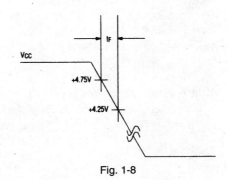

Fig. 1-8

Microprocessor-supervisory circuit titles and descriptions

Continued

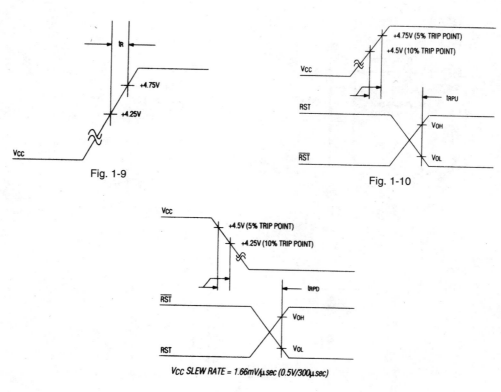

Fig. 1-9

Fig. 1-10

Fig. 1-11

Microprocessor monitor

Figures 1-1, 1-2, and 1-3 show the block diagram, pin configuration and pin description, respectively, for the MAX1232. The IC is similar to that described for Figs. 1-K and 1-L, except that only ⅒ power is consumed. Figures 1-4 and 1-5 show the basic connections and waveforms, respectively, for the pushbutton reset. Figures 1-6 and 1-7 show the basic connections and waveforms, respectively, for watchdog timing. Notice that the software routine that strobes ST is critical. The code must be in a section of software that executes frequently enough so that the time between toggles is less than the watchdog timeout period, as shown in Fig. 1-7. Figures 1-8 and 1-9 show the power-down and power-up slew rate, respectively. Figures 1-10 and 1-11 show the V_{CC}-detect reset-output delays. MAXIM NEW RELEASES DATA BOOK, 1992, P. 5-11, 5-14, 5-15, 5-16.

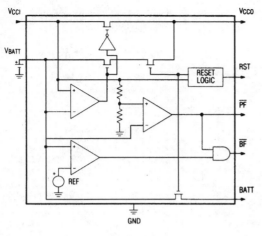

Fig. 1-12

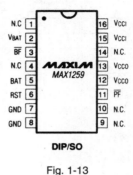

DIP/SO

Fig. 1-13

Battery manager
Figures 1-12 and 1-13 show the block diagram and pin configuration, respectively, for the MAX1259. The IC is similar to that described for Fig. 1-N, and switches to the backup battery when the primary power supply is interrupted. The input-output differential is 200 mV when supplying 250 mA from the primary power or 15 mA from the battery. The battery-failure output at pin 3 indicates when the battery is below +2 V. The power-fail output at pin 11 indicates when the primary power is low. MAXIM NEW RELEASES DATA BOOK, 1992, P. 5-17.

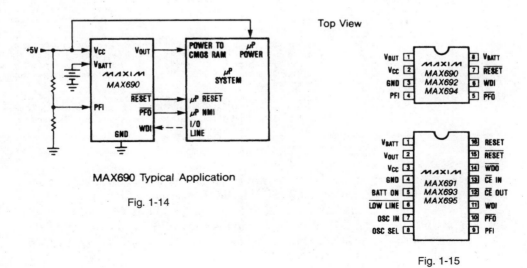

MAX690 Typical Application

Fig. 1-14

Top View

Fig. 1-15

Continued

NAME	PIN		FUNCTION
	MAX690/ 692/694	MAX691/ 693/695	
V$_{CC}$	2	3	The +5V input.
V$_{BATT}$	8	1	Backup battery input. Connect to Ground if a backup battery is not used.
V$_{OUT}$	1	2	The higher of V$_{CC}$ or V$_{BATT}$ is internally switched to V$_{OUT}$. Connect V$_{OUT}$ to V$_{CC}$ if V$_{OUT}$ and V$_{BATT}$ are not used.
GND	3	4	0V Ground reference for all signals.
$\overline{RESET}$	7	15	$\overline{RESET}$ goes low whenever V$_{CC}$ falls below either the reset voltage threshold or the V$_{BATT}$ input voltage. The reset threshold is typically 4.65V for the MAX690/691/694/695, and 4.4V for the MAX692 and MAX693. $\overline{RESET}$ remains low for 50ms after V$_{CC}$ returns to 5V, (except 200ms in MAX694/695). $\overline{RESET}$ also goes low for 50ms if the Watchdog Timer is enabled but not serviced within its timeout period. The $\overline{RESET}$ pulse width can be adjusted as shown in Table 1.
WDI	6	11	The watchdog input, WDI, is a three level input. If WDI remains either high or low for longer than the watchdog timeout period, $\overline{RESET}$ pulses low and $\overline{WDO}$ goes low. The Watchdog Timer is disabled when WDI is left floating or is driven to mid-supply. The timer resets with each transition at the Watchdog Timer Input.
PFI	4	9	PFI is the non-inverting input to the Power Fail Comparator. When PFI is less than 1.3V, $\overline{PFO}$ goes low. Connect PFI to GND or V$_{OUT}$ when not used. See Figure 1.
$\overline{PFO}$	5	10	$\overline{PFO}$ is the output of the Power Fail Comparator. It goes low when PFI is less than 1.3V. The comparator is turned off and $\overline{PFO}$ goes low when V$_{CC}$ is below V$_{BATT}$.
$\overline{CE}$ IN	—	13	The input to the $\overline{CE}$ gating circuit. Connect to GND or V$_{OUT}$ if not used.
$\overline{CE}$ OUT	—	12	$\overline{CE}$ OUT goes low only when $\overline{CE}$ IN is low and V$_{CC}$ is above the reset threshold (4.65V for MAX691 and MAX695, 4.4V for MAX693). See Figure 6.
BATT ON	—	5	BATT ON goes high when V$_{OUT}$ is internally switched to the V$_{BATT}$ input. It goes low when V$_{OUT}$ is internally switched to V$_{CC}$. The output typically sinks 25mA and can directly drive the base of an external PNP transistor to increase the output current above the 50mA rating of V$_{OUT}$.
$\overline{LOW\ LINE}$	—	6	$\overline{LOW\ LINE}$ goes low when V$_{CC}$ falls below the reset threshold. It returns high as soon as V$_{CC}$ rises above the reset threshold. See Figure 6, Reset Timing.
RESET	—	16	RESET is an active high output. It is the inverse of $\overline{RESET}$.
OSC SEL	—	8	When OSC SEL is unconnected or driven high, the internal oscillator sets the reset time delay and watchdog timeout period. When OSC SEL is low, the external oscillator input, OSC IN, is enabled. OSC SEL has a 3µA internal pullup. See Table 1.
OSC IN	—	7	When OSC SEL is low, OSC IN can be driven by an external clock to adjust both the reset delay and the watchdog timeout period. The timing can also be adjusted by connecting an external capacitor to this pin. See Figure 8. When OSC SEL is high or floating, OSC IN selects between fast and slow Watchdog timeout periods.
$\overline{WDO}$	—	14	The Watchdog Output, $\overline{WDO}$, goes low if WDI remains either high or low for longer than the Watchdog timeout period. $\overline{WDO}$ is set high by the next transition at WDI. If WDI is unconnected or at mid-supply, $\overline{WDO}$ remains high. $\overline{WDO}$ also goes high when $\overline{LOW\ LINE}$ goes low.

Fig. 1-16

Microprocessor supervisory circuit

Figures 1-14 and 1-15 show a typical application circuit and block diagram, respectively, for the MAX690. The IC is similar to that described for Fig. 1-M, but also includes watchdog and battery-backup protection, and has a 1.3-V threshold detector, as shown in Fig. 1-16. Use the equations shown in Fig. 1-M to calculate the values for the resistors at the PFI input (except use a threshold of 1.3 V, instead of 2.54 V). MAXIM NEW RELEASES DATA BOOK, 1992, P. 5-19, 5-22.

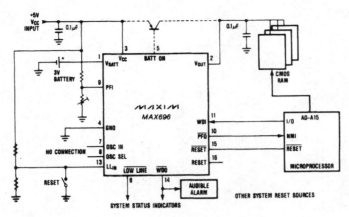

Fig. 1-17

NAME	PIN		FUNCTION
	MAX696	**MAX697**	
V_{CC}	3	3	The +5V input.
V_{BATT}	1	—	Backup battery input. Connect to Ground if a backup battery is not used.
V_{OUT}	2	—	The higher of V_{CC} or V_{BATT} is internally switched to V_{OUT}. Connect V_{OUT} to V_{CC} if V_{OUT} and V_{BATT} are not used.
GND	4	5	0V ground reference for all signals.
$\overline{RESET}$	15	15	$\overline{RESET}$ goes low whenever LL_{IN} falls below 1.3 volts or V_{CC} falls below the V_{BATT} input voltage. $\overline{RESET}$ remains low for 50ms after LL_{IN} goes above 1.3 volts. $\overline{RESET}$ also goes low for 50ms if the Watchdog Timer is enabled but not serviced within its timeout period. The $\overline{RESET}$ pulse width can be adjusted as shown in Table 1.
WDI	11	11	The watchdog input, WDI, is a three level input. If WDI remains either high or low for longer than the watchdog timeout period, $\overline{RESET}$ pulses low and $\overline{WDO}$ goes low. The Watchdog Timer is disabled when WDI is left floating or is driven to mid-supply. The timer resets with each transition at the Watchdog Timer Input.
PFI	9	9	PFI is the non-inverting input to the Power Fail Comparator. When PFI is less than 1.3V, $\overline{PFO}$ goes low. Connect PFI to GND or V_{OUT} when not used. See Figure 1.
$\overline{PFO}$	10	10	$\overline{PFO}$ is the output of the Power Fail Comparator. It goes low when PFI is less than 1.3V. The comparator is turned off and $\overline{PFO}$ goes low when V_{CC} is below V_{BATT}.
$\overline{CE\ IN}$	—	13	The input to the $\overline{CE}$ gating circuit. Connect to GND or V_{OUT} if not used.
$\overline{CE\ OUT}$	—	12	$\overline{CE\ OUT}$ goes low only when $\overline{CE\ IN}$ is low and LL_{IN} is above 1.3V. See Figure 5.
BATT ON	5	—	BATT ON goes high when V_{OUT} is internally switched to the V_{BATT} input. It goes low when V_{OUT} is internally switched to V_{CC}. The output typically sinks 7mA and can directly drive the base of an external PNP transistor to increase the output current above the 50mA rating of V_{OUT}.
$\overline{LOW\ LINE}$	6	6	$\overline{LOW\ LINE}$ goes low when LL_{IN} falls below 1.3 volts. It returns high as soon as LL_{IN} rises above 1.3 volts. See Figure 5, Reset Timing.
RESET	16	16	RESET is an active high output. It is the inverse $\overline{RESET}$.
OSC SEL	8	8	When OSC SEL is unconnected or driven high, the internal oscillator sets the reset time delay and watchdog timeout period. When OSC SEL is low, the external oscillator input, OSC IN, is enabled. OSC SEL has a 3µA internal pullup. See Table 1.
OSC IN	7	7	OSC IN sets the Reset delay timing and Watchdog timeout period when OSC SEL floats or is driven low. The timing can also be adjusted by connecting an external capacitor to this pin. See Figure 7. When OSC SEL is high, OSC IN selects between fast and slow Watchdog timeout periods.
$\overline{WDO}$	14	14	The Watchdog Output, $\overline{WDO}$, goes low if WDI remains either high or low for longer than the Watchdog timeout period. $\overline{WDO}$ is set high by the next transition at WDI. If WDI is unconnected or at mid-supply, $\overline{WDO}$ remains high. $\overline{WDO}$ also goes high when $\overline{LOW\ LINE}$ goes low.
NC	12	2	NO CONNECT. Leave this pin open.
LL_{IN}	13	4	LOW LINE INPUT. LL_{IN} is the CMOS input to a comparator whose other input is a precision 1.3 volt reference. The output is $\overline{LOW\ LINE}$ and is also connected to the reset pulse generator. See Figure 2.
TEST	—	1	Used during Maxim manufacture only. Always ground this pin.

Fig. 1-18

Microprocessor-supervisory circuit titles and descriptions

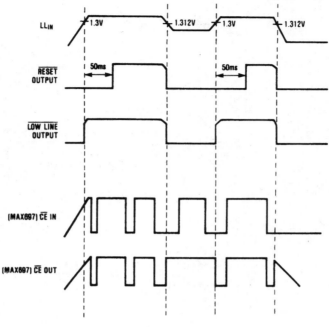

Fig. 1-19

Table 1. MAX696 and MAX697 Reset Pulse Width and Watchdog Timeout Selections

OSC SEL (Note 3)	OSC IN	WATCHDOG TIMEOUT PERIOD NORMAL	IMMEDIATELY AFTER RESET	RESET TIMEOUT PERIOD
Low	External Clock Input	1024 clks	4096 clks	512 clks
Low	External Capacitor	$\frac{400\text{ms}}{47\text{pf}} \times C$	$\frac{1.6\text{ sec}}{47\text{pf}} \times C$	$\frac{200\text{ms}}{47\text{pf}} \times C$
High/Floating	Low	100ms	1.6 sec	50ms
High/Floating	Floating	1.6 sec	1.6 sec	50ms

Note 1: When the MAX696/697 OSC SEL pin is low, OSC IN can be driven by an external clock signal, or an external capacitor can be connected between OSC IN and GND. The nominal internal oscillator frequency is 10.24kHz. The nominal oscillator frequency with external capacitor is $F_{OSC}(\text{Hz}) = \frac{184,000}{C_{OSC}(\text{pF})}$

Note 2: See Electrical Characteristics Table for minumum and maximum timing values.

Note 3: "HIGH" for the OSC SEL pin should be connected to V_{OUT}, not V_{CC} (on MAX696).

Fig. 1-20

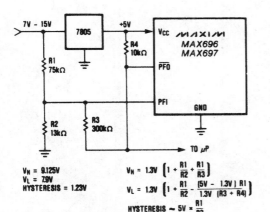

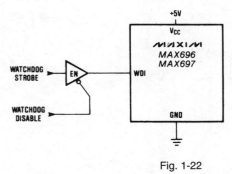

Fig. 1-22

Fig. 1-21

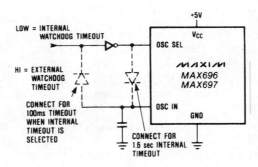

Fig. 1-23

Microprocessor supervisory circuit (added features)

Figures 1-17 and 1-18 show a typical application circuit and pin descriptions, respectively for a MAX696. The IC is similar to that described for Figs. 1-J through 1-N, but with added (or combined) features. Use the equations shown in Fig. 1-M to calculate the values for the resistors at the PFI and LLIN inputs (except use a threshold of 1.3 V, instead of 2.54 V). Figures 1-19, 1-20, 1-21, 1-22, and 1-23 show the reset timing, watchdog and reset timeout selection, calculations and connections for adding hysteresis, circuit for disabling the watchdog under program control, and circuit for selecting watchdog timeout, respectively. Table 1 in Fig. 1-18 refers to Fig. 1-20 in this book. Figure 1 in Fig. 1-18 refers to Fig. 1-17 in this book. Figure 5 in Fig. 1-18 refers to Fig. 1-19 in this book. Figure 7 in Fig. 1-18 refers to Fig. 1-21 in this book. Figure 2 in Fig. 1-18 is an internal function not shown here. MAXIM NEW RELEASES DATA BOOK, 1992, P. 5-36, 5-37, 5-41, 5-43.

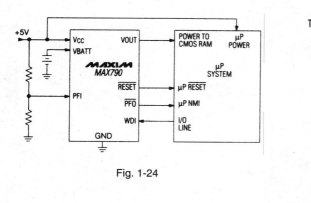

Fig. 1-24

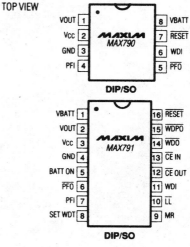

Fig. 1-25

High-performance supervisory circuit
Figures 1-24 and 1-25 show a typical application circuit and pin configurations, respectively, for the MAX790/791. The IC is similar to the MAX690 series (Figs. 1-14 and 1-15), but with several improvements, including 70-μA supply current, 10-ns CE propagation delay, 250-mA output current (V_{CC} mode), and 25-mA output current (V_{BATT} mode). The MAX790 is pin compatible with the MAX690. The reset output is guaranteed to be in the correct state for V_{CC} down to 1 V. MAXIM NEW RELEASES DATA BOOK, 1992, P. 5-53.

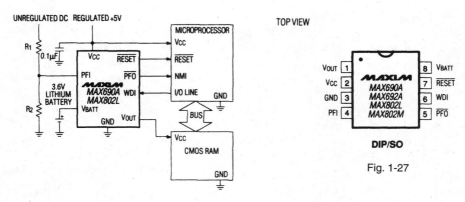

Fig. 1-26

Fig. 1-27

Digital/microprocessor supervisory circuits

Microprocessor supervisory circuit (high reliability)
Figures 1-26 and 1-27 show a typical application circuit and pin configurations, respectively, for the MAX690A/MAX802L. The ICs are similar to that described for Figs. 1-J through 1-N, but with added reliability and other reset thresholds. The MAX690A/MAX802L generate a reset pulse when the supply drops below 4.65 V, and the MAX692A/MAX802M generate a reset below 4.40 V. The threshold is 1.25 V for both power-fail warning and low-battery detection. The reset delay is 200 ms. The watchdog timer has a 1.6-s timeout. Quiescent current is 200 μA with normal power and 50 nA in battery-backup mode. The power-fail accuracy is ±2% for MAX802L/M. MAXIM HIGH-RELIABILITY DATA BOOK, 1993, P. 5-17.

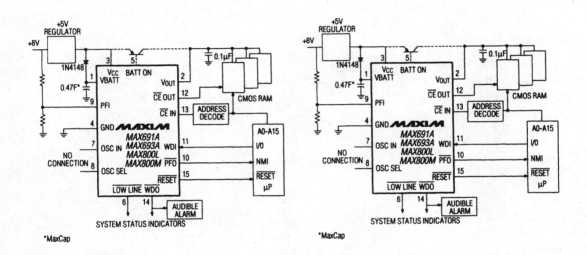

Fig. 1-28   Fig. 1-29

Microprocessor supervisory circuit (high reliability/added features
Figures 1-28 and 1-29 show a typical application circuit and pin configurations, respectively, for the MAX691A/693A/800L/800M. The ICs are similar to that described for Figs. 1-J through 1-M, but with more reliability and added features. The ICs improve performance with 35-μA supply current, 200-ms typical reset delay on power-up, and 6-ns chip-enable propagation delay. Other features include write protection of CMOS RAM or EEPROM, separate watchdog outputs, battery-backup switchover, and a reset output that is valid with V_{CC} down to 1 V. The MAX691A/MAX800L have a 4.65-V typical reset-threshold. The MAX693A/MAX800M reset threshold is 4.4 V (typical). The MAX800L/MAX800M is guaranteed to deliver power-fail accuracies to ±2%. MAXIM HIGH-RELIABILITY DATA BOOK, 1993, P. 5-19.

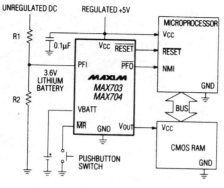

Fig. 1-30

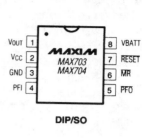

Fig. 1-31

Low-cost supervisory circuit with battery backup

Figures 1-30 and 1-31 show a typical application circuit and pin configurations, respectively, for the MAX703/MAX704. The ICs are similar to that described for Figs. 1-J through 1-M, but with generally lower cost. The ICs have a 200-ms reset pulse width, 200-µA quiescent current, 50-nA quiescent with battery backup, and a 1.25-V threshold detector for power-fail warning, low-battery detection, or for monitoring a power supply voltage other than +5 V. MAXIM HIGH-RELIABILITY DATA BOOK, 1993, P. 5-29.

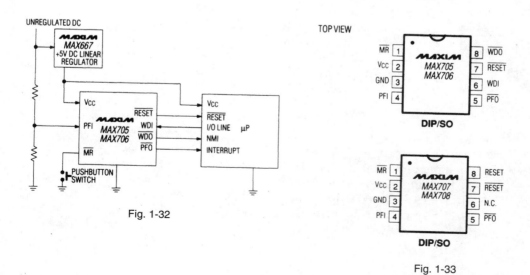

Fig. 1-32

Fig. 1-33

Digital/microprocessor supervisory circuits

Low-cost supervisory circuit (high reliability, manual reset)

Figures 1-32 and 1-33 show a typical application circuit and pin configurations, respectively, for the MAX705-708. The ICs are similar to that described for Figs. 1-J through 1-M, but with generally lower cost. The ICs have a 200-ms reset pulse width, 200-μA quiescent current, and a 1.25-V threshold detector for power-fail warning, low-battery detection, or for monitoring a power supply other than +5 V. The MAX705/MAX707 generate a reset pulse when the supply is below 4.65 V. The MAX706/708 generate a reset below 4.4 V. The MAX707/708 is the same as the MAX705/706, except an active-high reset is substituted for the watchdog timer. The watchdog is activated when the WDI input has not been toggled within 1.6 s. MAXIM HIGH-RELIABILITY DATA BOOK, 1993, P. 5-31.

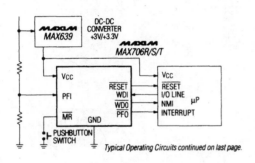

Fig. 1-34

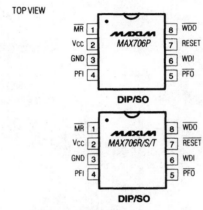

Fig. 1-35

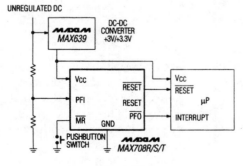

Fig. 1-36

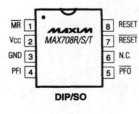

Fig. 1-37

Low-cost supervisory circuit with +3-V monitoring
Figures 1-34 and 1-35 show a typical application circuit and pin configurations, respectively, for the MAX706P/R/S/T. Figures 1-36 and 1-37 show a typical application circuit and pin configuration, respectively, for the MAX708R/S/T. The ICs are similar to that described for Figs. 1-J through 1-M, but with the capability of monitoring +3-V supply levels. The ICs have a 200-ms reset pulse width, 100-μA quiescent current, a watchdog with 1.6-s timeout, and a 1.25-V threshold detector for power-fail warning, low-battery detection, or for monitoring supply levels in the +3-V to +5-V range. The difference among the MAX706R,S, and T is the reset-threshold levels, which are 2.63 V, 2.93 V and 3.08 V, respectively. All have active-low reset output signals. The MAX706P reset output is active-high with a 2.63-V threshold. The MAX708R/S/T are identical to the corresponding MAX706, except that the MAX708 series does not have a watchdog, but provides both RESET and RESET outputs. The watchdog is activated when the WDI input has not been toggled within 1.6 s. All of the ICs have an active-low manual reset. MAXIM HIGH-RELIABILITY DATA BOOK, 1993, P. 5-33, 5-34.

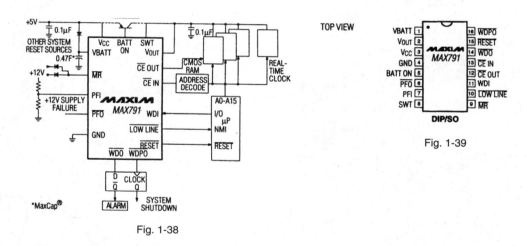

Fig. 1-38

Fig. 1-39

Microprocessor supervisory circuit with on-board chip-enable
Figures 1-38 and 1-39 show a typical application circuit and pin configuration, respectively, for the MAX791. The IC is similar to that described for Figs. 1-J through 1-M, but with on-board chip-enable. The IC has a 200-ms reset pulse width, 1-μA standby current, 250-mA output in V_{CC} mode, 25-mA output in battery-backup, 1.25-V threshold detector for power-fail warning, low-battery detection, or for monitoring a supply other than +5 V, a separate low-line comparator that compares V_{CC} to a threshold 150 mV above the reset threshold, and a pulsed watchdog output to give advance warning of impending WDO assertion caused by watchdog timeout. MAXIM HIGH-RELIABILITY DATA BOOK, 1993, P. 5-35.

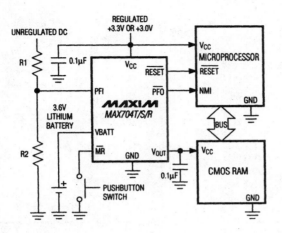

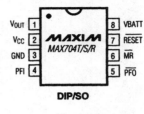

Fig. 1-41

Fig. 1-40

SUFFIX	TYPICAL RESET THRESHOLD (V)
T	3.06
S	2.91
R	2.61

Fig. 1-42

Low-cost supervisory circuit with battery backup (3.0 V/3.3 V)
Figures 1-40 and 1-41 show a typical application circuit and pin configuration, respectively, for the MAX704T/S/R. The ICs are similar to that described for Figs. 1-J through 1-M, but with generally lower cost, and for use with 3.0-V/3.3-V systems. The ICs have a 200-ms reset time delay, 50-µA quiescent current, 50-nA quiescent with battery backup, and a 1.25-V threshold detector for power-fail warning, low-battery detection, or for monitoring a supply other than 3.0 V or 3.3 V. The T, S, and R versions have different threshold levels, as shown in Fig. 1-42. MAXIM NEW RELEASES DATA BOOK, 1994, P. 5-43.

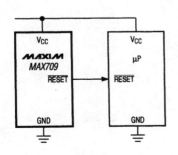

Fig. 1-43

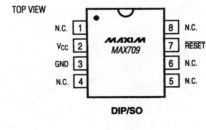

Fig. 1-44

Microprocessor-supervisory circuit titles and descriptions

PIN	NAME	FUNCTION
1, 4, 5, 6, 8	N.C.	No Connect. There is no internal connection to this pin.
2	V_{CC}	+5V, +3.3V, or +3V Supply Voltage
3	GND	Ground
7	RESET	Reset Output remains low while V_{CC} is below the reset threshold, and for 280ms after V_{CC} rises above the reset threshold.

Fig. 1-45

RESET THRESHOLD	
SUFFIX	VOLTAGE (V)
L	4.65
M	4.40
T	3.08
S	2.93
R	2.63

Fig. 1-46

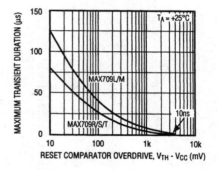

Fig. 1-47

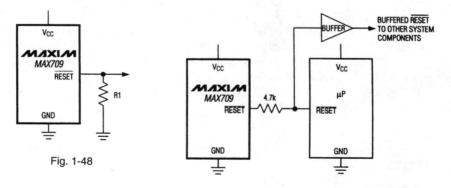

Fig. 1-48

Fig. 1-49

Power-supply monitor with reset

Figures 1-43, 1-44, and 1-45 show a typical application circuit, pin configuration, and pin descriptions, respectively, for the MAX709. The IC provides a system reset during power up, power down, and brownout conditions. When V_{CC} falls below the reset threshold (Fig. 1-46), $\overline{RESET}$ becomes low and holds the microprocessor for 140 ms (minimum) after V_{CC} rises above the threshold. The $\overline{RESET}$ output is guaranteed to be in the correct state with V_{CC} down to 1 V. The IC is relatively immune to short-duration negative-going V_{CC} transients (glitches), as shown in Fig. 1-47. Figure 1-48 shows the addition of R1 when the $\overline{RESET}$ output must be valid down to 0 V. The value of R1 should be about 100 kΩ, but it is not critical. Figure 1-49 shows the circuit for interfacing to microprocessors with bidirection reset pins (such as the Motorola 68HC11 series). MAXIM NEW RELEASES DATA BOOK, 1994, P. 5-67, 5-70, 5-71.

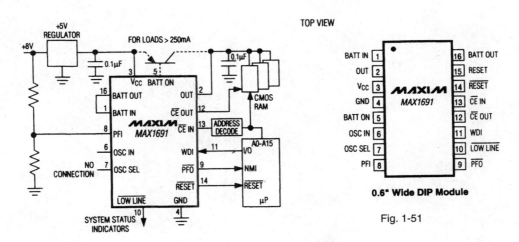

Fig. 1-50

Fig. 1-51

Microprocessor supervisor with internal backup battery

Figures 1-50 and 1-51 show a typical application circuit and pin configuration, respectively, for the MAX1691. The IC switches over to an internal backup battery to provide write protection and a watchdog function. The internal 3-V 125-mAh lithium battery connects to the supervisory circuit through external pin strapping (to minimize battery drain during shipping.) The IC is shipped in special nonconductive material. Storing the IC in conductive foam will discharge the internal battery. The power-OK/reset time delay is 200 ms, and standby current is 1 μA, with a typical 35-μA operating current. On-board gating of the chip-enable signals has a maximum delay of 10 ns. MAXIM NEW RELEASES DATA BOOK, 1995, P. 5-7.

=2=

Multiplexer/
switch circuits

This chapter is devoted to IC multiplexer (mux) and switch circuits. Although it appears that the only problem in selecting a suitable mux or switch IC is to choose one with the correct number of channels or switch positions, other problems affect circuit operation. We start with a summary of these problems.

Voltage extremes

Assume that a strain-gauge bridge-type transducer in an aircraft is connected to a multiplexed alarm system and a multiplexed computer input. The typical strain gauge is a resistive bridge with 10 V applied and produces 0 to 50 mV across the differential bridge arms, approximately 5 V above ground. The alarm circuit is fed by a mux that scans several such transducers, and the alarm actuates when the differential voltage of any channel exceeds the preset value.

The same differential voltage feeds an independent computer-driven mux that drives a cockpit display. Either mux must continue to operate if the other fails. Also, both mux ICs must block the effect of a failed transducer (where the transducer terminals could then source the full aircraft-battery voltage, 28 V). The alarm mux, computer mux, and transducers are interconnected by several connectors and many feet of unshielded wire, bundled in harnesses.

To further complicate the job of each mux, a pilot can remove or apply power to any transducer during flight. During maintenance the connecters might be disconnected and reconnected while power to the various modules is either on or off. Because mux inputs tie directly to the connector pins, the mux must cope with harsh voltage extremes. Here are some examples.

If one or more signal sources are energized while one mux is de-energized, that mux must not only avoid damage, but must also present high-impedance inputs that do not load the signal sources. The mux must maintain both forms of isolation, whether power is on or off, and during the transducer-failure modes that apply 28 V to the mux terminals.

When mux inputs are exposed to voltages that exceed the IC supplies (typically ±15 V), the mux must protect itself, as well as the op amp or other circuit connected to the output. There are two typical cases.

First, the mux inputs must not draw excessive current when voltage in the selected (on) channel exceeds the mux supply voltage. The mux must also limit V_{OUT} to a safe level under these conditions.

Second, when voltage in excess of the supply is applied to an off channel, the mux must limit its input current, while simultaneously blocking feedthrough to the output.

Mux protection can be included on-chip, or supplied externally. For example, the typical external protection scheme shown in Fig. 2-A provides a current-limiting resistor and two shunt diodes for each channel (one channel shown). When a positive voltage at any input attempts to exceed the 12-V zener voltage, the top diode turns on and clamps the mux input just above 12 V. A similar mechanism handles negative inputs below −12 V.

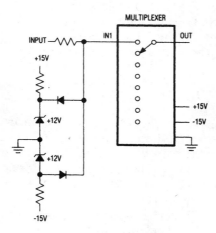

Fig. 2-A   External resistor/diode protection circuit.

The external protection of Fig. 2-A is good, but it can produce some design problems (such as high current during fault conditions, and diode-distributed leakage current in the signal paths during normal operation). If the Fig. 2-A circuit is used in the aircraft system just described, fault currents resulting from the turn-off of one mux could completely disrupt signals at the other mux.

Figure 2-B shows a variation of the external-protection circuit. This approach uses the built-in PN junctions of the CMOS mux IC in place of external diodes, and adds an external resistor to each channel. The resistor value must provide adequate protection, but too high a value allows unacceptable voltage errors caused

Fig. 2-B   External resistor protection.

by leakage current from the mux—especially at high temperatures (where leakage usually increases). Lower-value resistors minimize the error, but can allow higher fault currents, and might not provide adequate protection.

In general, external protection for IC mux and switch circuits is not recommended. The exception is where leakage is extremely low (in the nA and pA range, at most). A few application circuits using external protection are provided in the remainder of this chapter.

Remember the following points when using any form of external resistors with a mux or switch IC. The error voltage depends on total current through the resistor, including (for example) input bias current from the op amp following the mux. Also, the resistor combines with parasitic shunt capacitance to form a low-pass RC filter that might limit bandwidth. Of greater importance, the on-channel error, caused by fault current flowing in an off channel, could affect the design.

Multiplexer/switch IC testing/troubleshooting

The most practical in-circuit test of an IC mux/switch is to check the basic function. If the IC fails to perform this function, the IC is suspect. Of course, it is assumed that you will check all voltages and grounds to the IC (as covered in Chapter 1), and that all external components (and your wiring!) are good. After the basic function has been checked, test all of the applicable IC characteristics. The following paragraphs describe the most important characteristics for mux/switch ICs.

Basic function tests

Figures 2-C and 2-D show the typical application circuit and pin configuration/logic, respectively, for the MAX326/27. Test this ultra-low-leakage CMOS analog switch by applying and removing +5 V to the channel-control pins, and checking

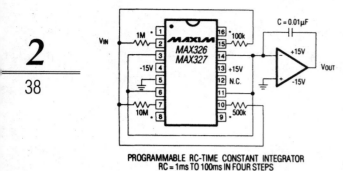

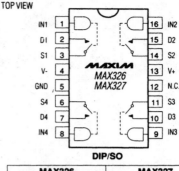

PROGRAMMABLE RC-TIME CONSTANT INTEGRATOR
RC = 1ms TO 100ms IN FOUR STEPS

*NOTE: PINS 1, 8, 9, AND 16 ARE CHANNEL-CONTROL PINS.

Fig. 2-C   MAX326/27 typical application circuit.

DIP/SO

MAX326		MAX327	
LOGIC	**SWITCH**	**LOGIC**	**SWITCH**
0	ON	1	ON
1	OFF	0	OFF

Fig. 2-D   MAX326/27 pin config-
uration/logic.

that the corresponding switches open and close. Use an ohmmeter connected
across the switch pins.

Off-channel leakage tests

Figure 2-E is a test circuit for measuring off-channel input voltage versus the resulting input-leakage current. Figure 2-F shows typical test results. As shown, the
input leakage during fault conditions is less than 20 nA. The voltmeter at the output shows the effect of off-channel leakage on the selected channel. The effect with
100-kΩ input resistors (less than 0.1 mV) is hardly noticeable.

V_{OUT} is limited by internal clamps to about 3 V less than the supply rails, and
ranges between ±12 V with ±15-V supplies. V_{OUT} collapses to 0 V when the power
is off.

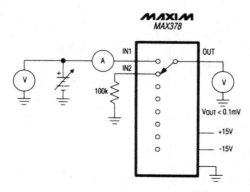

Fig. 2-E   Input-leakage test circuit.

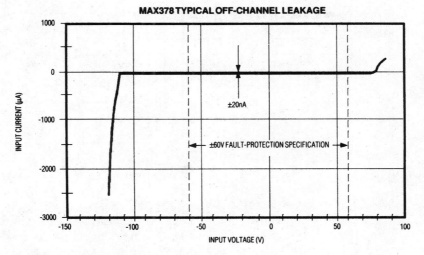

Fig. 2-F   Off-channel leakage.

Switching-time tests

Figure 2-G shows a test circuit and waveforms for measuring switching time for the MAX326/27. As a point of reference, t_{ON} should be between 500 and 1000 ns, with t_{OFF} between 50 and 500 ns. Figure 2-H shows how switching speeds are affected by supply voltages.

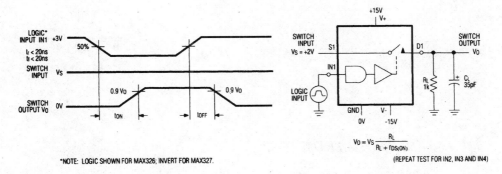

Fig. 2-G   Switching-time test circuit and waveforms.

POWER SUPPLY (V)	tON (μs)	tOFF (ns)
±15	0.5	50
±10	1	80
±5	2.5	200
+10	2.5	200
+15	1.5	100

Fig. 2-H Effect of supply voltages on switching speeds.

On-resistance tests

Figure 2-I shows a basic test circuit for measuring on-resistance, $r_{DS(ON)}$, for the MAX326/27. Figure 2-J shows how $r_{DS(ON)}$ changes with various analog inputs and supply combinations. Although specified at TTL threshold levels, the logic threshold is about 1.5 V ± 0.2 V. The IC switches properly with CMOS input levels from −15 V to +15 V. Never allow logic levels to exceed supply voltages in any circuit!

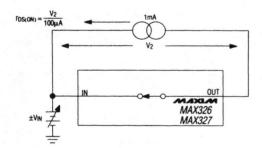

Fig. 2-I On-resistance vs. analog-signal level supply voltage.

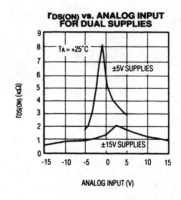

Fig. 2-J On-resistance test results.

On-leakage-current tests

Figure 2-K shows a basic test circuit for measuring on-leakage-current $I_{D(ON)}$ for the MAX326/27. Figure 2-L shows how $I_{D(ON)}$ changes with temperature.

Off-leakage-current tests

Figure 2-M shows a basic test circuit for measuring off-leakage-current $I_{S(OFF)}$ and $I_{D(OFF)}$ for the MAX326/27. Figure 2-N shows how $I_{S(OFF)}$ changes with temperature.

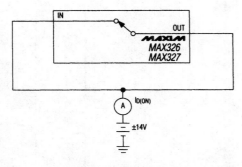

Fig. 2-K   On-leakage-current test circuit.

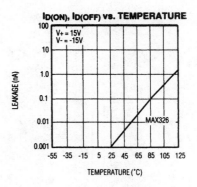

Fig. 2-L   On-leakage-current
test results.

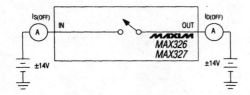

Fig. 2-M   Off-leakage-current test circuit.

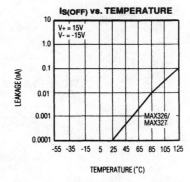

Fig. 2-N   Off-leakage-current
test results.

Basic mux function tests

Figures 2-O and 2-P shows the typical application circuit and pin configuration, respectively, for the MAX328/29. Figures 2-Q and 2-R shows the logic truth tables for the MAX328 and MAX329, respectively. This ultra-low-leakage CMOS analog mux is tested by applying and removing +5 V to the control pins, and checking that the corresponding switches open and close. Simple sine waves can be used at the analog inputs, and the op-amp output can be measured on a scope.

Access-time tests

Figure 2-S shows a test circuit for measuring access time versus logic level (high) for the MAX328/29. Notice that a pulse signal is required at the analog inputs.

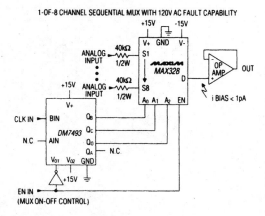

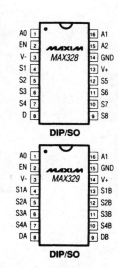

Fig. 2-O MAX 328/29 typical application circuit.

TOP VIEW

Fig. 2-P MAX328/329 pin configurations.

TRUTH TABLE - MAX328

A₂	A₁	A₀	EN	ON SWITCH
X	X	X	0	NONE
0	0	0	1	1
0	0	1	1	2
0	1	0	1	3
0	1	1	1	4
1	0	0	1	5
1	0	1	1	6
1	1	0	1	7
1	1	1	1	8

Fig. 2-Q MAX328 logic truth table.

TRUTH TABLE - MAX329

A₁	A₀	EN	ON SWITCH
X	X	0	NONE
0	0	1	1
0	1	1	2
1	0	1	3
1	1	1	4

Note: Logic "0" = $V_{AL} \leq 0.8V$, Logic "1" = $V_{AH} \geq 2.4V$

Fig. 2-R MAX329 logic truth table.

Multiplexer/switch circuits

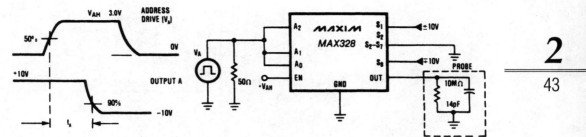

Fig. 2-S Access-time test circuit and waveforms.

Typical access time (or transition time) for the MAX328/39M is 1 μs maximum, with 1.5 μs maximum for the MAX328/29C/E.

Break-before-make delay

Figure 2-T shows a test circuit for measuring break-before-make delay for the MAX328/29. Again, a pulse is required at the analog inputs. Typical break-before-make delay is 0.2 μs.

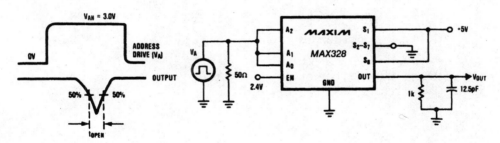

Fig. 2-T Break-before-make delay (t_{OPEN}).

Enable-delay tests

Figure 2-U shows a test circuit for measuring enable time for the MAX328/29. Typical enable turn-on time is 1 μs for the MAX328/29M and 1.5 μs for the MAX328/29C/E.

Charge-injection tests

Figure 2-V shows a test circuit for measuring charge injection for the MAX301/3/5. Notice that charge injection (Q) is measured in coulombs (C). A pulse is applied to the control pin and the difference in output voltage is noted. Q is calcu-

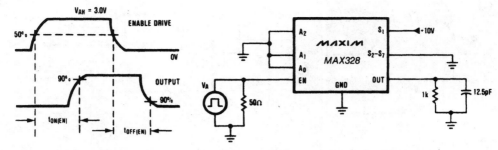

Fig. 2-U   Enable delay ($t_{ON(EN)}$, $t_{OFF(EN)}$).

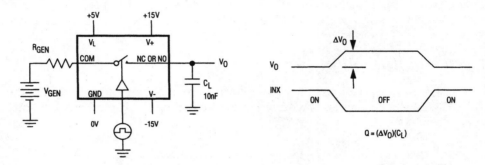

Fig. 2-V   Charge-injection test circuit.

lated when the difference in output voltage is divided by the capacitance at the output. Typical charge injection is 10 to 15 pC.

Off-isolation tests

Figure 2-W shows a test circuit for measuring the isolation between channels (when the channel switch is open) for the MAX301/3/5. Typical isolation is 72 dB.

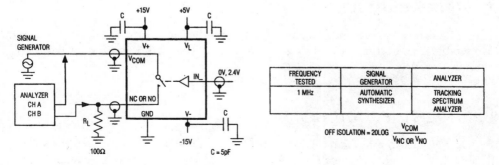

FREQUENCY TESTED	SIGNAL GENERATOR	ANALYZER
1 MHz	AUTOMATIC SYNTHESIZER	TRACKING SPECTRUM ANALYZER

$$\text{OFF ISOLATION} = 20 \text{LOG} \frac{V_{COM}}{V_{NC} \text{ OR } V_{NO}}$$

Fig. 2-W   Off-isolation.

Crosstalk tests

Figure 2-X shows a test circuit for measuring the isolation between channels (or the lack of crosstalk when the channel switches are closed) for the MAX301/3/5. Typical crosstalk isolation is 90 dB.

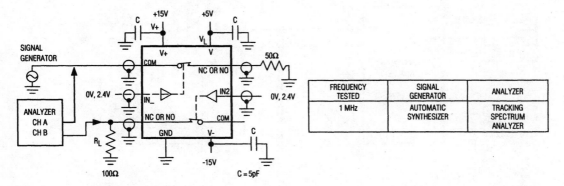

FREQUENCY TESTED	SIGNAL GENERATOR	ANALYZER
1 MHz	AUTOMATIC SYNTHESIZER	TRACKING SPECTRUM ANALYZER

Fig. 2-X Crosstalk test circuit.

Channel capacitance

Figures 2-Y and 2-Z show test circuits for measuring channel capacitance with the channel switch open and closed. Typical capacitance is 12 pF with the channel open and 39 pF when the channel is closed.

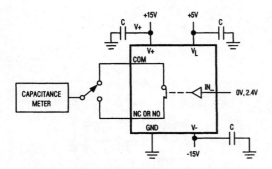

Fig. 2-Y Channel-off capacitance.

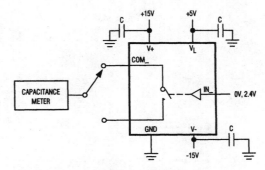

Fig. 2-Z Channel-on capacitance.

Multiplexer/switch circuit titles and descriptions

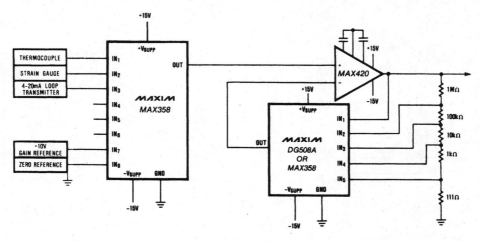

Fig. 2-1

Data acquisition front end

Figure 2-1 shows a typical data-acquisition system using a MAX358 multiplexer. Because the mux is driving a high-impedance input, the error is a function of IC resistance $r_{DS(ON)}$ times the mux leakage current $I_{D(ON)}$, and the amplifier bias current I_{BIAS}, or:

$$V_{ERROR} = r_{DS(ON)} \times I_{D(ON)} + I_{BIAS} \text{ (MAX420)}$$
$$= 1.5 \text{ k}\Omega \times (2 \text{ nA} + 30 \text{ pA})$$
$$= 3.05\text{-}\mu\text{V maximum error}$$

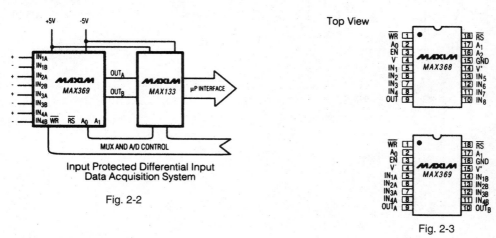

**Input Protected Differential Input
Data Acquisition System**

Fig. 2-2

Fig. 2-3

TRUTH TABLE — MAX368

A₂	A₁	A₀	EN	$\overline{WR}$	$\overline{RS}$	ON SWITCH
Latching						
X	X	X	X	⌐	1	Maintains previous switch condition
Reset						
X	X	X	X	X	0	NONE (latches cleared)
Transparent Operation						
X	X	X	0	0	1	NONE
0	0	0	1	0	1	1
0	0	1	1	0	1	2
0	1	0	1	0	1	3
0	1	1	1	0	1	4
1	0	0	1	0	1	5
1	0	1	1	0	1	6
1	1	0	1	0	1	7
1	1	1	1	0	1	8

Fig. 2-4

TRUTH TABLE — MAX369

A₁	A₀	EN	$\overline{WR}$	$\overline{RS}$	ON SWITCH
Latching					
X	X	X	⌐	1	Maintains previous switch condition
Reset					
X	X	X	X	0	NONE (latches cleared)
Transparent Operation					
X	X	0	0	1	NONE
0	0	1	0	1	1
0	1	1	0	1	2
1	0	1	0	1	3
1	1	1	0	1	4

Fig. 2-5

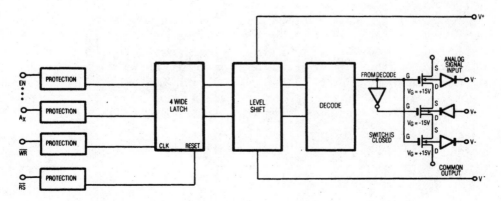

Fig. 2-6

Multiplexer/switch circuit titles and descriptions

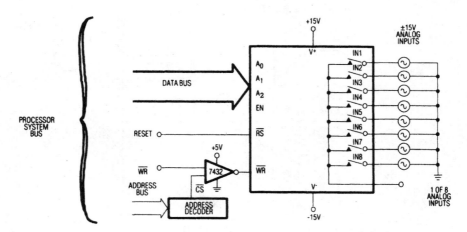

Fig. 2-7

Frequency	100kHz	500kHz	1MHz
One Channel Driven	74dB	72dB	66dB
All Channels Driven	64dB	48dB	44dB

Fig. 2-8

Frequency	100kHz	500kHz	1MHz
One Channel Driven	74dB	72dB	66dB
All Channels Driven	64dB	48dB	44dB

Fig. 2-9

Input-protected differential-input data-acquisition system
Figures 2-2 and 2-3 show a typical application circuit and pin configurations, respectively, for the MAX368/69. Figures 2-4 and 2-5 show the logic truth tables, respectively, for the MAX368 and MAX369. As shown in Fig. 2-6, these fault-protected analog mux ICs have a latch function so that both write $\overline{WR}$ and reset $\overline{RS}$ signals are required. Figure 2-7 shows the IC connected with a bus interface. Figure 2-8 shows the typical off-isolation rejection ratio. Figure 2-9 shows typical crosstalk-rejection ratio. MAXIM NEW RELEASES DATA BOOK, 1992, P. 1-31, 1-37, 1-40, 1-44, 1-46, 1-47.

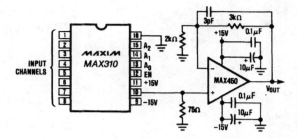

No Insertion Loss, 8 Channel Mux

Fig. 2-10

CMOS RF/video multiplexer
Figure 2-10 shows a typical application circuit and pin configurations for the MAX310/11. The key feature of the IC is extremely high off-isolation at high frequencies. The isolation of each off channel to the output is guaranteed to be −66 dB at 5 MHz. The input signal range is +12 V to −15 V with ±15-V supplies. Power consumption is typically 1.1 mW. All control inputs are fully compatible with TTL and CMOS. Decoding is in standard BCD. An enable input is provided to simplify cascading of the ICs. The ICs will operate with power supply combinations that total less than 36 V ($V+ - V-$), including single-supply operation at +12 V, +15-V, and +28 V with $V-$ connected to ground. MAXIM HIGH-RELIABILITY DATA BOOK, 1993, P. 1-1.

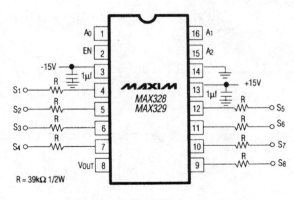

Fig. 2-11

Multiplexer/switch circuit titles and descriptions

Fault-tolerant mux

Figure 2-11 shows how the MAX328/29 (Figs. 2-O, 2-P) can be converted to a fault-tolerant mux. The internal diodes limit the voltage at the input to ±15.7 V (±15-V supplies). No external diodes are required. The resistors limit power dissipation to 0.28 W when a 120-Vac fault occurs. MAXIM HIGH-RELIABILITY DATA BOOK, 1993, P. 1-6.

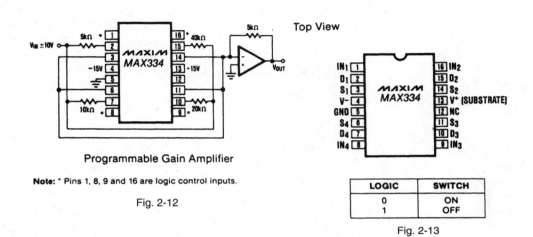

Programmable Gain Amplifier

Note: * Pins 1, 8, 9 and 16 are logic control inputs.

Fig. 2-12

Top View

LOGIC	SWITCH
0	ON
1	OFF

Fig. 2-13

Programmable-gain amplifier

Figures 2-12 and 2-13 show a typical application circuit and pin configuration/ logic, respectively, for the MAX334, which is a direct replacement for the Siliconix DG271 and HI-201HS. The IC has guaranteed break-before-make switching, t_{ON} of 100 ns, t_{OFF} of 50 ns, and channel-on resistance of 50 Ω (max). Split supplies (±5 V to ±15 V) or single positive supplies (+5 V to +30 V) can be used without affecting switching speed or the CMOS/TTL logic-compatible inputs. MAXIM HIGH-RELIABILITY DATA BOOK, 1993, P. 1-17.

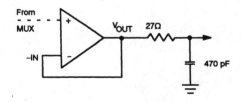

Fig. 2-14

Isolating capacitive loads
Figure 2-14 shows a circuit used to isolate a mux output that feeds into a large capacitive load. MAXIM HIGH-RELIABILITY DATA BOOK, 1993, P. 1-46.

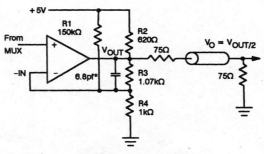

*Chosen to compensate for stray input capacitance.

Fig. 2-15

Minimizing phase distortion
Figure 2-15 shows a circuit used to minimize phase distortion at a mux output. MAXIM HIGH-RELIABILITY DATA BOOK, 1993, P. 1-46.

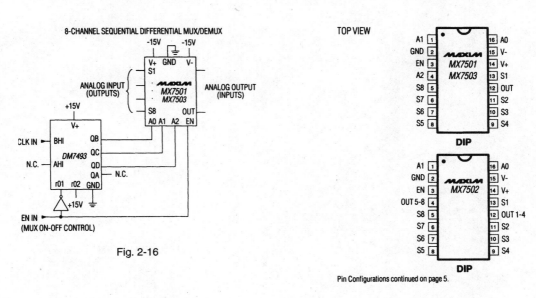

Fig. 2-16

Pin Configurations continued on page 5.

Fig. 2-17

Multiplexer/switch circuit titles and descriptions

8-channel sequential differential mux/demux
Figures 2-16 and 2-17 show a typical application circuit and pin configurations, respectively, for the MAX7501/02/03. In this circuit, the IC acts as a multiplexer or demultiplexer by inter-changing the analog inputs and outputs. Each channel is sampled in sequence by signals from a DM7493 under the control of an external clock. MAXIM HIGH-RELIABILITY DATA BOOK, 1993, P. 1-47.

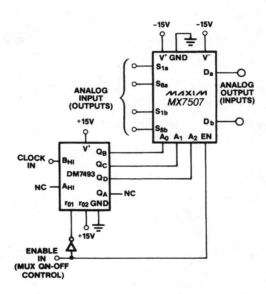

8 Channel Sequential Differential MUX/DEMUX

Fig. 2-18

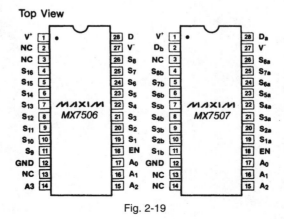

Fig. 2-19

16-channel sequential differential mux/demux

Figures 2-18 and 2-19 show a typical application circuit and pin configurations, respectively, for the MAX7506/07. This circuit is similar to that shown in Fig. 2-17, but with 16-channel operation. MAXIM HIGH-RELIABILITY DATA BOOK, 1993, P. 1-49.

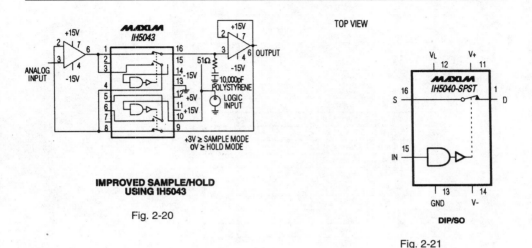

IMPROVED SAMPLE/HOLD USING IH5043

Fig. 2-20

Fig. 2-21

Sample-hold circuit

Figures 2-20 and 2-21 show a typical application circuit and pin configuration/ switching-state, respectively, for the IH5040-IH5045/47. These general-purpose CMOS analog switches are latch-up proof, with 1-nA leakage current and less than 1-μA quiescent current. MAXIM HIGH-RELIABILITY DATA BOOK, 1993, P. 1-153.

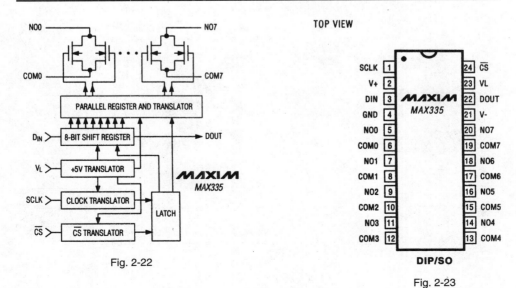

Fig. 2-22

DIP/SO

Fig. 2-23

Multiplexer/switch circuit titles and descriptions

Continued

PIN	NAME	FUNCTION
1	SCLK	Serial clock input
2	V+	Positive supply voltage
3	DIN	Serial data input
4	GND	Ground
5	NOØ	Switch 0
6	COMØ	Switch 0
7	NO1	Switch 1
8	COM1	Switch 1
9	NO2	Switch 2
10	COM2	Switch 2
11	NO3	Switch 3
12	COM3	Switch 3
13	COM4	Switch 4
14	NO4	Switch 4
15	COM5	Switch 5
16	NO5	Switch 5
17	COM6	Switch 6
18	NO6	Switch 6
19	COM7	Switch 7
20	NO7	Switch 7
21	V-	Negative supply voltage
22	DOUT	Serial data output
23	V_L	Logic supply/Reset
24	$\overline{CS}$	Chip select

Fig. 2-24

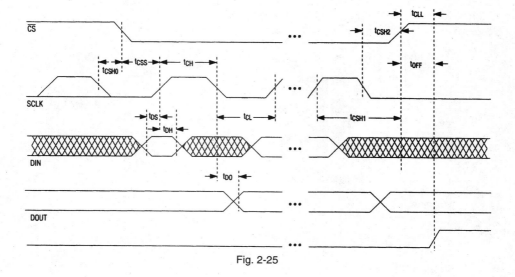

Fig. 2-25

Multiplexer/switch circuits

Continued

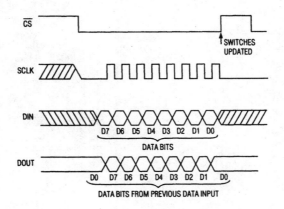

Fig. 2-26

DATA BITS								FUNCTION
D7	D6	D5	D4	D3	D2	D1	D0	
0	X	X	X	X	X	X	X	Switch 7 open (off)
1	X	X	X	X	X	X	X	Switch 7 closed (on)
X	0	X	X	X	X	X	X	Switch 6 open
X	1	X	X	X	X	X	X	Switch 6 closed
X	X	0	X	X	X	X	X	Switch 5 open
X	X	1	X	X	X	X	X	Switch 5 closed
X	X	X	0	X	X	X	X	Switch 4 open
X	X	X	1	X	X	X	X	Switch 4 closed
X	X	X	X	0	X	X	X	Switch 3 open
X	X	X	X	1	X	X	X	Switch 3 closed
X	X	X	X	X	0	X	X	Switch 2 open
X	X	X	X	X	1	X	X	Switch 2 closed
X	X	X	X	X	X	0	X	Switch 1 open
X	X	X	X	X	X	1	X	Switch 1 closed
X	X	X	X	X	X	X	0	Switch 0 open
X	X	X	X	X	X	X	1	Switch 0 closed

X = Don't care

Fig. 2-27

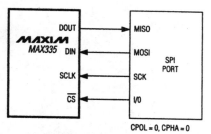

CPOL = 0, CPHA = 0

THE DOUT-MISO CONNECTION IS NOT REQUIRED FOR WRITING TO THE
MAX335, BUT MAY BE USED FOR DATA-ECHO PURPOSES.

Fig. 2-28

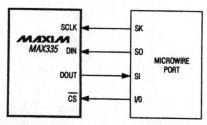

THE DOUT-SI CONNECTION IS NOT REQUIRED FOR WRITING TO THE
MAX335, BUT MAY BE USED FOR DATA-ECHO PURPOSES.

Fig. 2-29

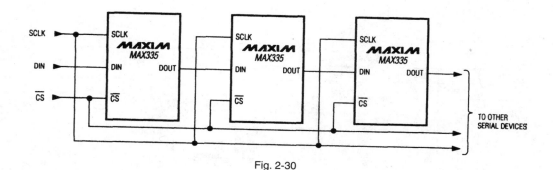

Fig. 2-30

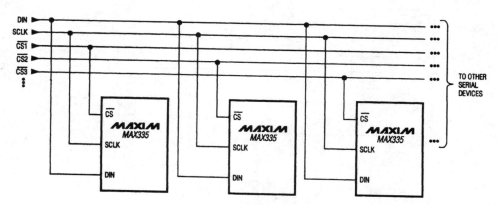

Fig. 2-31

Serial-controlled 8-channel SPST switch (MAX335)
Figures 2-22, 2-23, and 2-24 show the functional diagram, pin configuration, and pin descriptions, respec-tively, for the MAX335. Figures 2-25 and 2-26 show the timing, and three-wire interface timing, respectively. Figure 2-27 shows the serial-interface switch programming. This IC has eight separately controlled single-pole-single-throw (SPST) switches, all of which conduct equally in either direction, with an on-resistance (100 Ω) that is constant over the analog signal range. The switches can continuously operate with supplies from ±4.5 V to ±20 V, and handle rail-to-rail analog signals. Upon power-up, all switches are off, and the internal serial and parallel shift registers are reset to zero. The IC is equivalent to two DG211 quad switches, but controlled by a serial interface. The interface is compatible with both Motorola SPI (Fig. 2-28) and Microwire (Fig. 2-29). Functioning as a shift register, this serial interface allows data (at DIN) to be clocked in sync with the rising edge of clock (SCLK). The shift register output (D_{OUT}) enables several MAX335 ICs to be daisy-chained (Fig. 2-30), or the ICs can be connected as an addressable serial interface (Fig. 2-31). Digital-feedthrough transients are typically 10 mVp-p when a 100-pF capacitance is used at the switch channels. MAXIM NEW RELEASES DATA BOOK, 1995, P. 1-59, 1-64, 1-65, 1-66, 1-67, AND 1-68.

8-×-1 multiplexer (MAX335)
To use the MAX335 as an 8-×-1 mux, tie all drains together (COM0 to COM7). The mux inputs then source each switch (N00 to N07). Input a single 0-V to +3-V pulse at DIN. As the input is clocked through the register by SCLK, each switch will sequence-on (one at a time).

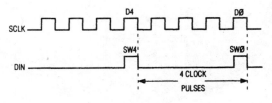

Fig. 2-32

4-2 differential mux (MAX335)
To use the MAX335 as a 4-×-2 differential mux, tie COM0 through COM3 together and COM4 through COM7 together. Differential inputs will be at the source inputs as follows: (N00, N04), (N01, N05), (N02, N06), and (N03, N07). Figure 2-32 shows the serial input control at DIN required to turn on two switches to form a differential mux. $\overline{CS}$ is held low for four clock pulses; the first pulse is clocked into

the fifth switch position as the second pulse is clocked into the first switch position. $\overline{CS}$ is pulled high to update the switches; then CS is pulled low, and SCLK advances pulses to S1 and S5 positions, where $\overline{CS}$ is pulled high to update, etc.

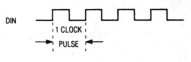

Fig. 2-33

SPDT switches (MAX335)

To use the MAX as a single-pole double-throw (SPDT) switch, tie COM0 to N01 so that N00 and COM1 are now inputs and COM0/N01 is the common output. Up to four SPDT switches can be made from each MAX335. Multiples of four or more can be made by daisy-chaining the ICs. Figure 2-33 shows the serial input control for SPDT switch operation. Again, $\overline{CS}$ is held low to clock in pulses, and pulled high to update; $\overline{CS}$ is also held low to shift pulses, then pulled high to update, etc.

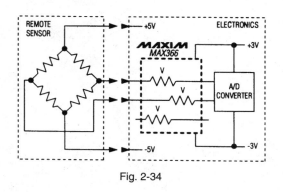

Fig. 2-34

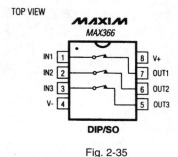

Fig. 2-35

Continued

TOP VIEW

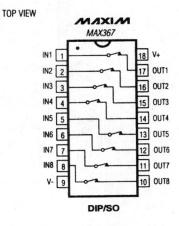

Fig. 2-36

Signal-line circuit protectors (fault-protected switches)
Figures 2-34 and 2-35 show a typical application circuit and pin configuration, respectively, for the MAX366. Pin configurations for the MAX337 are shown in Fig. 2-36. These ICs are multiple, two-terminal circuit protectors. Placed in series with signal lines, each two-terminal device protects sensitive circuit components from damaging voltage near and beyond the normal supply range. The ICs are used at interfaces where sensitive circuits connect to the external world, and potentially damaging voltage (up to ±35 V beyond the supply rails) might be encountered during power-up, power-down, or fault conditions. The ICs can be used to protect either analog or digital signals using unipolar (5 V to 44 V) or bipolar (±5 V to ±22 V) supplies. The ICs are essentially fault-protected switches that are always on when power is applied. On resistance is 100 Ω (max.), and leakage is less than 1 nA at +25°C. When signal voltages exceed, or are within 1.5 V of the supply voltages, or when power is off, the two-terminal resistance increases and becomes a virtual open circuit. This ensures low current during fault conditions. The protected side of the switch maintains the correct polarity and clamps about 1.5 V below the supply rail. There are no glitches or polarity reversals going into or coming out of a fault condition. ESD (electrostatic discharge) protection is greater than 2 kV. MAXIM NEW RELEASES DATA BOOK, 1995, P. 1-109, 1-110.

=3=

Interface circuits

This chapter is devoted to IC interface circuits (drivers/receivers and transceivers) used in connecting digital devices to the outside world. Although it appears that the only problem in selecting a suitable interface IC is to choose one with the correct number of channels, other problems affect circuit operation. The following section starts with a summary of these problems.

Cable length

The old RS-232C interface specification calls for a maximum cable length of 50 feet. The newer EIA-232D specification calls for a length of cable that produces a capacitance of 2500 pF. This change from feet of cable to pF of load capacitance recognizes one of the problems in RS-232C. Not only does RS-232C ignore the effects of cable capacitance, but it contains a 50-ft distance limit that is often ignored by designers.

The EIA-232D specification considers cable length indirectly, through the effect of load capacitance, but does not specify a maximum length. Because the capacitance of inexpensive cable can range from 12 pF/ft for a single twisted pair to 30 pF/ft for low-noise shielded, multiple-twisted-pair cable, there is some confusion.

Cable capacitance is important because EIA-232D transmission are ac signals. Higher capacitance demands higher peak currents from the transmitter, resulting in higher average supply current for a given data rate, as shown in Fig. 3-A. Also, the cable impedance forms an ac divider with the transmitter output impedance. Higher cable capacitance lowers the divider shunt component. In turn, this reduces signal amplitude at the receiver end of the cable, as shown in Fig. 3-B. This signal loss becomes a problem when the receiver-end voltage falls below the specified 5-V minimum input level (required by most digital equipment).

Finally, cable capacitance limits the slew rate available from a given transmitter, as shown in Fig. 3-C. The slew rate determines the transition time between the +3-V and −3-V signal levels. (Slew rate is discussed further in Chapter 6.) Transition time limits the maximum data rate according to the specifications in RS-232C, EIA-232D, and CCITT V.28, as shown in Fig. 3-D.

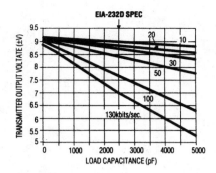

Fig. 3-A MAX220 supply current vs. output load capacitance (2 transmitters driven).

Fig. 3-B MAX220 output voltage vs. output load capacitance (two transmitters driven).

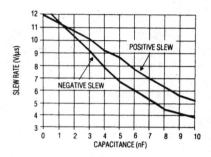

Fig. 3-C Slew-rate vs. output capacitance (MAX220/232A/233A/242/243).

	RS-232C	**EIA-232D**	**CCITT V.28**
Slew Rate, Maximum	30 V/μs	30 V/μs	30 V/μs
Transition Time, t$_T$, Maximum			
<40 bits/sec.	1ms	1ms	—
<30bits/sec. (V.28 only)	—	—	1ms
40bits/sec. < t$_T$ < 8kbits/sec.	4% of a unit interval	4% of a unit interval	3% of a unit interval
> 8kbits/sec.	4% of a unit inverval	5μs	3% of a unit interval

Fig. 3-D Comparison of transition times.

The simplest way to calculate the maximum allowable cable length for an EIA-232D interface IC (transceiver) is to divide the 2500-pF load specification by the capacitance per foot specified for the cable. Then, for cable lengths up to that limit, you can rest assured that the transceiver will operate properly and provide the maximum specified data rate.

Figure 3-E shows how the data rate can be extended past the specification limits. Longer cable adds to the line capacitance, lengthening the signal transition time, thus limiting the data rate. However, you can implement cable lengths and data rates well beyond the limits implied by the EIA-232D specification.

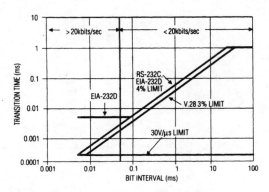

Fig. 3-E   Maximum transition time vs. bit interval.

Low power consumption and shutdown

Many types of portable equipment communicate over EIA-232D lines—from barcode readers to underwater data loggers. Minimum power consumption is a common goal in the design of such equipment because most portable devices are battery operated. Low power consumption in EIA-232D transceivers is important because dissipation in these circuits becomes a growing percentage of the total as you reduce the operating current in a CMOS system. (CMOS is used in most portable equipment because of the low power consumption compared to bipolars.)

The obvious solution for low power consumption is to shut down all or part of the circuits whenever possible. But it is not that simple. In some cases, the EIA-232D transceiver link must be active continuously. For such continuous operation, it is best to choose a transceiver with the lowest possible quiescent current. In those cases where operation is intermittent and the entire circuit can be shutdown, choose a transceiver with the lowest shutdown-mode current.

Figure 3-F shows a comparison of power consumption among several Maxim interface devices. Although the MAX222 and MAX242 draw the same current as the MAX232A/233A/243 (13.4 mA for the conditions listed) for continuous operation, the MAX222/242 consumes less average current if shut down for any duty cycle (as shown by the slanting line). On the other hand, the MAX 220 (which cannot be shut down) has the advantage over the MAX222/242 when data must be transmitted for more than two-thirds (67%) of the time (as shown by the dashed vertical line). Notice that this comparison is restricted to data rates below the MAX220 limit of 40 kbits/second. To sum up, the best choice for power consump-

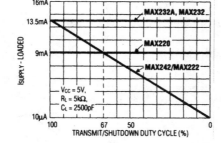

Fig. 3-F Comparison of power consumption.

tion in EIA-232D devices depends on the application and the ratio of transmitting time to shutdown time. For reference, the MAX220 has a 0.5-mA quiescent current, and the MAX222/242 draws 1 μA in the shutdown mode.

Line isolation

In telecom systems, the service provider incorporates line isolation to protect the network against unorthodox connections made by users. At the other end of the line, EIA-232D connections used in medical patient-monitoring equipment, data loggers, and supervisory computers (to name a few devices) must be isolated to protect the equipment from hardware failures. Isolation not only provides safety, but it can improve system performance.

As an example, EIA-232D links between a computer in one building and terminals in another building can show ground-current noise if the building earth-ground connections are at different potentials (as is usually the case). Isolation with a rating of 100 V can solve this problem. In severe industrial environments, a full UL-listed isolation barrier of 1500 V (or more) might be required.

The transmission of digital data, while maintaining an electrical barrier, usually involves transformers and opto-isolators, as shown in 6a of Fig. 3-G. The transformer supplies power to the other side of the barrier, and the opto-couplers handle data transmission across the barrier. However, there is a problem with this approach. The LEDs in opto-isolators (especially high-speed opto-isolators) require more current than normal logic circuits can provide. As a result, you must connect the outputs in parallel or add buffer ICs to get adequate drive current. Also, the isolated power supply must be fairly large because standard EIA-232D chips require a supply voltage of ±12 V, as well as +5 V.

One approach to solving these power and data-transfer problems is to use matching interface chip sets, such as the MAX250/MAX251, shown in 6a of Fig. 3-G. These two ICs include circuitry for two EIA-232D transmitters and two receivers, circuitry for generating isolated power-supply voltage from the main (non-isolated) +5-V supply, and interface circuitry for driving and receiving sig-

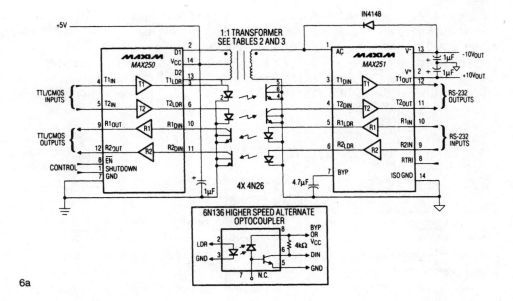

6a

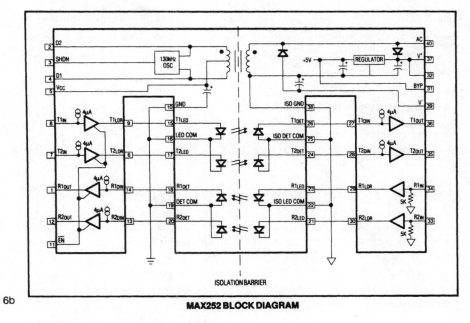

6b

MAX252 BLOCK DIAGRAM

Fig. 3-G   Isolated EIA-232D interface.

nals from the external opto-isolators. You only need to supply the isolators, a 1:1 transformer, and a few passive components to complete an isolated dual-transceiver EIA-232D port. (Such circuits are covered in this chapter.)

The alternate approach is to use a single-chip interface, such as the MAX252 shown in 6b of Fig. 3-G. This IC includes all of the circuitry in a standard 40-pin DIP, but provides the full UL-recognized 1500-V isolation barrier. (This approach is also covered in this chapter.)

Comparison of standards

Figure 3-H shows a comparison of common interface standards.

	EIA-232D	EIA-423A	EIA-422A	RS-485
Mode of Operation	Single Ended	Single Ended	Differential	Differential
Allowed # of Tx and Rx per Data Line	1 Tx, 1 Rx	1 Tx, 1 Rx	1 Tx, 10 Rx	32 Tx, 32Rx
Cable Length, Maximum	Load-Dependent	4kft	4kft	4kft
Data Rate, Maximum	20kbits/sec.	100kbits/sec.	10Mbits/sec.	10Mbits/sec.
Driver Output Range, Loaded (0V Offset):				
Minimum	±5V	±3.6V	±2V	±1.5V
Maximum	±15V	±5.4V	±5V	±5V
Driver Short-Circuit Current, Maximum	500mA	150mA	150mA	250mA
TX Load Impedance	3kΩ to 7kΩ	450Ω	100Ω	54Ω
Instantaneous Slew Rate	< 30V/µs	—	—	—
Rx Input Sensitivity	±3V	±200mV	±200mV	±200mV
Rx Input Resistance, Minimum	3kΩ to 7kΩ	4kΩ	4kΩ	12kΩ
Rx Input Range	±25V	±12V	±7V	-7V to +12V

Fig. 3-H Comparison of interface standards.

Interface IC testing/troubleshooting

The most practical in-circuit test of an interface IC is to check the basic function. If the IC fails to perform this function, the IC is suspect. Of course, it is assumed that you will check all voltages and grounds to the IC (as covered in Chapter 1), and that all external components (and your wiring!) are good. After the basic function has been checked, test all of the applicable IC characteristics. The following paragraphs describe the most important characteristics for interface ICs.

Basic function test

Figure 3-I shows the typical application circuit and pin configuration for the MAX220. This EIA-232D interface (driver/receiver) operates from 5 V, and is also

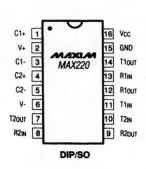

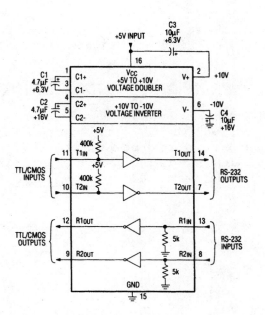

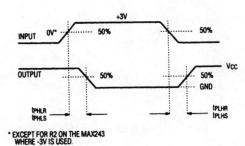

Fig. 3-I   MAX220 typical application circuit and pin configuration.

suited for use in systems designed to V.28/V.24 specifications. The basic function is tested by comparing waveforms at the input and output of all four channels. In theory, the waveforms should be identical, except for possible delay. Figures 3-J and 3-K show typical delay timing for both the transmitter and receiver. For reference, the MAX220 receiver delay is 0.6 μs (typical) and 3 μs (maximum). The transmitter delay is 4 μs (typical) and 10 μs (maximum).

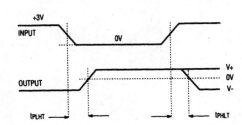

Fig. 3-J typical delay timing for transmitter.

Fig. 3-K   Typical delay timing for receiver.

If one or more (but not all) of the four channels shows no output, severe distortion, or excessive delay, the IC is suspect. If all four channels are defective, and +5 V is at pin 16, check the four capacitors (C1 through C4). These capacitors are used in the voltage-doubling (charge pump) and voltage-inverting functions within the IC. If the capacitors are good, and +5-V power is applied, but any or all of the channels are defective, suspect the IC.

Enable and disable timing tests

Figure 3-L shows the typical application circuit and pin configuration for the MAX242. This IC is similar to the MAX220, except that the receiver channels can be enabled and disabled, and all four channels can be shut down by external signals. Figures 3-M and 3-N show test circuits and waveforms for receiver-channel enable/disable, and transmitter-channel disable timing.

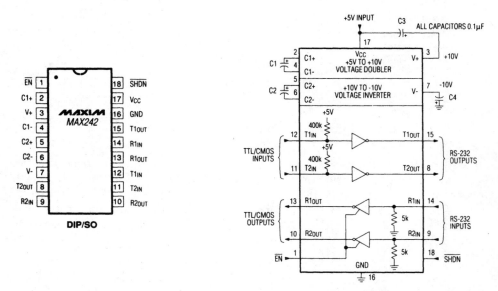

Fig. 3-L   MAX242 typical application circuit and pin configuration.

Shutdown timing tests

Figure 3-O shows a shutdown test circuit for the MAX230/35/36. In this case, the shutdown current measured at the V_{CC} terminal is being measured. For reference, the shutdown current is 1 μA (typical) and 10 μA (maximum) for the ICs involved.

Transition slew-rate tests

Figure 3-P shows a transition slew-rate test circuit for the MAX200/11/13. (Slew rate is covered in Chapter 6.) For reference, the slew rate is 5.5 V/μs (typical) and 30 V/μs (maximum) for the ICs involved.

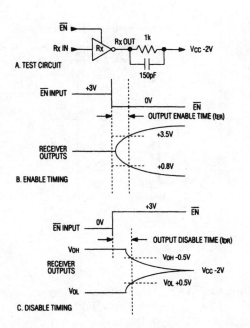

A. TEST CIRCUIT

B. ENABLE TIMING

C. DISABLE TIMING

Fig. 3-M Enable and disable timing for receiver output.

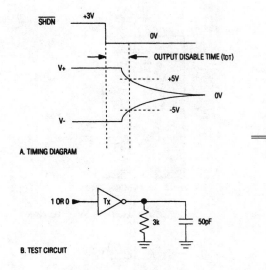

A. TIMING DIAGRAM

B. TEST CIRCUIT

Fig. 3-N Disable timing for trans-mitter output.

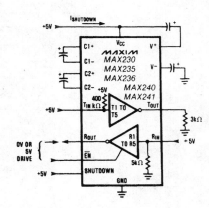

Fig. 3-O Shutdown current test circuit.

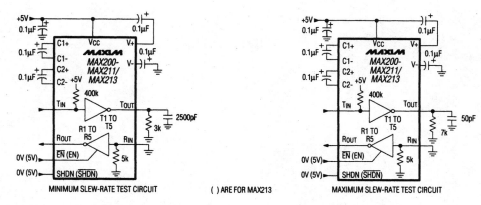

Fig. 3-P Transition slew-rate test circuit.

Loopback tests

Figure 3-Q shows a loopback test circuit for the MAX3241. Figures 3-R and 3-S show the loopback test results at 120 kbps and 240 kbps, respectively. Notice that the input is applied at the transmitter input, and that both the transmitter output (connected to receiver input) and the receiver output waveforms are monitored.

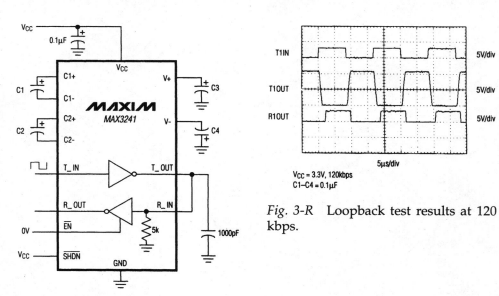

Fig. 3-Q Loopback test circuit.

Fig. 3-R Loopback test results at 120 kbps.

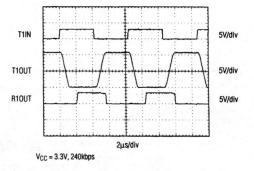

T1IN   5V/div

T1OUT   5V/div

R1OUT   5V/div

2μs/div

V_{CC} = 3.3V, 240kbps

Fig. 3-S   Loopback test results at 240 kbps.

Interface circuit titles and descriptions

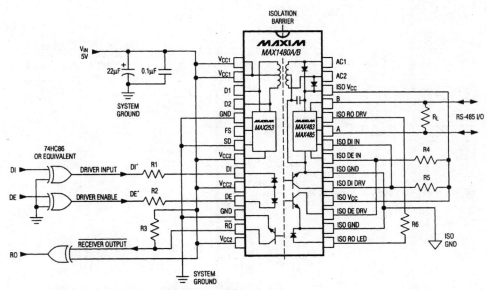

Table 1. Pull-Up and LED Drive Resistors

Part Number	R1 (Ω)	R2 (Ω)	R3 (Ω)	R4 (Ω)	R5 (Ω)	R6 (Ω)
MAX1480A	200	200	360	3k	360	200
MAX1480B	200	510	3k	2.2k	3k	200

Fig. 3-1

PIN	NAME	FUNCTION
1, 2	V$_{CC1}$	Logic-Side (non-isolated side) +5V Supply Voltage. Internally connected. Tie to V$_{CC2}$ for normal operation.
8, 10, 14	V$_{CC2}$	Logic-Side (non-isolated side) +5V Supply Voltages. Must be connected together, not internally connected. Tie to V$_{CC1}$ for normal operation.
3, 4	D1, D2	Internal Connections. Leave these pins unconnected.
5, 12	GND	Logic-Side Grounds. Must be tied together; not internally connected.
6	FS	Frequency Switch. If FS = V$_{CC}$ or open, switch frequency = 350kHz; if FS = 0V, switch frequency = 200kHz.
7	SD	Shutdown. Ground for normal operation. When high, the power oscillator is disabled.
9	DI	Driver Input. With DE' high, a low on DI' forces output A low and output B high. Similarly, a high on DI' forces output A high and output B low. Drives internal LED cathode through a resistor. (See Table 1 of Figure 2.)
11	DE	Driver Enable. The driver outputs, A and B, are enabled by bringing DE' high. The driver outputs are high impedance when DE' is low. If the driver outputs are enabled, the parts function as line drivers. While the driver outputs are high impedance, the chips can function as line receivers. Drives internal LED cathode through a resistor. (See Table 1 of Figure 2.)
13	$\overline{RO}$	Receiver Output. If A > B by 200mV, $\overline{RO}$ will be low; if A < B by 200mV, $\overline{RO}$ will be high. Open collector; must have pull-up to V$_{CC}$.
15	ISO RO LED	Isolated Receiver Output LED. If A > B by 20mV, ISO RO LED will be high; if A < B by 200mV, ISO RO LED will be low. Drives internal LED anode through a resistor. (See Table 1 of Figure 2.)
16, 20	ISO GND	Isolated Grounds. Must be tied together; not internally connected.
17	ISO DE DRV	Isolated Driver-Enable Drive. The driver outputs, A and B, are enabled by bringing DE' high. The driver outputs are high impedance when DE' is low. If the driver outputs are enabled, the parts function as line drivers. While the driver outputs are high impedance, the chips can function as line receivers. Open-collector output; must have pull-up to ISO V$_{CC}$ and be tied to ISO DE IN for normal operation.
18, 26	ISO V$_{CC}$	Isolated Supply Voltages. Must be tied together; not internally connected.
19	ISO DI DRV	Isolated Driver-Input Drive. With DE' high, a low on DI' forces output A low and output B high. Similarly, a high on DI' forces output A high and output B low. Open-collector output; must have pull-up to ISO V$_{CC}$ and be tied to ISO DI IN for normal operation.
21	ISO DE IN	Isolated Driver-Enable Input. Tie to ISO DE DRV for normal operation.
22	ISO DI IN	Isolated Driver-Input Input. Tie to ISO DI DRV for normal operation.
23	A	Noninverting Driver Output and Noninverting Receiver Input
24	ISO RO DRV	Isolated Receiver-Output Drive. Tie to ISO RO LED through a resistor for normal operation. (See Table 1 of Figure 2.)
25	B	Inverting Driver Output and Inverting Receiver Input
27, 28	AC2, AC1	Internal Connections; leave these pins unconnected.

Note: See Typical Application Circuit (Figure 2) for DE' and DI' pin descriptions.

Fig. 3-2

INPUTS		OUTPUTS	
DE'	DI'	B	A
1	1	0	1
1	0	1	0
0	X	High-Z	High-Z

X = Don't care
High-Z = High impedance

Fig. 3-3

Continued

INPUTS		OUTPUT
DE´	A-B	R̄O̅
0	≥ +0.2V	1
0	≤ -0.2V	0
0	Inputs open	1

Fig. 3-4

Isolated RS-485/RS-422 data interface
Figure 3-1 shows the typical application circuit for the MAX1480A/B. These ICs are complete, electrically isolated RS-485/RS-422 data-communications interfaces. Transceivers, opto-couplers, and a transformer provide a complete interface in one 28-pin package. Figure 3-2 shows the pin descriptions. Figures 3-3 and 3-4 show the function tables. The MAX1408B has reduced-slew-rate drivers that minimize EMI (electromagnetic interference) and reduce reflections caused by improperly terminated cables, allowing error-free data transitions at data rates up to 250 kbps. The MAX1480 driver slew rate is not limited, allowing transmission rates up to 2.5 Mbps. The ICs draw 28 mA of quiescent current. The MAX1408B provides a low-power shutdown mode that consumes only 0.2 μA. The drivers are short-circuit current-limited and are protected against excessive power dissipation by thermal shutdown circuits that place the driver outputs into a high-impedance state. The receiver input has a fail-safe feature that guarantees a logic-high output if the input is open circuit. The ICs typically can withstand 1600 Vrms for one minute or 2000 Vrms for one second. MAXIM NEW RELEASES DATA BOOK, 1995, P. 2-13, 2-15, 2-17.

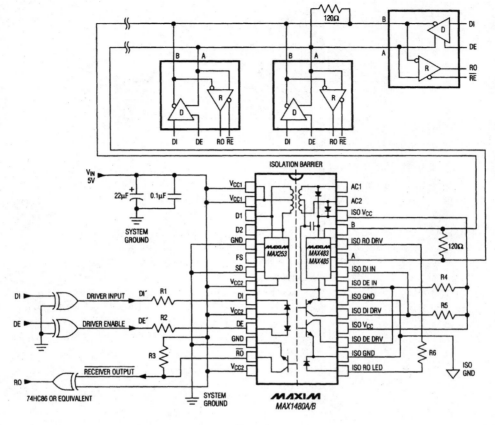

Fig. 3-5

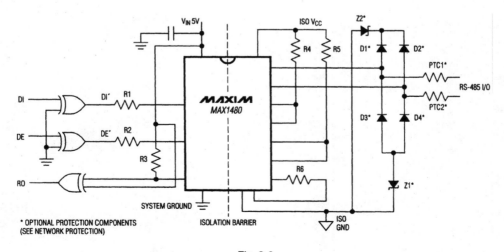

Fig. 3-6

Interface circuits

Vin 5V   TOP VIEW

PIN 1   PIN 28

MAXIM
MAX1480

FS

DI   PIN 1

DE

R1

R2

R3

RO

74HC86

SYSTEM GND

ISOLATION
BARRIER

ISO GND

Z2*   Z1*

D1*   D4*
D3*
D2*

PTC1*
RS-485 I/O
PTC2*

R6

R5

R4

* Optional Protection Components
(see Network Protection)

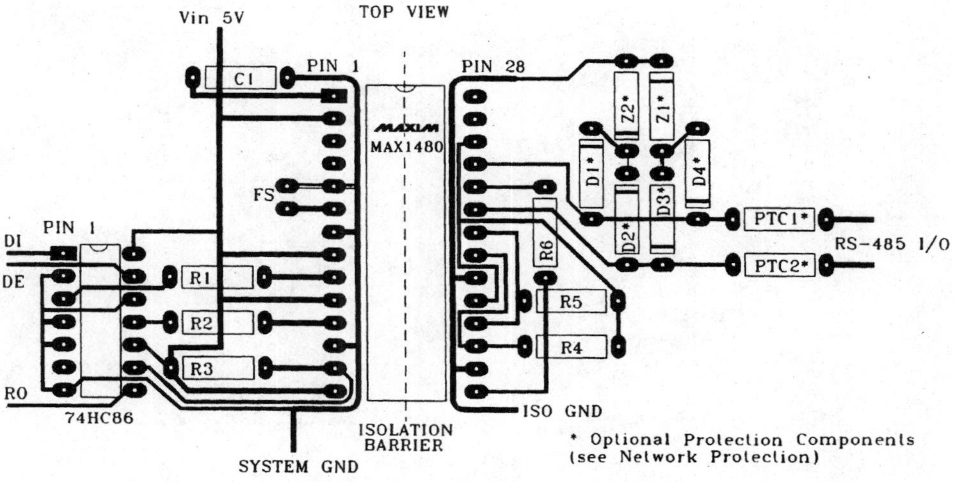

Fig. 3-7

Typical RS-485/RS422 network

Figure 3-5 shows the MAX1480A/B connected in a typical RS-485/RS422 network. Figures 3-6 and 3-7 show the layout schematic and PC layout. To minimize reflections, terminate the line at both ends with its characteristic impedance, and keep the stub lengths off the main line as short as possible. The slew-rate-limited MAX1408B is more tolerant of imperfect termination and stubs off the main line. MAXIM NEW RELEASES DATA BOOK, 1995, P. 2-19, 2-21.

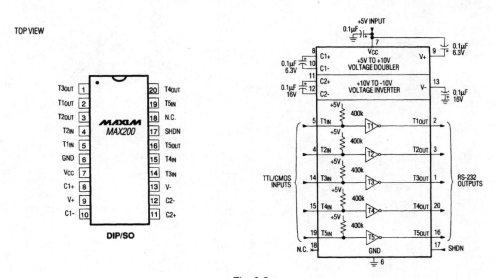

TOP VIEW

+5V INPUT
0.1µF

7
8   Vcc   V+   9   0.1µF
C1+   +5V TO +10V   6.3V
0.1µF   10 C1-   VOLTAGE DOUBLER
6.3V
11
0.1µF   12 C2+   +10V TO -10V   V-   13 0.1µF
16V   C2-   VOLTAGE INVERTER   16V

T3OUT 1   20 T4OUT
T1OUT 2   19 T5IN
T2OUT 3   18 N.C.
T2IN   4   MAXIM   17 SHDN
T1IN   5   MAX200 16 T5OUT
GND   6   15 T4IN
Vcc   7   14 T3IN
C1+   8   13 V-
V+   9   12 C2-
C1-   10   11 C2+

DIP/SO

+5V 400k
5 T1IN   T1   T1OUT 2
+5V 400k
4 T2IN   T2   T2OUT 3
+5V 400k
TTL/CMOS 14 T3IN   T3   T3OUT 1 RS-232
INPUTS   OUTPUTS
+5V 400k
15 T4IN   T4   T4OUT 20
+5V 400k
19 T5IN   T5   T5OUT 16

N.C. 18   17 SHDN
GND
6

Fig. 3-8

Interface circuit titles and descriptions

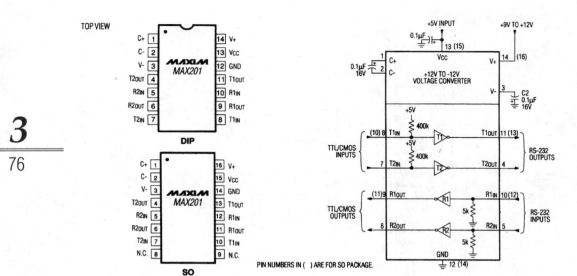

Fig. 3-9

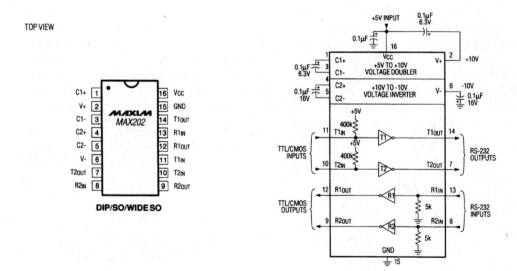

Fig. 3-10

TOP VIEW

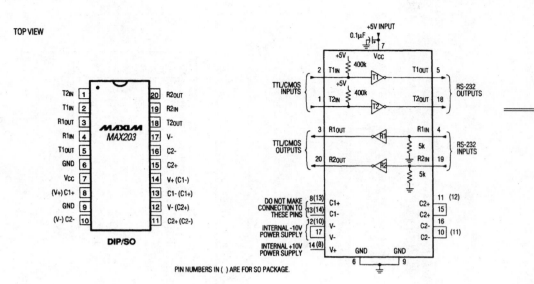

PIN NUMBERS IN ( ) ARE FOR SO PACKAGE.

Fig. 3-11

TOP VIEW

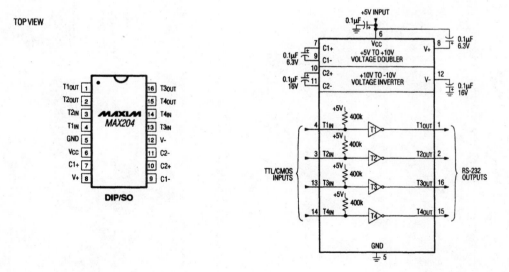

Fig. 3-12

TOP VIEW

DIP

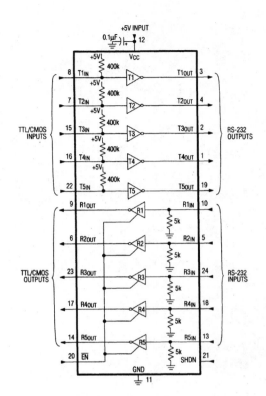

Fig. 3-13

TOP VIEW

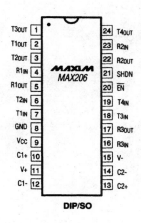

DIP/SO

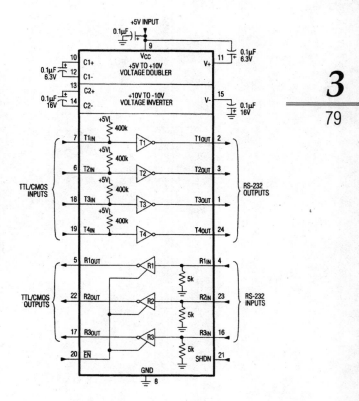

Fig. 3-14

3

80

TOP VIEW

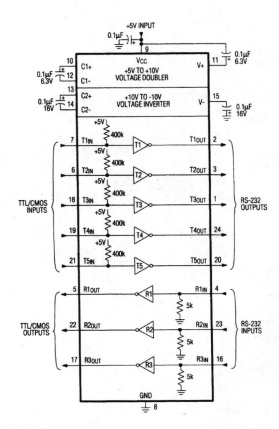

Fig. 3-15

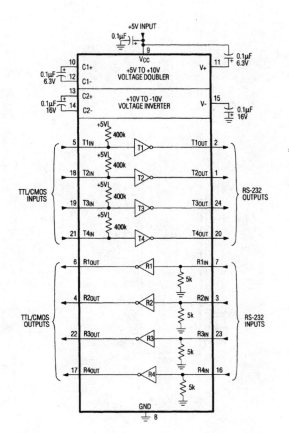

Fig. 3-16

TOP VIEW

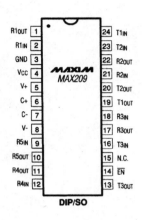

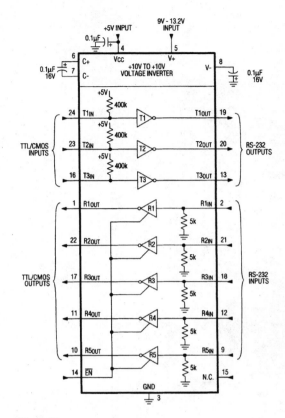

Fig. 3-17

TOP VIEW

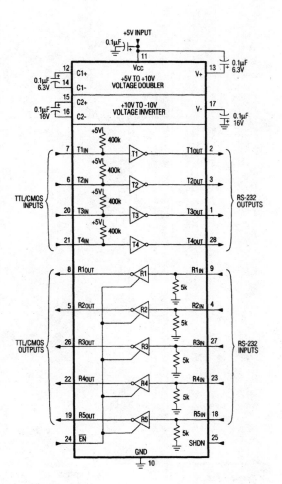

Fig. 3-18

TOP VIEW

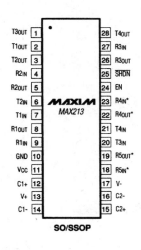

SO/SSOP

*ACTIVE IN SHUTDOWN

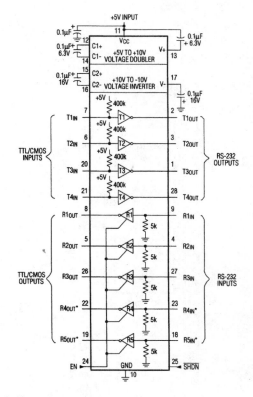

Fig. 3-19

Part Number	Power-Supply Voltage (V)	No. of RS-232 Drivers	No. of RS-232 Receivers	No. of Receivers Active in Shutdown	No. of External Capacitors (0.1μF)	Low-Power Shutdown/ TTL Three-State
MAX200	+5	5	0	0	4	Yes/No
MAX201	+5 and +9.0 to +13.2	2	2	0	2	No/No
MAX202	+5	2	2	0	4	No/No
MAX203	+5	2	2	0	None	No/No
MAX204	+5	4	0	0	4	No/No
MAX205	+5	5	5	0	None	Yes/Yes
MAX206	+5	4	3	0	4	Yes/Yes
MAX207	+5	5	3	0	4	No/No
MAX208	+5	4	4	0	4	No/No
MAX209	+5 and +9.0 to +13.2	3	5	0	2	No/Yes
MAX211	+5	4	5	0	4	Yes/Yes
MAX213	+5	4	5	2	4	Yes/Yes

Fig. 3-20

SHDN	OPERATION STATUS	TRANSMITTERS T1-T5
0	Normal Operation	All Active
1	Shutdown	All High-Z

Fig. 3-21

SHDN	$\overline{EN}$	OPERATION STATUS	TRANSMITTERS T1-T5	RECEIVERS R1-R5
0	0	Normal Operation	All Active	All Active
0	1	Normal Operation	All Active	All High-Z
1	0	Shutdown	All High-Z	All High-Z

Fig. 3-22

$\overline{SHDN}$	EN	OPERATION STATUS	TRANSMITTERS	RECEIVERS	
			T1-T4	R1-R3	R4, R5
0	0	Shutdown	All High-Z	High-Z	High-Z
0	1	Shutdown	All High-Z	High-Z	Active*
1	0	Normal Operation	All Active	High-Z	High-Z
1	1	Normal Operation	All Active	Active	Active

* Active = active with reduced performance.

Fig. 3-23

PARAMETER	CONDITION	EIA/TIA-232E, V.28 SPECIFICATION
Driver Output Voltage 0 Level 1 Level Output Level, Max	$3k\Omega$ to $7k\Omega$ load $3k\Omega$ to $7k\Omega$ load No load	+5.0V to +15V -5.0V to -15V ±25V
Data Rate	$3k\Omega \leq R_L \leq 7k\Omega$, $C_L \leq 2500pF$	Up to 20kbits/sec
Receiver Input Voltage 0 Level 1 Level Input Level, Max		+3.0V to +15V -3.0V to -15V ±25V
Instantaneous Slew Rate, Max	$3k\Omega \leq R_L \leq 7k\Omega$, $C_L \leq 2500pF$	30V/µs
Driver Output Short-Circuit Current, Max		100mA
Transition Rate on Driver Output	V.28	1ms or 3% of the period
	EIA/TIA-232E	4% of the period
Driver Output Resistance	-2V < V_{OUT} < +2V	300Ω

Fig. 3-24

Interface circuit titles and descriptions

PIN	CONNECTION	
1	Received Line Signal Detector, sometimes called Carrier Detect (DCD)	Handshake from DCE
2	Receive Data (RD)	Data from DCE
3	Transmit Data (TD)	Data from DTE
4	Data Terminal Ready	Handshake from DTE
5	Signal Ground	Reference point for signals
6	Data Set Ready (DSR)	Handshake from DCE
7	Request to Send (RTS)	Handshake from DTE
8	Clear to Send (CTS)	Handshake from DCE
9	Ring Indicator	Handshake from DCE

Fig. 3-25

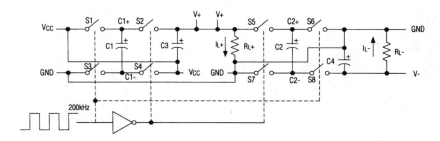

Fig. 3-26

Transceivers with external capacitors (5 V)

Figures 3-8 through 3-19 show typical application circuits and pin configurations for the MAX200-MAX211/MAX213. Figure 3-20 shows the selection table for these ICs. Figures 3-21, 3-22, and 3-23 show the control-pin configurations for the MAX200, MAX205/06/11, and MAX213, respectively. Figure 3-24 shows a summary of EIA/TIA-232E, V.28 specifications. Figure 3-25 shows DB9 cable connections commonly used for EIA/TIA-232E and V.24 asynchronous interfaces. These ICs are designed for RS-232 and V.28 communication interfaces, where ±12 V supplies are not available. On-board charge pumps (Fig. 3-26) convert the +5-V input to the ±10 V needed for RS-232 output levels. The MAX201 and MAX209 operate from +5 V and +12 V, and contain a +12-V to −12-V charge-pump voltage converter. The drivers and receivers of these ICs meet all EIA/TIA-232E and CCITV V.28 specifications at data rates of 20 kbits/second. The drivers (except MAX202, MAX203) maintain the ±5-V EIA/TIA-232E output signal levels at data rates in excess of 120 kbits/second, when loaded in accordance with the specification. The MAX200/205/206/211 have a 5-μW shutdown mode to conserve energy. The MAX213 has an active-low shutdown and an active-high

Interface circuits

receiver-enable control. Two receivers in the MAX213 are active, allowing ring indicator (Ri) to be monitored easily using only 75 μW of power. MAXIM NEW RELEASES DATA BOOK, 1995, P. 2-23, 2-28, 2-29, 2-30, 2-31, 2-32, 2-33, 2-34, 2-35, 2-36, 2-37, 2-38, 2-39, 2-40.

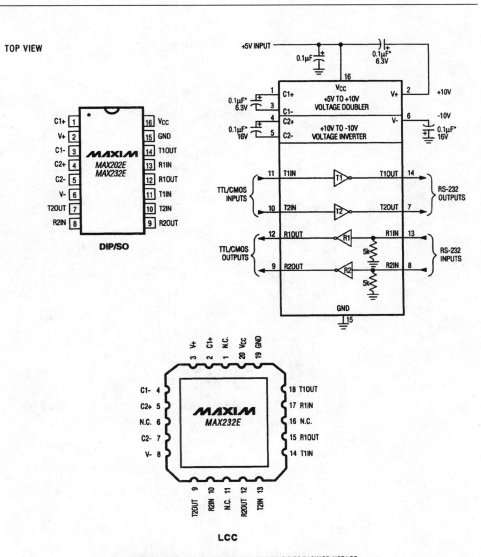

NOTE: PIN NUMBERS ON TYPICAL OPERATING CIRCUIT REFER TO DIP/SO PACKAGE, NOT LCC.
* 1.0μF CAPACITORS, MAX232E ONLY.

Fig. 3-27

3

88

TOP VIEW

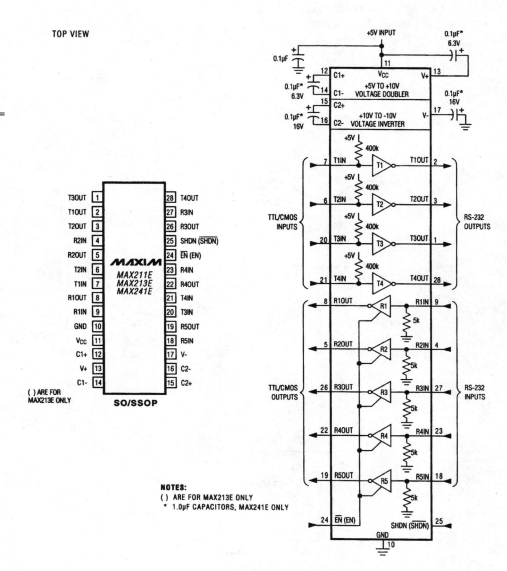

Fig. 3-28

ESD-protected transceivers (5 V)

Figure 3-27 shows the typical application circuits and pin configurations for the MAX202E,232E. Figure 3-28 shows the MAX211E, 213E, 241E. These ICs are designed to operate in harsh environments, but still meet EIA/TIA-232E specifications. Each transmitter output and receiver input is protected against ±15-kV electrostatic discharge (ESD). MAXIM NEW RELEASES DATA BOOK, 1995, P. 2-54, 2-55.

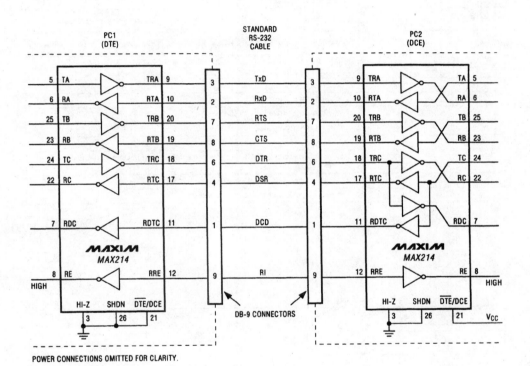

Fig. 3-29

3

90

PIN	NAME	FUNCTION
1, 2	C2+, C2-	Terminals for negative charge-pump capacitor
3	HI-Z	RS-232 receiver impedance control. Take high to disconnect the termination resistor.
4	N.C.	No connect—not internally connected
5, 24, 25	TA, TC, TB	TTL/CMOS driver A, C, B inputs
6, 8, 22, 23	RA, RE, RC, RB	TTL/CMOS receiver A, E, C, B outputs
7	RDC	TTL/CMOS DTE receiver output D for $\overline{DTE}$/DCE = 0V, or TTL/CMOS DCE receiver output C for $\overline{DTE}$/DCE = +5V
9, 18, 20	TRA, TRC, TRB	RS-232 DTE driver output for $\overline{DTE}$/DCE= 0V, or RS-232 DCE receiver input for $\overline{DTE}$/DCE = +5V
10, 17, 19	RTA, RTC, RTB	RS-232 DTE receiver input for $\overline{DTE}$/DCE = 0V, or RS-232 DCE driver output for $\overline{DTE}$/DCE = +5V
11	RDTC	RS-232 DTE receiver input D for $\overline{DTE}$/DCE = 0V, or RS-232 DCE driver output C for $\overline{DTE}$/DCE = +5V
12	RRE	RS-232 receiver input
13	GND	Ground
14	V-	-2V$_{CC}$ voltage generated by the charge pump
15	V+	+2V$_{CC}$ voltage generated by the charge pump
16	V$_{CC}$	+4.5V to +5.5V supply voltage
21	$\overline{DTE}$/DCE	Data terminal equipment (DTE) and data circuit-terminating equipment (DCE) control pin. DCE active high and DTE active low.
26	SHDN	Shutdown control; shutdown high, normal operation low
27, 28	C1+, C1-	Terminals for positive charge-pump capacitor

Fig. 3-30

TTL/CMOS I/O LABEL	MAX214 PIN	FUNCTION	MAX214 PIN	RS-232 I/O LABEL	DB-25 PIN	INPUT THRESHOLD
Transmitter (TxD)	5		9	TxD	2	
Receiver (RxD)	6		10	RxD	3	+
Request to Send (RTS)	25		20	RTS	4	
Clear to Send (CTS)	23		19	CTS	5	-
Data Terminal Ready (DTR)	24		18	DTR	6	
Data Set Ready (DSR)	22		17	DSR	20	-
Detector Carrier Data (DCD)	7		11	DCD	8	+
Ring Indicator (RI)	8		12	RI	22	+

Fig. 3-31

Interface circuits

MAX214 PIN	FUNCTION	MAX214 PIN	RS-232 I/O LABEL	DB-25 PIN	INPUT THRESHOLD
5	⊳	10	RxD	3	
6	⊲	9	TxD	2	+
25	⊳	19	CTS	5	
23	⊲	20	RTS	4	-
24		17	DSR	20	
		11	DCD	8	
22		18	DTR	6	-
7					
8	⊲	12	RI	22	+

Fig. 3-32

CONTROL INPUTS			RS-232 PINS		
SHUTDOWN	HI-Z	DTE/DCE	TRA, TRB, TRC	RTA, RTB, RTC, RDTC	RRE
0	0	0	Transmit Mode	Receive Mode/5kΩ	Receive Mode/5kΩ
0	0	1	Receive Mode/5kΩ	Transmit Mode	Receive Mode/5kΩ
0	1	0	Transmit Mode	Receive Mode/HI-Z	Receive Mode/HI-Z
0	1	1	Receive Mode/HI-Z	Transmit Mode	Receive Mode/HI-Z
1	0	0	Disabled/HI-Z	Slow Receive/HI-Z	Slow Receive/HI-Z
1	0	1	Slow Receive/HI-Z	Disabled/HI-Z	Slow Receive/HI-Z
1	1	0	Disabled/HI-Z	Slow Receive/HI-Z	Slow Receive/HI-Z
1	1	1	Slow Receive/HI-Z	Disabled/HI-Z	Slow Receive/HI-Z

Fig. 3-33

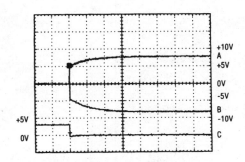

A = TRANSMITTER OUTPUT HIGH, +5V/div
B = TRANSMITTER OUTPUT LOW, +5V/div
C = SHDN INPUT, +5V/div

HORIZONTAL = 200µs

Fig. 3-34

Interface circuit titles and descriptions

Programmable DTE/DCE transceiver

Figure 3-29 shows a typical application circuit (where two PCs have both DTE and DCE operation) for the MAX214. Figures 3-30, 3-31, 3-32, and 3-33 show the pin descriptions. Figure 3-34 shows the scope display for transmitter outputs when exiting shutdown. This IC provides a software-configurable DTE (data terminal equipment) or DCE (data circuit-terminating equipment) port RS232 interface. Either DTE or DCE is selected using the $\overline{\text{DTE}/\text{DCE}}$ pin (21). This IC eliminates the need to swap cables when switching between DTE and DCE configurations. MAXIM NEW RELEASES DATA BOOK, 1995, P. 2-69, 2-70, 2-72, 2-73, 2-74.

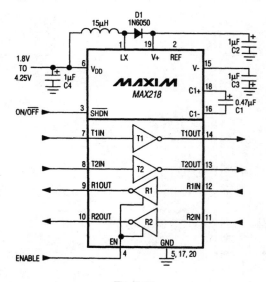

Fig. 3-35

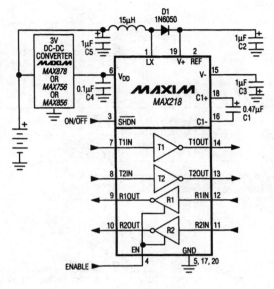

Fig. 3-36

PIN	NAME	FUNCTION
1	LX	Inductor/Diode Connection Point
2	REF	Internal Reference Bypass Node. Normally left open. Bypass to GND (between pin 2 and pin 5) with 0.1µF if V_{DD} is noisy.
3	$\overline{SHDN}$	Shutdown Control. Connect to V_{DD} for normal operation. Connect to GND to shut down the power supply and to disable the drivers. Receiver status is not changed by this control.
4	EN	Receiver Output Enable Control. Connect to V_{DD} for normal operation. Connect to GND to force the receiver outputs into high-Z state.
5, 17, 20	GND	Ground. Connect all GND pins to ground.
6	V_{DD}	Supply Voltage Input; 1.8V to 4.25V. Bypass to GND with at least 1µF. See *Capacitor Selection* section.
7, 8	T1IN, T2IN	Transmitter Inputs
9, 10	R1OUT, R2OUT	Receiver Outputs; swing between GND and V_{DD}.
11, 12	R2IN, R1IN	Receiver Inputs
13, 14	T2OUT, T1OUT	Transmitter Outputs; swing between V+ and V-.
15	V-	Negative Supply generated on-board
16, 18	C1-, C1+	Terminals for Negative Charge-Pump Capacitor
19	V+	Positive Supply generated on-board

Fig. 3-37

Interface circuit titles and descriptions

MANUFACTURER	PART NUMBER	PHONE	FAX
Inductors			
Murata	LQH4N150K-TA	USA (814) 237-1431 Japan (075) 951-9111	USA (814) 238-0490 Japan (075) 955-6526
Sumida	CD43150	USA (708) 956-0666 Japan (03) 3607-5111	USA (708) 956-0702 Japan (03) 3607-5428
TDK	NLC453232T-150K	USA (708) 803-6100 Japan (03) 3278-5111	USA (708) 803-6296 Japan (03) 3278-5358
Diodes—Surface Mount			
Central Semiconductor	CMPSH-3, Schottky	USA (516) 435-1110	USA (516) 435-1824
Motorola	MMBD6050LT1, Silicon	USA (408) 749-0510	USA (408) 991-7420
Philips	PMBD6050, Silicon	USA (401) 762-3800	USA (401) 767-4493
Diodes—Through-Hole			
Motorola	1N6050, Silicon 1N5817, Schottky	USA (408) 749-0510	USA (408) 991-7420

Fig. 3-38

SHDN	EN	RECEIVER OUTPUT	DRIVER OUTPUT	DC-DC CONVERTER	SUPPLY CURRENT
L	L	High-Z	High-Z	OFF	Minimum
L	H	Enabled	High-Z	OFF	Minimum
H	L	High-Z	Enabled	ON	Normal
H	H	Enabled	Enabled	ON	Normal

Fig. 3-39

Dual transceiver (1.8 V to 4.25 V)
Figure 3-35 shows a typical application circuit for single-supply operation of the MAX218. Figure 3-36 shows the connections for operation from unregulated and regulated supplies. Figure 3-37 shows the pin descriptions. Figure 3-38 shows suggested component suppliers. This IC runs on two alkaline (NiCd or NiMH) cells to provide full EIA/TIA-232E and V.28/V.24 communications interface. True RS-232 and EIA/TIA-562 voltage levels are maintained with a wide +1.8-V to +4.25-V operating range. A shutdown mode reduces current to 1 μA. The receivers can be enabled or disabled under logic control. Figure 3-39 shows the shutdown and enable/disable mode logic. The data rate is guaranteed at 120 kbps. MAXIM NEW RELEASES DATA BOOK, 1995, P. 2-79, 2-80, 2-81, 2-82.

TOP VIEW

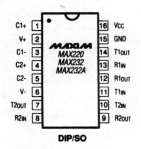

CAPACITANCE (µF)					
DEVICE	C1	C2	C3	C4	C5
MAX220	4.7	4.7	10	10	4.7
MAX232	1.0	1.0	1.0	1.0	1.0
MAX232A	0.1	0.1	0.1	0.1	0.1

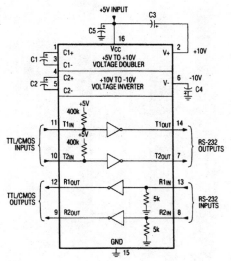

Fig. 3-40

TOP VIEW

( ) ARE FOR MAX222 ONLY.

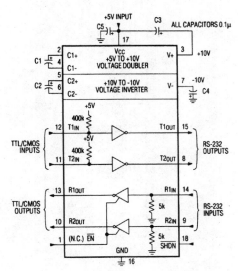

Fig. 3-41

Interface circuit titles and descriptions

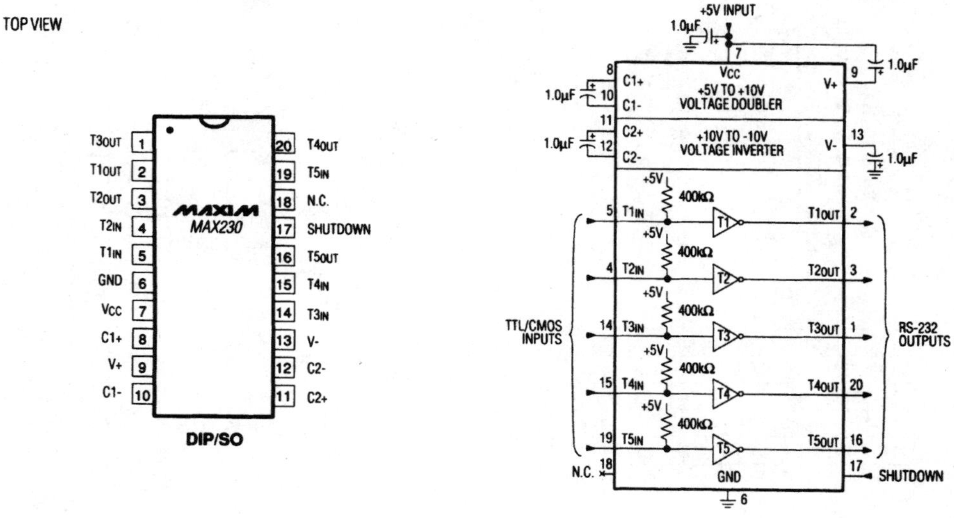

TOP VIEW

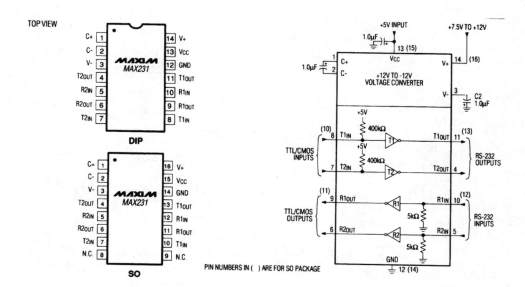

Fig. 3-42

TOP VIEW

PIN NUMBERS IN ( ) ARE FOR SO PACKAGE

Fig. 3-43

Continued

TOP VIEW

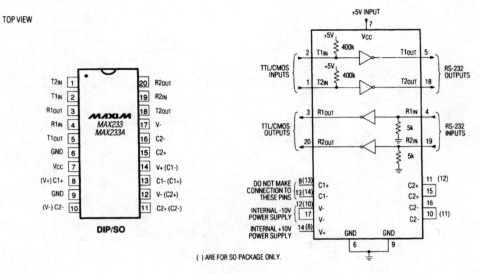

Fig. 3-44

TOP VIEW

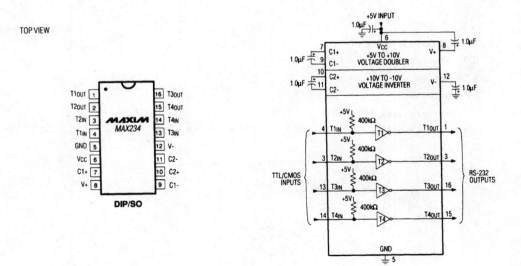

Fig. 3-45

TOP VIEW

SO

MAX225 Functional Description

5 Receivers

5 Transmitters

2 Control Pins

   1 Receiver Enable ($\overline{ENR}$)

   1 Transmitter Enable ($\overline{ENT}$)

PINS ($\overline{ENR}$, GND, V_{CC}, T_5OUT) ARE INTERNALLY CONNECTED.
CONNECT EITHER OR BOTH EXTERNALLY. T_5OUT IS A SINGLE DRIVER.

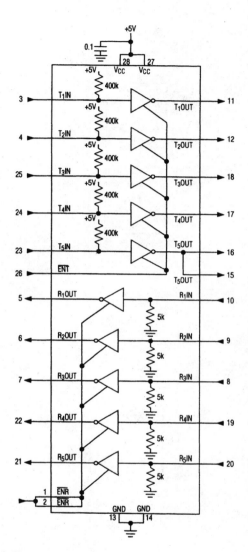

Fig. 3-46

TOP VIEW

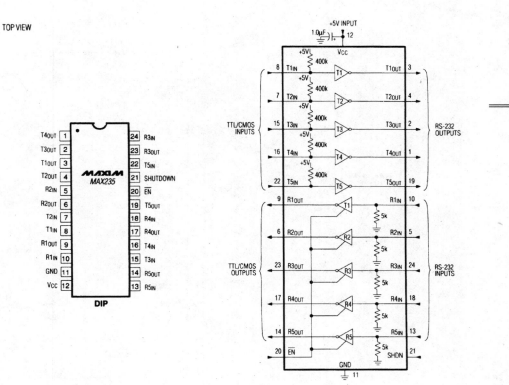

Fig. 3-47

TOP VIEW

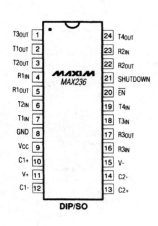

DIP/SO

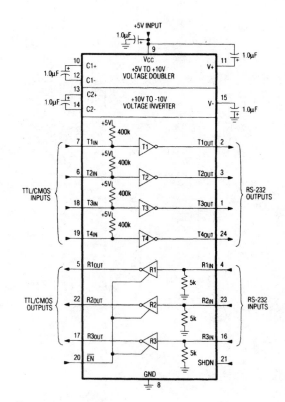

Fig. 3-48

TOP VIEW

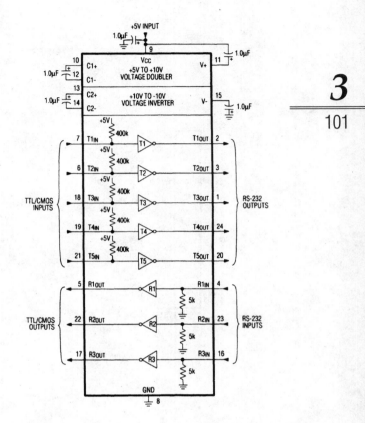

Fig. 3-49

TOP VIEW

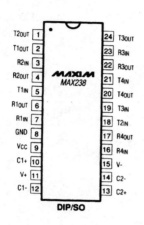

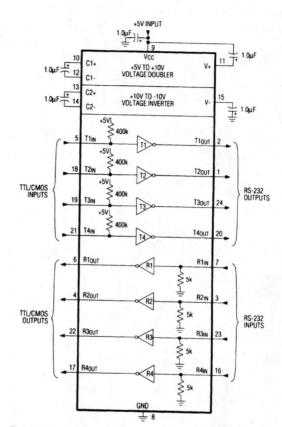

Fig. 3-50

Continued

TOP VIEW

DIP/SO

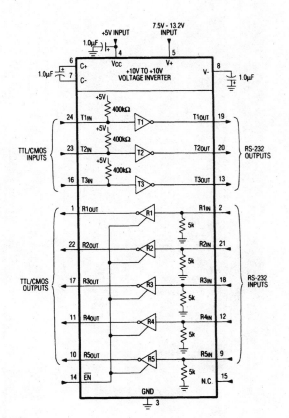

Fig. 3-51

3

104

TOP VIEW

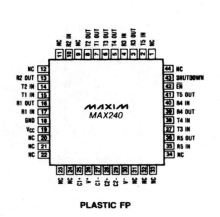

PLASTIC FP

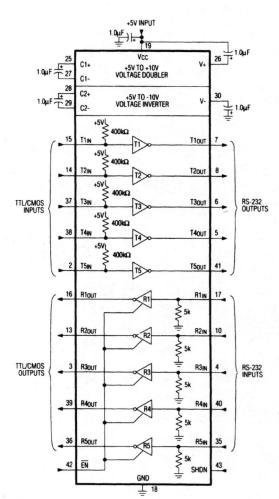

Fig. 3-52

Continued

TOP VIEW

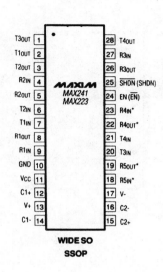

WIDE SO

SSOP

* R4 AND R5 IN MAX223 REMAIN ACTIVE IN SHUTDOWN

NOTE: PIN LABELS IN ( ) ARE FOR MAX241

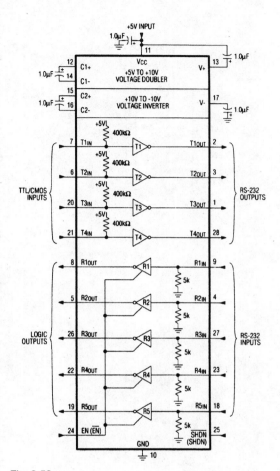

Fig. 3-53

TOP VIEW

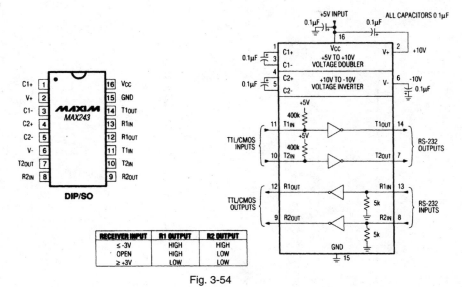

RECEIVER INPUT	R1 OUTPUT	R2 OUTPUT
≤ -3V	HIGH	HIGH
OPEN	HIGH	LOW
≥ +3V	LOW	LOW

Fig. 3-54

TOP VIEW

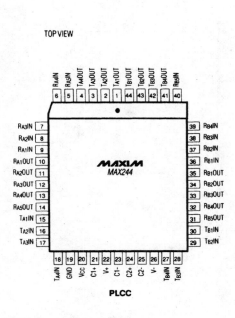

PLCC

MAX244 Functional Description

10 Receivers Always Active
  5 A-Side Receivers
  5 B-Side Receivers

8 Transmitters
  4 A-Side Transmitters
  4 B-Side Transmitters

No Control Pins

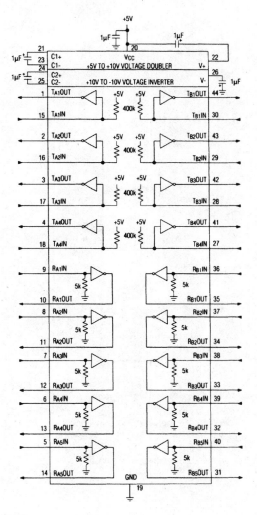

Fig. 3-55

TOP VIEW

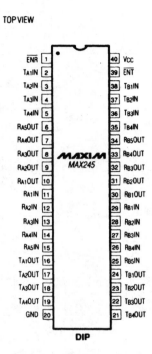

DIP

MAX245 Functional Description

10 Receivers
   5 A-Side Receivers (RA5 always active)
   5 B-Side Receivers (RB5 always active)

8 Transmitters
   4 A-Side Transmitters
   4 B-Side Transmitters

2 Control Pins
   1 Receiver Enable (ENR)
   1 Transmitter Enable (ENT)

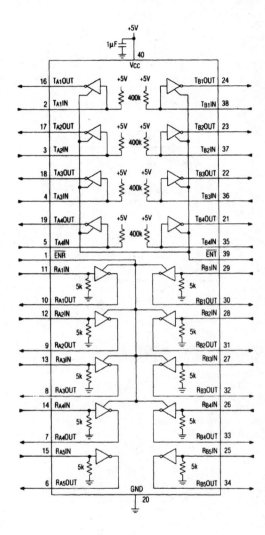

Fig. 3-56

TOP VIEW

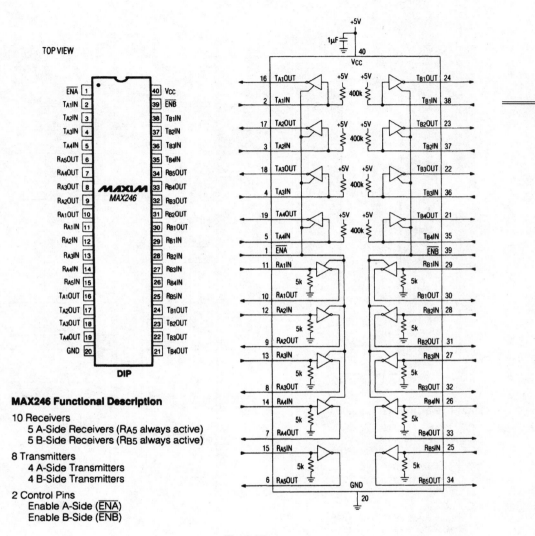

DIP

MAX246 Functional Description

10 Receivers
    5 A-Side Receivers (RA5 always active)
    5 B-Side Receivers (RB5 always active)

8 Transmitters
    4 A-Side Transmitters
    4 B-Side Transmitters

2 Control Pins
    Enable A-Side (ENA)
    Enable B-Side (ENB)

Fig. 3-57

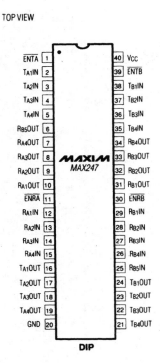

TOP VIEW

DIP

MAX247 Functional Description

9 Receivers
  4 A-Side Receivers
  5 B-Side Receivers (R$_{B5}$ always active)

8 Transmitters
  4 A-Side Transmitters
  4 B-Side Transmitters

4 Control Pins
  Enable Receiver A-Side ($\overline{ENRA}$)
  Enable Receiver B-Side ($\overline{ENRB}$)
  Enable Transmitter A-Side ($\overline{ENTA}$)
  Enable Transmitter B-Side ($\overline{ENTB}$)

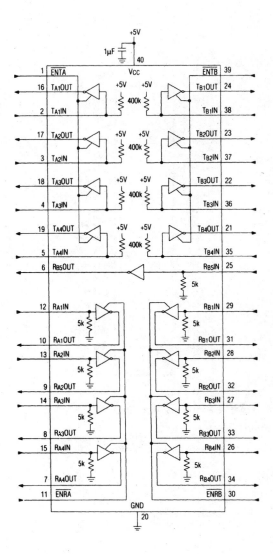

Fig. 3-58

TOP VIEW

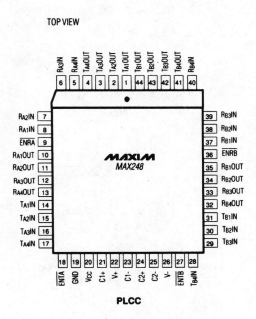

PLCC

MAX248 Functional Description

8 Receivers
    4 A-Side Receivers
    4 B-Side Receivers

8 Transmitters
    4 A-Side Transmitters
    4 B-Side Transmitters

4 Control Pins
    Enable Receiver A-Side ($\overline{ENRA}$)
    Enable Receiver B-Side ($\overline{ENRB}$)
    Enable Transmitter A-Side ($\overline{ENTA}$)
    Enable Transmitter B-Side ($\overline{ENTB}$)

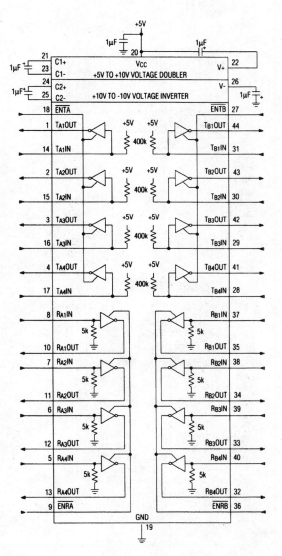

Fig. 3-59

Interface circuit titles and descriptions

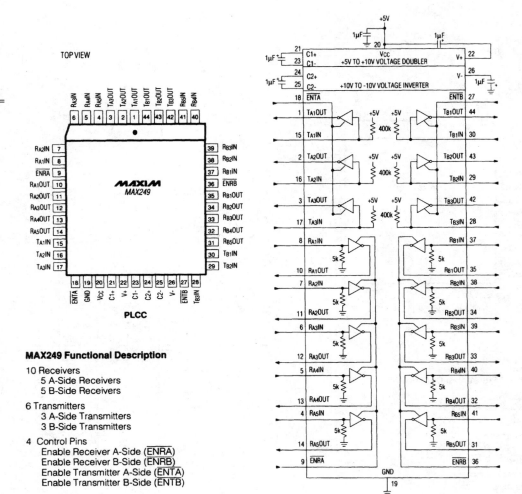

TOP VIEW

PLCC

MAX249 Functional Description

10 Receivers
    5 A-Side Receivers
    5 B-Side Receivers

6 Transmitters
    3 A-Side Transmitters
    3 B-Side Transmitters

4 Control Pins
    Enable Receiver A-Side ($\overline{\text{ENRA}}$)
    Enable Receiver B-Side ($\overline{\text{ENRB}}$)
    Enable Transmitter A-Side ($\overline{\text{ENTA}}$)
    Enable Transmitter B-Side ($\overline{\text{ENTB}}$)

Fig. 3-60

PART	TEMP. RANGE	PIN-PACKAGE
MAX220CPE	0°C to +70°C	16 Plastic DIP
MAX220CSE	0°C to +70°C	16 Narrow SO
MAX220CWE	0°C to +70°C	16 Wide SO
MAX220C/D	0°C to +70°C	Dice*
MAX220EPE	-40°C to +85°C	16 Plastic DIP
MAX220ESE	-40°C to +85°C	16 Narrow SO
MAX220EWE	-40°C to +85°C	16 Wide SO
MAX220EJE	-40°C to +85°C	16 CERDIP
MAX220MJE	-55°C to +125°C	16 CERDIP

Ordering Information continued at end of data sheet.
Contact factory for dice specifications.

Fig. 3-61

ENT	$\overline{\text{ENR}}$	OPERATION STATUS	TRANSMITTERS	RECEIVERS
0	0	Normal Operation	All Active	All Active
0	1	Normal Operation	All Active	All 3-State
1	0	Shutdown	All 3-State	All Low-Power Receive Mode
1	1	Shutdown	All 3-State	All 3-State

Fig. 3-62

ENT	$\overline{\text{ENR}}$	OPERATION STATUS	TRANSMITTERS		RECEIVERS	
			TA1-TA4	TB1-TB4	RA1-RA5	RB1-RB5
0	0	Normal Operation	All Active	All Active	All Active	All Active
0	1	Normal Operation	All Active	All Active	RA1-RA4 3-State RA5 Active	RB1-RB4 3-State RB5 Active
1	0	Shutdown	All 3-State	All 3-State	All Low Power Receiver Mode	All Low Power Receiver Mode
1	1	Shutdown	All 3-State	All 3-State	RA1-RA4 3-State RA5 Low-Power Receiver Mode	RB1-RB4 3-State RA5 Low-Power Receiver Mode

Fig. 3-63

$\overline{\text{ENA}}$	$\overline{\text{ENB}}$	OPERATION STATUS	TRANSMITTERS		RECEIVERS	
			TA1-TA4	TB1-TB4	RA1-RA5	RB1-RB5
0	0	Normal Operation	All Active	All Active	All Active	All Active
0	1	Normal Operation	All Active	All 3-State	All Active	RB1-RB4 3-State RB5 Active
1	0	Shutdown	All 3-State	All Active	RA1-RA4 3-State RA5 Active	All Active
1	1	Shutdown	All 3-State	All 3-State	RA1-RA4 3-State RA5 Low-Power Receiver Mode	RB1-RB4 3-State RA5 Low-Power Receiver Mode

Fig. 3-64

Interface circuit titles and descriptions

3

113

ENTA̅	ENTB̅	ENRA̅	ENRB̅	OPERATION STATUS	MAX247 TA1–TA4	TB1–TB4	RA1–RA4	RB1–RB5
					TRANSMITTERS		**RECEIVERS**	
					MAX248 TA1–TA4	TB1–TB4	RA1–RA4	RB1–RB4
					MAX249 TA1–TA3	TB1–TB3	RA1–RA5	RB1–RB5
0	0	0	0	Normal Operation	All Active	All Active	All Active	All Active
0	0	0	1	Normal Operation	All Active	All Active	All Active	All 3-State, except RB5 stays active on MAX247
0	0	1	0	Normal Operation	All Active	All Active	All 3-State	All Active
0	0	1	1	Normal Operation	All Active	All Active	All 3-State	All 3-State, except RB5 stays active on MAX247
0	1	0	0	Normal Operation	All Active	All 3-State	All Active	All Active
0	1	0	1	Normal Operation	All Active	All 3-State	All Active	All 3-State, except RB5 stays active on MAX247
0	1	1	0	Normal Operation	All Active	All 3-State	All 3-State	All Active
0	1	1	1	Normal Operation	All Active	All 3-State	All 3-State	All 3-State, except RB5 stays active on MAX247
1	0	0	0	Normal Operation	All 3-State	All Active	All Active	All Active
1	0	0	1	Normal Operation	All 3-State	All Active	All Active	All 3-State, except RB5 stays active on MAX247
1	0	1	0	Normal Operation	All 3-State	All Active	All 3-State	All Active
1	0	1	1	Normal Operation	All 3-State	All Active	All 3-State	All 3-State, except RB5 stays active on MAX247
1	1	0	0	Shutdown	All 3-State	All 3-State	Low-Power Receive Mode	Low-Power Receive Mode
1	1	0	1	Shutdown	All 3-State	All 3-State	Low-Power Receive Mode	All 3-State, except RB5 stays active on MAX247
1	1	1	0	Shutdown	All 3-State	All 3-State	All 3-State	Low-Power Receive Mode
1	1	1	1	Shutdown	All 3-State	All 3-State	All 3-State	All 3-State, except RB5 stays active on MAX247

Fig. 3-65

Multi-channel drivers/receivers (5 V)
Figures 3-40 through 3-60 show typical application circuits and pin configurations for the MAX220-MAX249. Figure 3-61 shows the selection table for these ICs. Figures 3-62 through 3-65 show the control pin configurations. These ICs are line drivers/receivers intended for all EIA/TIA-232E and V.28/V.24 communications interfaces—especially for those applications where ±12 V is not available. The low-power shutdown mode reduces power dissipation to less than 5 μW. MAXIM NEW RELEASES DATA BOOK, 1995, P. 2-85, 2-96, 2-97, 2-100 THROUGH 2-117.

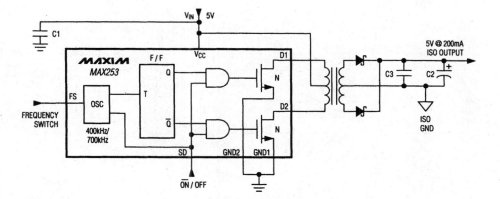

Fig. 3-66

PIN	NAME	FUNCTION
1	D1	Open drain of N-channel transfomer drive 1.
2	GND1	Ground. Connect both GND1 and GND2 to ground.
3	FS	Frequency switch. If FS = V_{CC} or open, switch frequency = 350kHz; if FS = 0V, switch frequency = 200kHz.
4	SD	Shutdown. Ground for normal operation, tie high for shutdown.
5	N.C.	Not internally connected.
6	V_{CC}	+5V supply voltage.
7	GND2	Ground. Connect both GND1 and GND2 to ground.
8	D2	Open drain of N-channel transformer drive 2.

Fig. 3-67

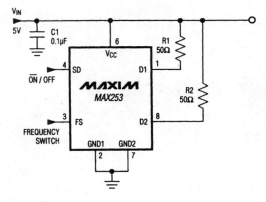

Fig. 3-68

Interface circuit titles and descriptions

Continued

CHARACTERISTIC		+5V to ±10V	+5V to +5V	+3.3V to +5V	+5V to +24V	+5V to ±5V; ±12V
Figure		9a	2, 3, 5, 6	4, 7	8	10
Turns Ratio		1CT*:1	1CT:1.3CT	1CT:2.1CT	1CT:5CT	1CT:1.5CT:3CT
Typical Windings	Primary	44CT	44CT	28CT	44CT	44CT
	Secondary	44	56CT	56CT	220CT	66CT, 132CT
Primary ET Product	FS Low	18.3V-µs	18.3V-µs	12V-µs	18.3V-µs	18.3V-µs
	FS High	11V-µs	11V-µs	7.2V-µs	11V-µs	11V-µs

*CT = Center Tapped

Fig. 3-69

TRANSFORMERS	TRANSFORMER CORES	OPTOCOUPLERS
BH Electronics Phone: (507) 532-3211 FAX: (507) 532-3705	Philips Components Phone: (407) 881-3200 FAX: (407) 881-3300	Quality Technology Phone: (408) 720-1440 FAX: (408) 720-0848
Coilcraft Phone: (708) 639-6400 FAX: (708) 639-1469	Magnetics Inc. Phone: (412) 282-8282 FAX: (412) 282-6955	Sharp Electronics Phone: (206) 834-2500 FAX: (206) 834-8903
Coiltronics Phone: (407) 241-7876 FAX: (407) 241-9339	Fair-Rite Products Phone: (914) 895-2055 FAX: (914) 895-2629	Siemens Components Phone: (408) 777-4500 FAX: (408) 777-4983

Fig. 3-70

PRODUCTION METHOD	CAPACITORS
Surface Mount	Matsuo 267 series (low ESR) USA Phone: (714) 969-2491, FAX: (714) 960-6492 Sprague Electric Co. 595D/293D series (very low ESR) USA Phone: (603) 224-1961, FAX: (603) 224-1430 Murata Erie Ceramic USA Phone: (800) 831-9172, FAX: (404) 436-3030
High-Performance Through Hole	Sanyo OS-CON series (very low ESR) USA Phone: (619) 661-6835, FAX: (619) 661-1055 Japan Phone: 81-7-2070-1005, FAX: 81-7-2070-1174
Through Hole	Nichicon PL series (low ESR) USA Phone: (708) 843-7500, FAX: (708) 843-2798 Japan Phone: 81-7-5231-8461, FAX: 81-7-5256-4158

* Nihon Inter Electronics Corp.
USA Phone: (805) 867-2555
FAX: (805) 867-2556
Japan Phone: 81-3-3494-7411
FAX: 81-3-3494-7414

Fig. 3-71

DATA RATE	FULL DUPLEX RS-485 IC	HALF DUPLEX RS-485 IC	OPTOCOUPLER FOR DI / RO	OPTOCOUPLER FOR DE
250kbps	MAX488/MAX489	MAX483/MAX487	PC417*	PC357T*
2.5Mbps	MAX490/MAX491	MAX481/MAX485	PC410*	PC357T

* PC-Series Optocouplers, Sharp Electronics
 USA Phone: (206) 834-2500
 FAX: (206) 834-8903
Sharp Electronics, Europe GmbH
 Germany Phone: (040) 2376-0
 FAX: (040) 230764

Fig. 3-72

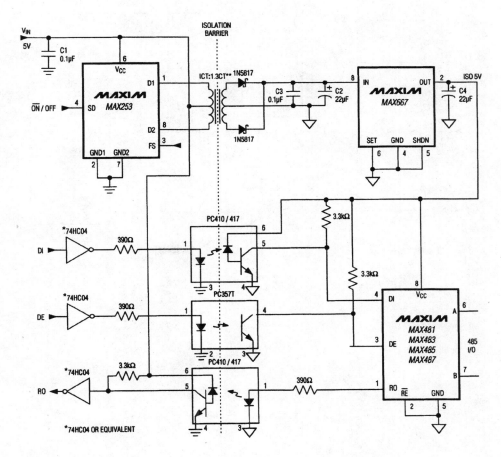

Fig. 3-73

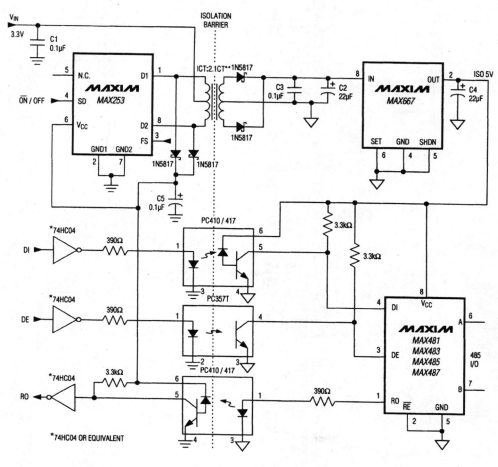

Fig. 3-74

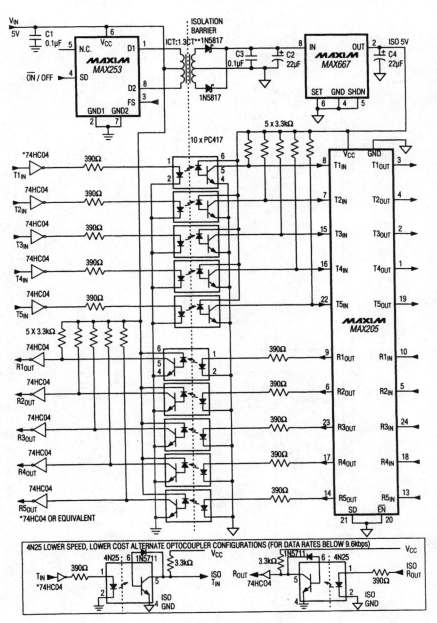

Fig. 3-75

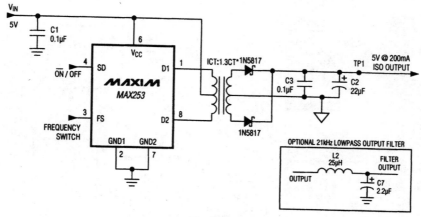

Fig. 3-76

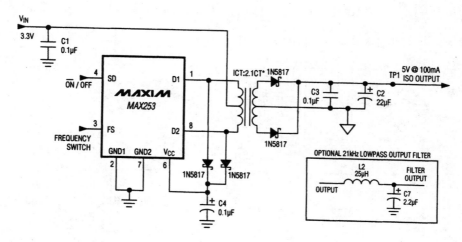

Fig. 3-77

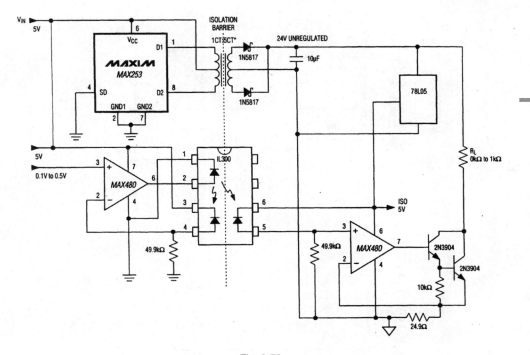

Fig. 3-78

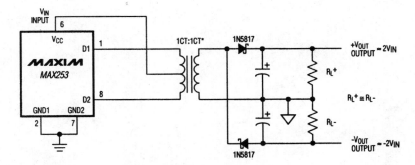

Fig. 3-79

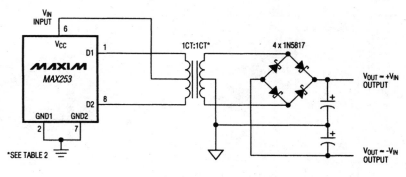

Fig. 3-80

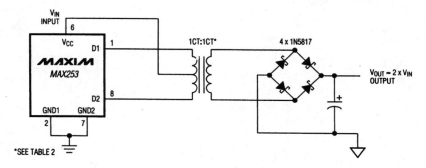

Fig. 3-81

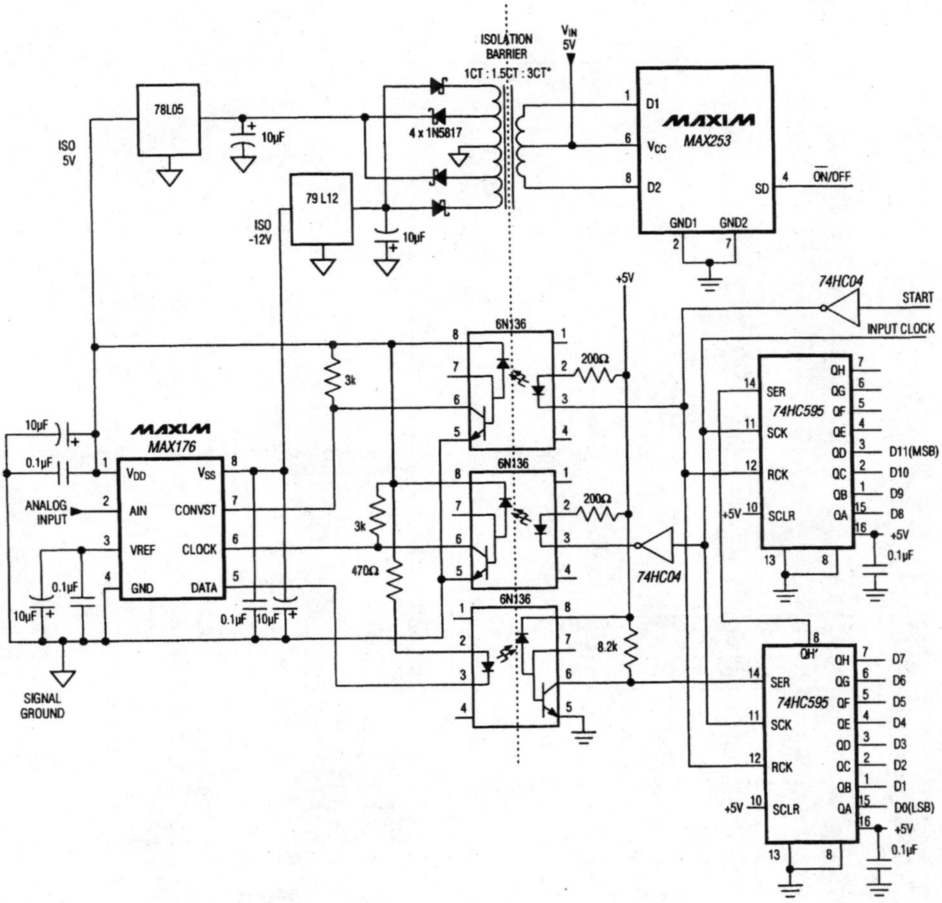

Fig. 3-82

Transformer driver for isolated RS485 interface
Figures 3-66, 3-67, and 3-68 show the block diagram, pin descriptions, and test circuit for the MAX253. Figures 3-69 through 3-72 show recommended component suppliers. This IC is a monolithic oscillator/driver specifically designed to provide isolated power for an isolated RS485 or RS-232 data interface. The IC drives a center-tapped transformer primary from a 5-V or 3.3-V supply. The secondary can be wound to provide any isolated voltage needed at power levels up to 1 W. The oscillator runs at double the output frequency, driving a toggle flip-flop to ensure a 50% duty cycle to each of the switches. The current is 0.4 µA in

shutdown. Figures 3-73 through 3-82 show typical circuits for a variety of applications. MAXIM NEW RELEASES DATA BOOK, 1995, P. 2-125, 2-136, 2-127, 2-128, 2-129, 2-130, 2-131, 2-132, 2-133, 2-134, 2-135.

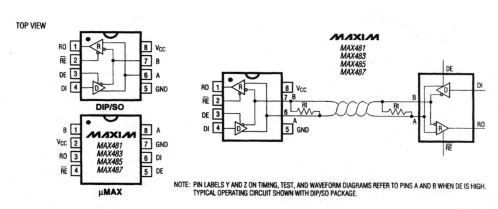

NOTE: PIN LABELS Y AND Z ON TIMING, TEST, AND WAVEFORM DIAGRAMS REFER TO PINS A AND B WHEN DE IS HIGH. TYPICAL OPERATING CIRCUIT SHOWN WITH DIP/SO PACKAGE.

Fig. 3-83

PIN					NAME	FUNCTION
MAX481/MAX483/ MAX485/MAX487		MAX488/ MAX490		MAX489/ MAX491		
DIP/SO	μMAX	DIP/SO	μMAX	DIP/SO		
1	3	2	4	2	RO	Receiver Output: If A > B by 200mV, RO will be high; If A < B by 200mV, RO will be low.
2	4	—	—	3	$\overline{RE}$	Receiver Output Enable. RO is enabled when $\overline{RE}$ is low; RO is high impedance when $\overline{RE}$ is high.
3	5	—	—	4	DE	Driver Output Enable. The driver outputs, Y and Z, are enabled by bringing DE high. They are high impedance when DE is low. If the driver outputs are enabled, the parts function as line drivers. While they are high impedance, they function as line receivers if $\overline{RE}$ is low.
4	6	3	5	5	DI	Driver Input. A low on DI forces output Y low and output Z high. Similarly, a high on DI forces output Y high and output Z low.
5	7	4	6	6, 7	GND	Ground
—	—	5	7	9	Y	Noninverting Driver Output
—	—	6	8	10	Z	Inverting Driver Output
6	8	—	—	—	A	Noninverting Receiver Input and Noninverting Driver Output
—	—	8	2	12	A	Noninverting Receiver Input
7	1	—	—	—	B	Inverting Receiver Input and Inverting Driver Output
—	—	7	1	11	B	Inverting Receiver Input
8	2	1	3	14	Vcc	Positive Supply: 4.75V ≤ Vcc ≤ 5.25V
—	—	—	—	1, 8, 13	N.C.	No Connect—not internally connected

Fig. 3-84

Interface circuits

INPUTS			OUTPUTS	
$\overline{RE}$	DE	DI	Z	Y
X	1	1	0	1
X	1	0	1	0
0	0	X	Hi-Z	Hi-Z
1	0	X	Hi-Z*	Hi-Z*

X = Don't care
Hi-Z = High impedance
* Shutdown mode for MAX481/MAX483/MAX487

Fig. 3-85

INPUTS			OUTPUT
$\overline{RE}$	DE	A-B	RO
0	0	$\geq$ +0.2V	1
0	0	$\leq$ -0.2V	0
0	0	Inputs open	1
1	0	X	Hi-Z*

X = Don't care
Hi-Z = High impedance
* Shutdown mode for MAX481/MAX483/MAX487

Fig. 3-86

PART NUMBER	HALF/FULL DUPLEX	DATA RATE (Mbps)	SLEW-RATE LIMITED	LOW-POWER SHUTDOWN	RECEIVER/ DRIVER ENABLE	QUIESCENT CURRENT	NUMBER OF TRANSMITTERS ON BUS	PIN COUNT
MAX481	Half	2.5	No	Yes	Yes	300	32	8
MAX483	Half	0.25	Yes	Yes	Yes	120	32	8
MAX485	Half	2.5	No	No	Yes	300	32	8
MAX487	Half	0.25	Yes	Yes	Yes	120	128	8
MAX488	Full	0.25	Yes	No	No	120	32	8
MAX489	Full	0.25	Yes	No	Yes	120	32	14
MAX490	Full	2.5	No	No	No	300	32	8
MAX491	Full	2.5	No	No	Yes	300	32	14

Fig. 3-87

Interface circuit titles and descriptions

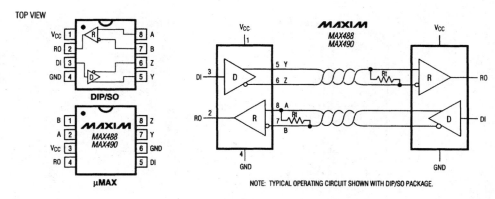

Fig. 3-88

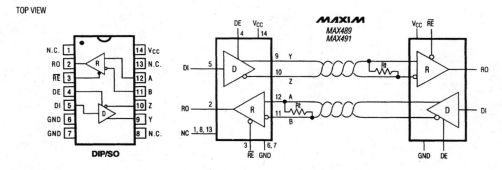

Fig. 3-89

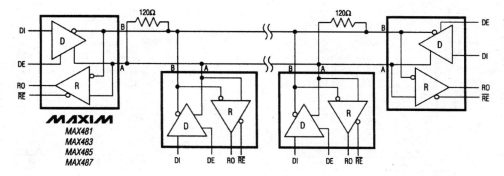

Fig. 3-90

Continued

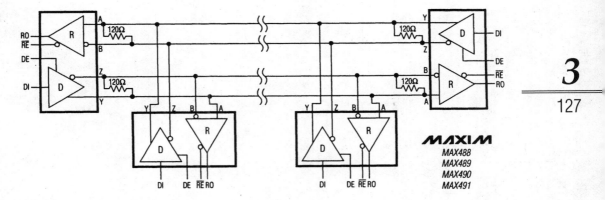

NOTE: $\overline{RE}$ AND DE ON MAX489/MAX491 ONLY.

Fig. 3-91

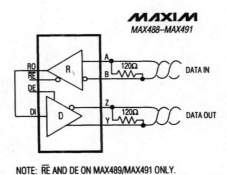

NOTE: $\overline{RE}$ AND DE ON MAX489/MAX491 ONLY.

Fig. 3-92

Low-power RS-485/RS-422 transceivers
Figure 3-83 shows the typical application circuit and pin configurations for the MAX481/83/85/87. Figures 3-84, 3-85, and 3-86 show the pin functions. Figure 3-87 is the selection table. Figures 3-88 and 3-89 show typical application circuits and pin configurations for the MAX488 through MAX491. Figure 3-90 shows the MAX481/83/85/87 in a typical RS-485 network. Figure 3-91 shows the MAX488 through MAX491 connected in a full-duplex RS485 network. Figure 3-92 shows the MAX488 through MAX491 connected as a line repeater. MAXIM NEW RELEASE DATA BOOK, 1995, P. 2-159, 2-165, 2-166, 2-167, 2-168, 2-171, 2-172.

3

127

Interface circuit titles and descriptions

4

Bridge-based circuits

This chapter is devoted to bridge-based circuits. Such circuits are often used with instrumentation amplifiers. The basic testing and troubleshooting techniques for amplifiers are covered in Chapter 6, and are not repeated here.

Figure 4-A lists performance data for some specific instrumentation amplifiers. Figures 4-B and 4-C summarize some options for dc bridge signal conditioning. The characteristics listed in Figs. 4-A, 4-B, and 4-C, such as CMRR (common-mode rejection ratio) are covered in Chapter 6.

PARAMETER	LTC1100	LT1101	LT1102	LTC1043 (USING LTC1050 AMPLIFIER)
Offset	10μV	160μV	500μV	0.5μV
Offset Drift	100nV/°C	2μV/°C	2.5μV/°C	50nV/°C
Bias Current	50pA	8nA	50pA	10pA
Noise (0.1Hz–10Hz)	2μVp-p	0.9μV	2.8μV	1.6μV
Gain	100	10,100	10,100	Resistor Programmable
Gain Error	0.03%	0.03%	0.05%	Resistor Limited 0.001% Possible
Gain Drift	4ppm/°C	4ppm/°C	5ppm/°C	Resistor Limited <1ppm/°C Possible
Gain Non-Linearity	8ppm	8ppm	10ppm	Resistor Limited 1ppm Possible
CMRR	104dB	100dB	100dB	160dB
Power Supply	Single or Dual, 16V Max	Single or Dual, 44V Max	Dual, 44V Max	Single, Dual 18V Max
Supply Current	2.2mA	105μA	5mA	2mA
Slew Rate	1.5V/μs	0.07V/μs	25V/μs	1mV/ms
Bandwidth	8kHz	33kHz	220kHz	10Hz

Fig. 4-A Performance data for typical instrumentation amplifiers.

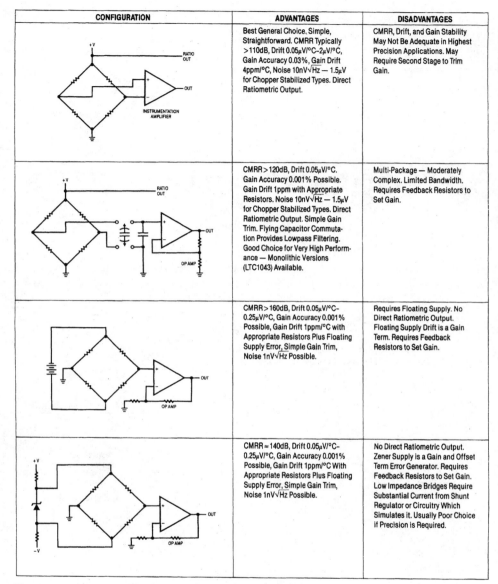

CONFIGURATION	ADVANTAGES	DISADVANTAGES
	Best General Choice. Simple, Straightforward. CMRR Typically >110dB, Drift 0.05µV/°C–2µV/°C, Gain Accuracy 0.03%, Gain Drift 4ppm/°C, Noise 10nV√Hz — 1.5µV for Chopper Stabilized Types. Direct Ratiometric Output.	CMRR, Drift, and Gain Stability May Not Be Adequate in Highest Precision Applications. May Require Second Stage to Trim Gain.
	CMRR > 120dB, Drift 0.05µV/°C. Gain Accuracy 0.001% Possible. Gain Drift 1ppm with Appropriate Resistors. Noise 10nV√Hz — 1.5µV for Chopper Stabilized Types. Direct Ratiometric Output. Simple Gain Trim. Flying Capacitor Commutation Provides Lowpass Filtering. Good Choice for Very High Performance — Monolithic Versions (LTC1043) Available.	Multi-Package — Moderately Complex. Limited Bandwidth. Requires Feedback Resistors to Set Gain.
	CMRR > 160dB, Drift 0.05µV/°C– 0.25µV/°C, Gain Accuracy 0.001% Possible, Gain Drift 1ppm/°C with Appropriate Resistors Plus Floating Supply Error, Simple Gain Trim, Noise 1nV√Hz Possible.	Requires Floating Supply. No Direct Ratiometric Output. Floating Supply Drift is a Gain Term. Requires Feedback Resistors to Set Gain.
	CMRR ≈ 140dB, Drift 0.05µV/°C– 0.25µV/°C, Gain Accuracy 0.001% Possible, Gain Drift 1ppm/°C With Appropriate Resistors Plus Floating Supply Error, Simple Gain Trim, Noise 1nV√Hz Possible.	No Direct Ratiometric Output. Zener Supply is a Gain and Offset Term Error Generator. Requires Feedback Resistors to Set Gain. Low Impedance Bridges Require Substantial Current from Shunt Regulator or Circuitry Which Simulates it. Usually Poor Choice if Precision is Required.

Fig. 4-B   Options for dc-bridge signal conditioning.

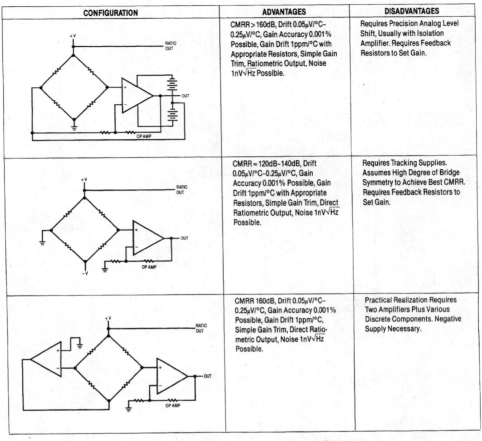

CONFIGURATION	ADVANTAGES	DISADVANTAGES
	CMRR > 160dB, Drift 0.05μV/°C–0.25μV/°C, Gain Accuracy 0.001% Possible, Gain Drift 1ppm/°C with Appropriate Resistors, Simple Gain Trim, Ratiometric Output, Noise 1nV√Hz Possible.	Requires Precision Analog Level Shift, Usually with Isolation Amplifier. Requires Feedback Resistors to Set Gain.
	CMRR ≈ 120dB–140dB, Drift 0.05μV/°C–0.25μV/°C, Gain Accuracy 0.001% Possible, Gain Drift 1ppm/°C with Appropriate Resistors, Simple Gain Trim, Direct Ratiometric Output, Noise 1nV√Hz Possible.	Requires Tracking Supplies. Assumes High Degree of Bridge Symmetry to Achieve Best CMRR. Requires Feedback Resistors to Set Gain.
	CMRR 160dB, Drift 0.05μV/°C–0.25μV/°C, Gain Accuracy 0.001% Possible, Gain Drift 1ppm/°C, Simple Gain Trim, Direct Ratiometric Output, Noise 1nV√Hz Possible.	Practical Realization Requires Two Amplifiers Plus Various Discrete Components. Negative Supply Necessary.

Fig. 4-C   Additional options for dc-bridge signal conditioning.

Bridge-based circuit titles and descriptions

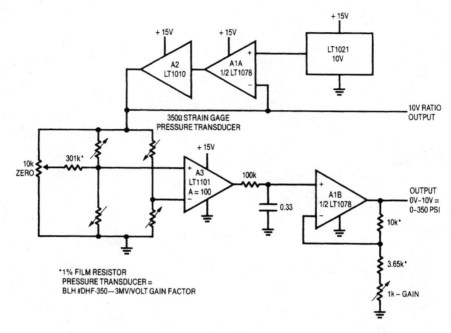

Fig. 4-1

Bridge-based instrumentation amplifier
Figure 4-1 shows signal conditioning for a 350-Ω transducer bridge. The specified strain-gauge pressure transducer produces 3-mV output per volt of bridge excitation. The LT1021 reference, buffered by A1A and A2, drives the bridge. This potential also supplies the ratio output, permitting ratiometric operation of a monitoring analog-to-digital converter (Chapter 8). Amplifier A3 extracts the bridge differential output at a gain of 100, with additional trimmed gain supplied by A1B. The circuit shown can be adjusted for a precise 10-V output at full-scale pressure. To trim the circuit, apply zero pressure to the transducer, adjust the 10-kΩ pot until the output just comes off 0 V. Then apply full-scale pressure and trim the 1-kΩ adjustment. Repeat the procedure until both points are fixed. LINEAR TECHNOLOGY, APPLICATION NOTE 43, P. 5.

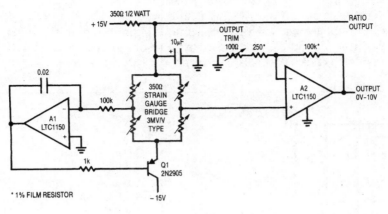

Fig. 4-2

Servo-controlling bridge drive
Figure 4-2 shows a way to reduce errors because of the bridge common-mode
output voltage. A1 biases Q1 to servo the bridge left midpoint to zero under all
operating conditions. The 350-Ω resistor ensures that A1 will find a stable
operating point with 10 V of drive delivered to the bridge. This allows A2 to take
a single-ended measurement, eliminating all common-mode voltage errors.
LINEAR TECHNOLOGY, APPLICATION NOTE 43, P. 5.

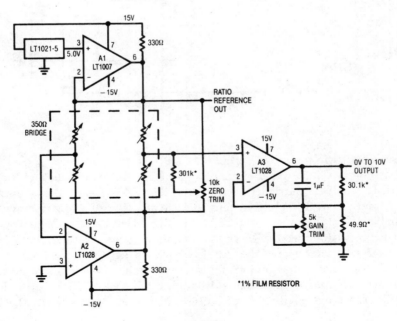

Fig. 4-3

Low-noise bridge amplifier with common-mode suppression
Figure 4-3 shows a circuit that is similar to that of Fig. 4-2, except that low-noise bipolar amplifiers are used. This circuit trades slightly higher dc offset drift for lower noise; it is a good candidate for stable resolution of small, slowly varying measurements. LINEAR TECHNOLOGY, APPLICATION NOTE 43, P. 6.

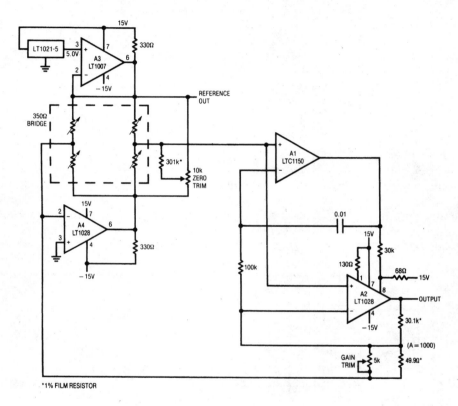

Fig. 4-4

Low-noise, chopper-stabilized bridge amplifier
Figure 4-4 shows a circuit that is similar to that of Fig. 4-3, except that A1 is chopper stabilized. This reduces the offset error even further. A1 measures the dc error at the A2 inputs and biases the A1 offset pins to force an offset of a few microvolts. The offset-pin bias at A2 is arranged so that A1 will always be able to find the servo point. The 0.01-μF capacitor rolls off A1 at low frequencies, with A2 handling the high-frequency signals. Returning the A2 feedback string to the bridge midpoint eliminates the A4 offset contribution. If this was not done, A4 would require a similar offset-correction loop. The circuit has a drift of less than 0.05-μV/°C, 1-nV/Hz noise, and CMRR exceeding 160 dB. LINEAR TECHNOLOGY, APPLICATION NOTE 43, P. 6.

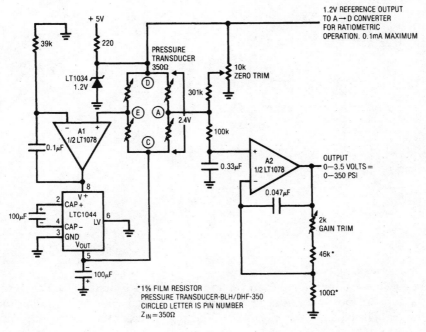

Fig. 4-5

Single-supply bridge amplifier with common-mode suppression
Figure 4-5 shows a circuit that is similar to that of Fig. 4-4, except that a single power supply is required. A2 biases the LTC1044 positive-to-negative converter. The LTC1044 output pulls the bridge output negative, causing the A1 input to balance at 0 V. This local loop permits a single-ended amplifier (A2) to extract the bridge output signal. The 100-kΩ/033-μF RC filter minimizes noise. The A2 gain is set to provide the desired output scale factor. Because bridge drive is taken from the LT1034 reference, the A2 output is not affected by supply shifts. The LT1034 output is available for ratio operation. Although the supply is 5 V, the transducer "sees" only 2.4 V of drive. This reduced drive results in lower transducer outputs for a given measurement value, effectively magnifying amplifier offset-drift terms. The limit on the available bridge drive is set by the CMOS LTC1044 output impedance. LINEAR TECHNOLOGY, APPLICATION NOTE 43, P. 7.

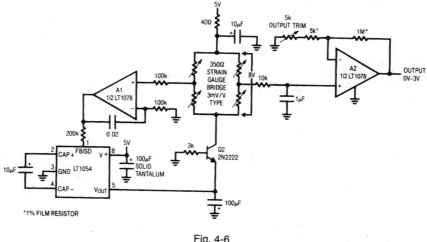

Fig. 4-6

High-resolution single-supply bridge amplifier
Figure 4-6 shows a high-resolution version of the Fig. 4-5 circuit. The Fig. 4-6 circuit uses a bipolar positive-to-negative converter, which has much lower output impedance. The biasing used permits 8 V to appear across the bridge, requiring the 100-mA capability of the LT1054 to sink about 24 mA. This increased drive results in a more favorable transducer-gain per slope, increasing the signal-to-noise ratio. LINEAR TECHNOLOGY, APPLICATION NOTE 43, P. 7.

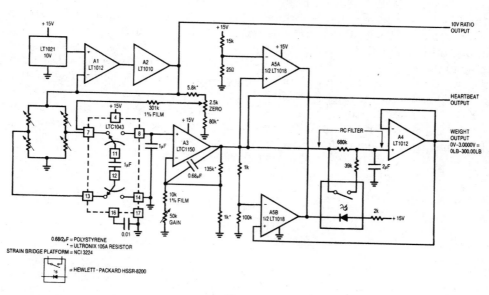

Fig. 4-7

Bridge-based circuits

High-precision weight scale

Figure 4-7 shows a switched-capacitor type of bridge circuit used in weight-scale applications. (The circuit is intended for weighing human subjects.) Resolution is 0.01 pound at 300.00 pounds, full scale. To trim the circuit, adjust the zero pot for 0 V out with no weight on the platform. Then set the gain adjustment for 3.0000-V output for a 300.00-pound platform weight. Repeat this procedure until both points are fixed. LINEAR TECHNOLOGY, APPLICATION NOTE 43, P. 8.

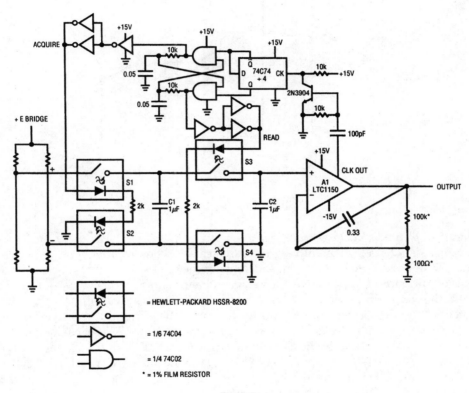

Fig. 4-8

Floating-input bridge

Figure 4-8 shows an optically coupled switched-capacitor bridge with a floating input. The common-mode rejection ratio at dc for the front-end exceeds 160 dB. The amplifier operates over a ±200-V common-mode range. Gain-accuracy and stability are limited only by external resistors. The offset drift is 0.05-μV/°C. The optical drive to the MOSFET eliminates the charge-injection problems that are common to FET switched-capacitor networks. LINEAR TECHNOLOGY, APPLICATION NOTE 43, P. 10.

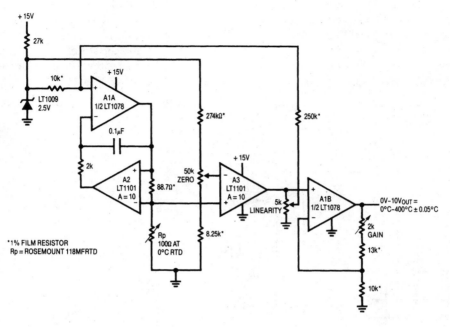

Fig. 4-9

Linearized platinum RTD resistance bridge

Figure 4-9 shows a bridge with a ground-referred RTD to improve noise rejection. The RTD leg is driven by a current source, with the opposite bridge branch voltage biased. The current drive allows the bridge voltage across the RTD to vary directly with the temperature-induced resistance shift. The difference between this potential and that of the opposing bridge leg forms the bridge output. To calibrate, substitute a precision decade box (General Radio 1432k) for R_p. Set the box to the 0°C value (100.00 Ω) and adjust the offset trim for a 0.00-V output. Then set the decade box for a 140°C output (154.26 Ω) and adjust the gain trim for a 3.500-V output reading. Finally, set the box to 249.0-Ω (400.00°C) and trim the linearity adjustment for a 10.000-V output. Repeat the sequence until all three points are fixed. The total error over the entire range will be within ±0.05°C. The resistance values given are for a nominal 100.00 Ω (0°C) sensor. Sensors deviating from this nominal value can be used by factoring in the deviation from 100.00 Ω. LINEAR TECHNOLOGY, APPLICATION NOTE 43, P. 11.

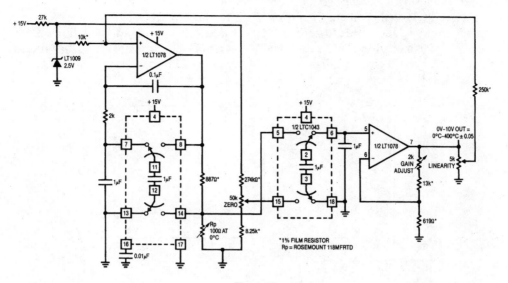

Fig. 4-10

Linearized platinum RTD resistance bridge (switched capacitor)
Figure 4-10 is a switched-capacitor version of the Fig. 4-9 circuit, with A2 and A3 replaced with an LTC1043. The differential-to-single-ended transitions in the current-source and bridge-output amplifier are performed by the LTC1043. The values shifts in the current source and output stage reflect the LTC1043 lack of gain. The primary trade-off between the two circuits is component count versus cost. LINEAR TECHNOLOGY, APPLICATION NOTE 43, P. 12.

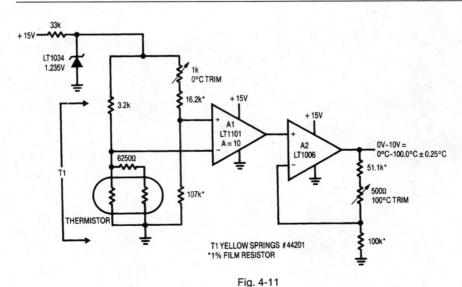

Fig. 4-11

Bridge-based circuit titles and descriptions

Linear-output thermistor bridge

Figure 4-11 uses a thermistor in one leg of the bridge circuit. A trim in the opposing leg sets the bridge output to zero at 0°C. A1 and A2 provide additional trimmed gain to furnish a calibrated output. Calibration is performed in similar fashion to the platinum RTD circuits (Figs. 4-9 and 4-10). LINEAR TECHNOLOGY, APPLICATION NOTE 43, P. 17.

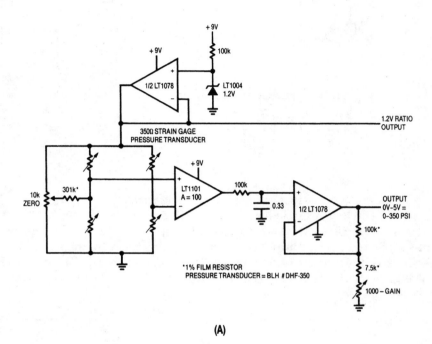

(A)

Fig. 4-12A

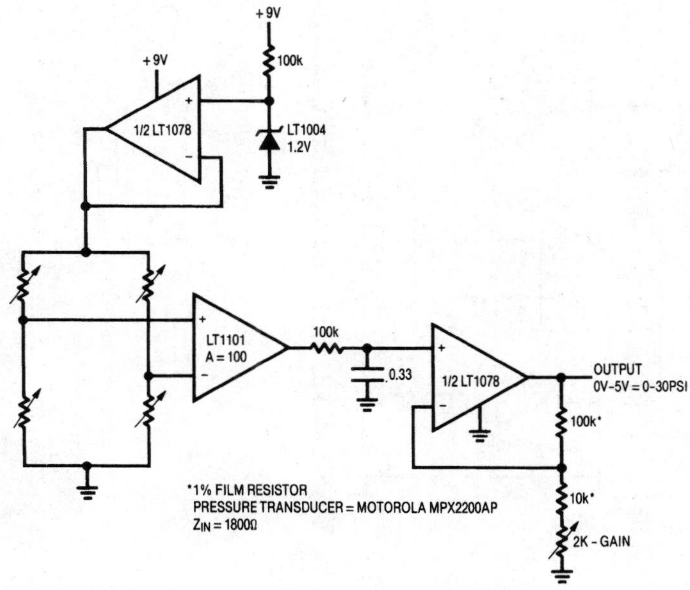

(B)

Fig. 4-12B

Low-power bridge
Figure 4-12A is identical to Fig. 4-1, except that the bridge excitation is reduced to 1.2 V. This cuts bridge current from about 30 mA to about 3.5 mA. A 0.01% reading of a 10-V powered 350-Ω strain-gauge bridge (Fig. 4-1) requires 3 μV of stable resolution. The Fig. 4-12A circuit requires only 360 nV. Figure 4-12B is similar, but the bridge current is reduced below 700 μA. LINEAR TECHNOLOGY, APPLICATION NOTE 43, P. 18.

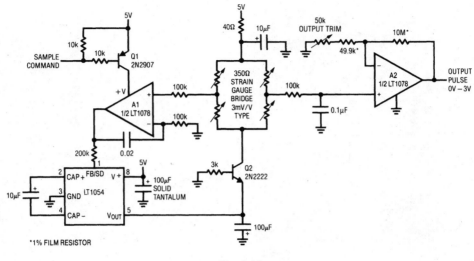

Fig. 4-13

Strobed-power strain-gauge bridge
Figure 4-13 (derived from Fig. 4-6) shows a way to reduce power without sacrificing bridge signal-output level. The technique is applicable where continuous output is not a requirement. The circuit is in a quiescent state for long periods with relatively brief on times. (A typical application would be where remote weight information in storage tanks is sampled once per week.) Quiescent current is about 150 µA, with on-state current typically 50 mA. LINEAR TECHNOLOGY, APPLICATION NOTE 43, P. 19.

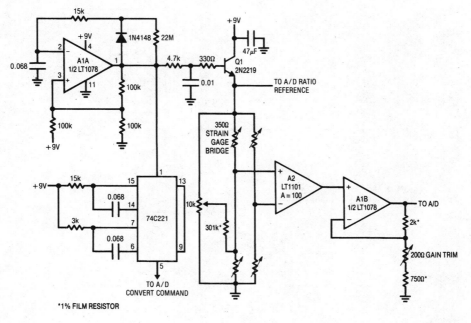

Fig. 4-14

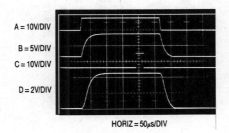

Fig. 4-15

Sampled-output bridge signal conditioner
Figure 4-14 shows a bridge signal conditioner with sampled output. (This is an extension of the Fig. 4-13 circuit.) Figure 4-15 shows the waveforms. Traces A, B, C, and D show the A1A output, Q1 emitter, 74C221 output, and A2 output, respectively. The strobing action is automated by a clocked sequence. Circuit on time is restricted to 250 μs, at a clock rate of about 2 Hz. This keeps average power consumption down to about 200 μA. Oscillator A1A produces a 250-μs clock pulse every 500 ms (trace A, Fig. 4-15). To calibrate, trim both zero and gain for the appropriate outputs. LINEAR TECHNOLOGY, APPLICATION NOTE 43, P. 20.

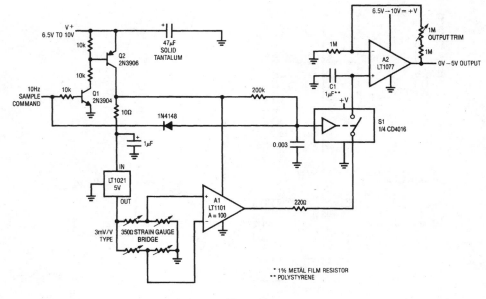

Fig. 4-16

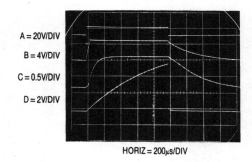

HORIZ = 200μs/DIV

Fig. 4-17

Continuous-output sampled bridge

Figure 4-16 shows a bridge signal conditioner with sampled continuous output. Figure 4-17 shows the waveforms. Traces A, B, C, and D show the Q2 collector, LT1021-5 output, A1 output, and S1 input, respectively. The output is made continuous by the addition of a sample-hold stage. Q2 is off when the sample command is low. Under these conditions, only A2 and S1 receive power, resulting in a current drain of less than 60 μA. When the sample command is pulsed high, the Q2 collector (trace A, Fig. 4-17) goes high, providing power to all other circuit elements. During the sampling phase, supply current approaches 20 mA, but a 10-Hz sampling rate cuts effective drain below 250 μA. Slower sampling rates will

Bridge-based circuits

further reduce drain, but the C1 droop rate (about 1 mV/100 ms) sets the accuracy constant. The 10-Hz rate provides adequate bandwidth for most transducers. For 3-mV/V-slope-factor transducers, the gain trim shown (1 M) allows proper calibration. It might be necessary to rescale the gain trim for other transducer types. The A2 output is accurate enough for 12-bit systems. Although the output is continuous, information is collected at a 10-Hz rate, and the Nyquist limit applies when interpreting the results. LINEAR TECHNOLOGY, APPLICATION NOTE 43, P. 21.

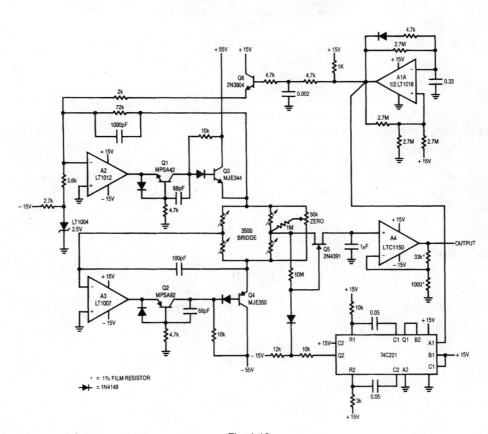

Fig. 4-18

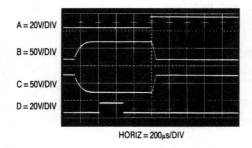

A = 20V/DIV

B = 50V/DIV

C = 50V/DIV

D = 20V/DIV

HORIZ = 200μs/DIV

Fig. 4-19

High-resolution continuous-output sampled bridge
Figure 4-18 shows a sampled bridge with continuous output and high resolution. Figure 4-19 shows the waveforms. Traces A, B, C, and D show the A1A output, Q3 emitter, comple-mentary output to A2 loop, and the 74C221 delayed pulse to Q5. The circuit applies 100 V across a 10-V, 350-Ω strain-gauge bridge for short periods of time. The high pulsed-voltage drive increases bridge output proportionally, without forcing excessive dissipation. The average bridge power is far below the normal 29 mA obtained with the usual 10-Vdc excitation. Combining the 10× higher bridge gain (300 mV full-scale versus the normal 30 mV) with a chopper-stabilized amplifier in the sample-hold output stage is the key to the high resolution. To calibrate, simply adjust the 50-kΩ zero control for zero output with the bridge at mechanical zero. Although the output is continuous, information is collected at a 1-Hz rate, and the Nyquist limit applies when interpreting results. LINEAR TECHNOLOGY, APPLICATION NOTE 43, P. 23.

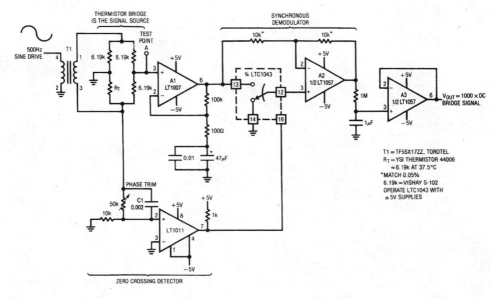

Fig. 4-20

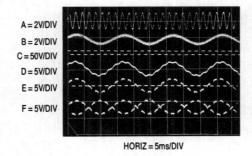

A = 2V/DIV
B = 2V/DIV
C = 50V/DIV
D = 5V/DIV
E = 5V/DIV
F = 5V/DIV

HORIZ = 5ms/DIV

Fig. 4-21

Ac-driven bridge/synchronous demodulator

Figure 4-20 shows a circuit that combines a bridge with a synchronous detector to obtain very high noise rejection capability. Figure 4-21 shows the waveforms. Traces A, B, C, D, E, and F show T1 input, A1 input, LTC1043 clock, A1 output, A2 positive input, and A2 output, respectively. An ac carrier excites the bridge and synchronizes the gain-stage demodulator. In this application, the signal source is a thermistor bridge that detects extremely small temperature shifts in a biochemical microcalorimetry-reaction chamber. To calibrate, adjust the phase pot so that C1 switches when the carrier crosses through zero. LINEAR TECHNOLOGY, APPLICATION NOTE 43, P. 24.

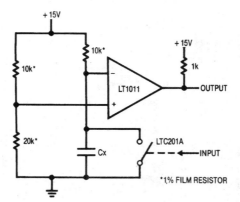

Fig. 4-22

Time-domain bridge

Figure 4-22 shows a bridge circuit that operates in the time domain, and is particularly suitable for capacitance measurement. With S1 closed, the comparator output is high. When S1 opens, capacitance C_x charges. When the C_x potential crosses the voltage established at the comparator noninverting input, the comparator output goes low. The elapsed time between the switch opening and the comparator going low is proportionate to the C_x value. The circuit is insensitive to supply and repetition rate variations, and can provide good accuracy if the time constants are kept much larger than comparator and switch delays. The LT1011 delay is about 200 ns, and the LTC201A delay is about 450 ns. To ensure 1% accuracy, the time constant of C_x and the series resistor should not go below 65 μs. Extremely low values of capacitance might be influenced by switch charge injection. In such cases, switching should be implemented by alternating the bridge drive between ground and +15 V. LINEAR TECHNOLOGY, APPLICATION NOTE 43, P. 27.

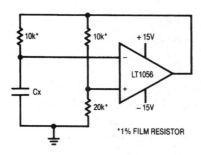

Fig. 4-23

Bridge-based circuits

Bridge oscillator with square-wave output
Figure 4-23 shows an oscillator circuit formed from the circuit of Fig. 4-22. In effect, the Fig. 4-23 circuit is a classic op-amp multivibrator (MV). The 10-kΩ/20-kΩ bridge leg provide switching-point hysteresis, with C_x charged through the remaining 10-kΩ resistor. When C_x reaches the switching point, the amplifier output changes state, abruptly reversing the sign of the positive output. The charging direction of C_x also reverses to sustain oscillation. Output frequency depends on bridge components (at frequencies that are low compared to amplifier delays). Amplifier input errors tend to cancel, and supply shifts are generally rejected. The duty cycle is influenced by output saturation and supply asymmetry. LINEAR TECHNOLOGY, APPLICATION NOTE 43, P. 27.

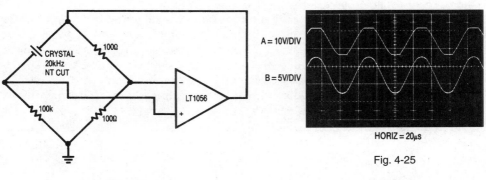

Fig. 4-24

Fig. 4-25

Quartz-stabilized bridge oscillator
Figure 4-24 shows a bridge-based oscillator, where one leg of the bridge is replaced with a resonant element. Figure 4-25 shows the waveforms. With the crystal removed, the circuit is a basic non-inverting gain-of-two amplifier with grounded input. Inserting the crystal closes a positive-feedback path at the crystal resonant frequency. The amplifier output (trace A, Fig. 4-25) swings in an attempt to maintain input balance. Excessive circuit gain prevents linear operation, and oscillations start when the amplifier repeatedly overshoots in an attempt to null the bridge. The high Q of the crystal is evident in the filtered waveform (trace B) at the amplifier positive input. LINEAR TECHNOLOGY, APPLICATION NOTE 43, P. 27.

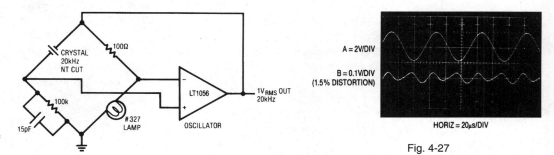

Fig. 4-26

Fig. 4-27

Quartz-stabilized bridge oscillator with sine-wave output
Figure 4-26 shows a crystal-controlled bridge-based oscillator, where one leg of the bridge is formed by a lamp (a classic circuit). Figure 4-27 shows the waveforms. As the oscillation amplitude builds, the lamp current increases, as does lamp resistance. This causes a reduction in gain, and the circuit finds a stable operating point. The 15-pF capacitor suppresses spurious oscillation. The amplifier output (trace A, Fig. 4-27) is a sine wave without about 1.5% distortion (trace B), caused primarily by common-mode swing. LINEAR TECHNOLOGY, APPLICATION NOTE 43, P. 28.

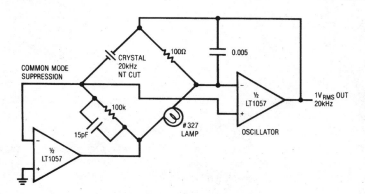

Fig. 4-28

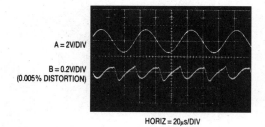

Fig. 4-29

Bridge-based circuits

Quartz-stabilized bridge oscillator with common-mode suppression
Figure 4-28 shows a bridge-oscillator circuit where the common-mode swing is suppressed by the addition of a second amplifier. This configuration forces the bridge midpoint to virtual ground by measuring the midpoint value and comparing it to ground. Because the bridge drive is complementary, the oscillator amplifier sees no common-mode swing, thus reducing distortion. Figure 4-29 shows less than 0.005% distortion (trace B) in the output (trace A) waveform. (Distortion measurements are covered in Chapter 6.) LINEAR TECHNOLOGY, APPLICATION NOTE 43, P. 28.

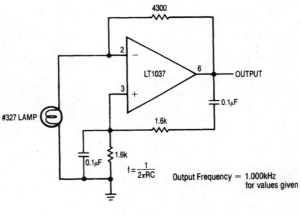

Fig. 4-30

Basic Wien-bridge sine-wave oscillator
Figure 4-30 shows a Wien-bridge sine-wave oscillator with a lamp in one leg of the bridge (similar to that of Figs. 4-26 and 4-28). LINEAR TECHNOLOGY, APPLICATION NOTE 43, P. 29.

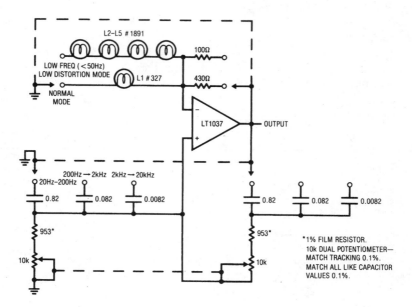

Fig. 4-31

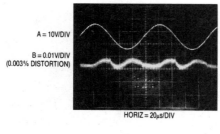

Fig. 4-32

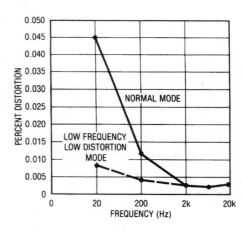

Fig. 4-33

Multi-range Wien-bridge sine-wave oscillator
Figure 4-31 shows a Wien-bridge sine-wave oscillator with a frequency range from
20 Hz to 20 kHz, in three decade ranges, with 0.25-dB amplitude flatness. Trace A
of Fig. 4-32 shows circuit output at 10 kHz. Trace B shows distortion (below
0.003%). Figure 4-33 plots distortion versus frequency. LINEAR TECHNOLOGY,
APPLICATION NOTE 43, P. 29, 30.

Bridge-based circuits

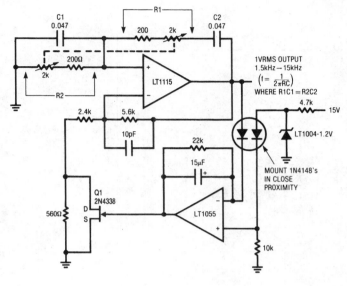

Fig. 4-34

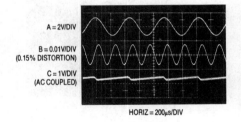

HORIZ = 200μs/DIV

Fig. 4-35

Wien-bridge sine-wave oscillator with electronic amplitude stabilization
Figure 4-34 shows a sine-wave oscillator that uses electronic stabilization in place
of the lamp. Figure 4-35 shows the waveforms. Trace A shows the output. Trace B
shows distortion, and trace C shows the rectifier peaking residue at the Q1 gate.
LINEAR TECHNOLOGY, APPLICATION NOTE 43, P. 30, 31.

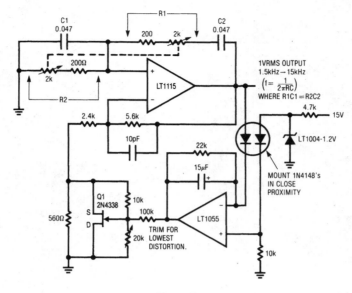

Fig. 4-36

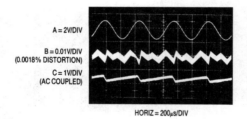

HORIZ = 200μs/DIV

Fig. 4-37

Wien-bridge oscillator with improved stabilization
Figure 4-36 shows an improved version of the Fig. 4-34 circuit. Figure 4-37 shows waveforms for the Fig. 4-36 circuit. The 20-kΩ trimmer in the Q1 circuit makes it possible to get 0.0018% distortion (trace B, Fig. 4-37). LINEAR TECHNOLOGY, APPLICATION NOTE 43, P. 31.

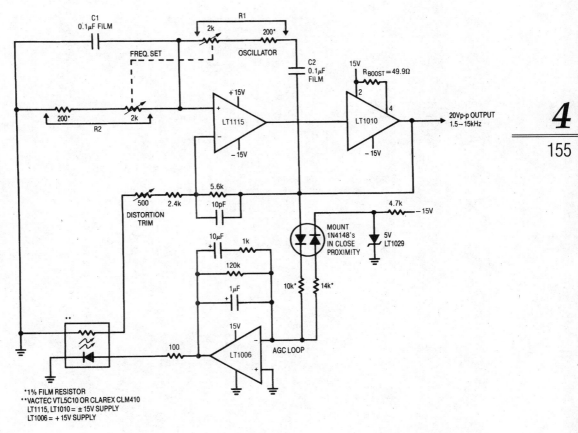

Fig. 4-38

Wien-bridge oscillator with automatic trim
Figure 4-38 shows another version of the Fig. 4-34 circuit, but with automatic trim for distortion. Q1 is replaced with an optically driven CdS photocell, and the LT1055 is replaced with a ground-sensing op amp operating in the single-supply mode. Distortion improves slightly to 0.0015%. LINEAR TECHNOLOGY, APPLICATION NOTE 43, P. 33.

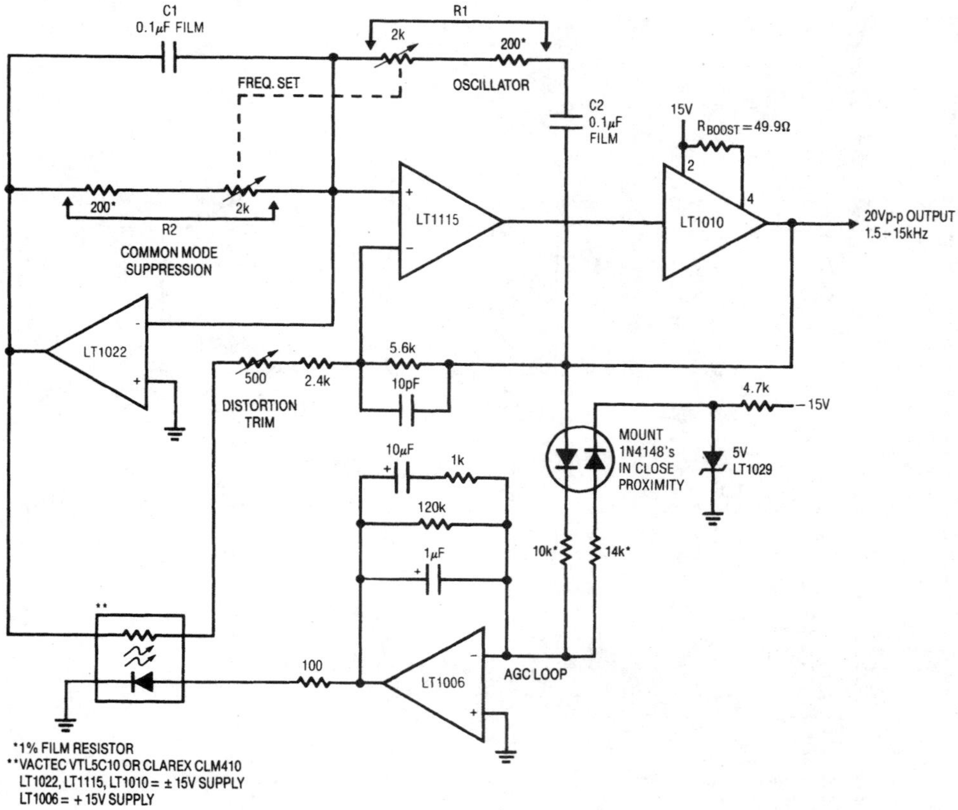

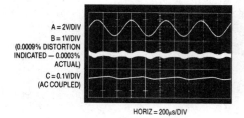

Fig. 4-39

C1
0.1μF FILM

R1

2k

200*

FREQ. SET

OSCILLATOR

C2
0.1μF
FILM

15V

R_BOOST = 49.9Ω

2

LT1010

4

20Vp-p OUTPUT
1.5–15kHz

200*

R2

2k

COMMON MODE
SUPPRESSION

LT1115

LT1022

500

2.4k

DISTORTION
TRIM

5.6k

10pF

MOUNT
1N4148's
IN CLOSE
PROXIMITY

4.7k

–15V

5V
LT1029

10μF

1k

120k

1μF

10k*

14k*

100

LT1006

AGC LOOP

*1% FILM RESISTOR
**VACTEC VTL5C10 OR CLAREX CLM410
LT1022, LT1115, LT1010 = ± 15V SUPPLY
LT1006 = + 15V SUPPLY

A = 2V/DIV

B = 1V/DIV
(0.0009% DISTORTION
INDICATED — 0.0003%
ACTUAL)

C = 0.1V/DIV
(AC COUPLED)

HORIZ = 200μs/DIV

Fig. 4-40

Bridge-based circuits

Wien-bridge oscillator with common-mode suppression
Figure 4-39 shows another version of the Fig. 4-34 circuit, but with automatic trim and common-mode suppression (similar to that of Fig. 4-28). The circuit output (trace A, Fig. 4-40) contains less than 0.0003% (3 ppm) distortion (trace B), with no visible correlation to gain-loop ripple residue (trace C). This level of distortion is below the uncertainty floor of most distortion analyzers. LINEAR TECHNOLOGY, APPLICATION NOTE 43, P. 32, 33.

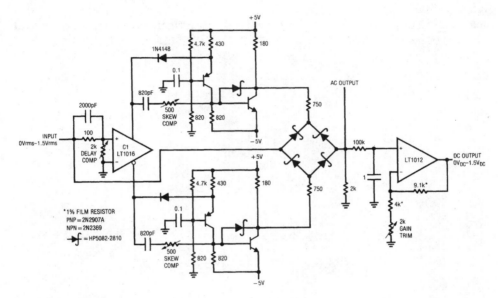

Fig. 4-41

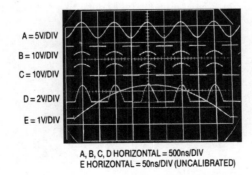

A, B, C, D HORIZONTAL = 500ns/DIV
E HORIZONTAL = 50ns/DIV (UNCALIBRATED)

Fig. 4-42

Diode bridge-based rectifier/AC-voltmeter

Figure 4-41 shows a circuit that provides both ac and dc outputs for inputs up to 2.5 MHz. Figure 4-42 shows the waveforms. Trace A shows the input sine wave, traces B and C are the switched corners of the bridge, trace D is the ac output, trace E is an expanded version of trace D. To calibrate, apply a 1-MHz to 2-MHz 1-V$_{p-p}$ sine wave and adjust the delay compensation so that bridge switching occurs when the sine crosses zero. This adjustment corrects for the small delays through the LT1016 and the level shifters. Next, adjust the skew-compensation pots for minimum distortion in the ac output signal (trace D). These trims shift the phase of the rising output edge of their respective level shifter. This allows skew in the complementary bridge-drive signals to be kept within 1 ns to 2 ns, minimizing output disturbances when switching occurs. A 100-mV sine input will produce a clean output with a dc output accuracy of better than 0.25%. LINEAR TECHNOLOGY, APPLICATION NOTE 43, P. 34.

=5=

High-speed (wideband) circuits

This chapter is devoted to high-speed (wideband) circuits. Such circuits are primarily some form of op amp, or make extensive use of op amps. The basic testing and troubleshooting techniques for op amps are covered in Chapter 6, and are not repeated here.

High-speed (wideband) circuit titles and descriptions

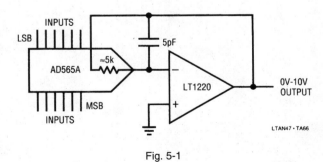

Fig. 5-1

Continued

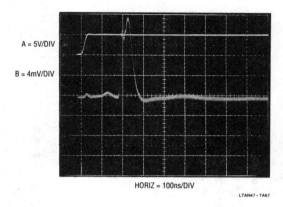

A = 5V/DIV

B = 4mV/DIV

HORIZ = 100ns/DIV

LTAN47 · TA67

Fig. 5-2

DAC amplifier
Figure 5-1 shows an amplifier that is suitable for the output of a fast 12-bit DAC (digital-to-analog converter, Chapter 8). Figure 5-2 shows clean 0.01% settling in 280 ns (trace B) to an all-bits-on input step (trace A). LINEAR TECHNOLOGY, APPLICATION NOTE 47, P. 32.

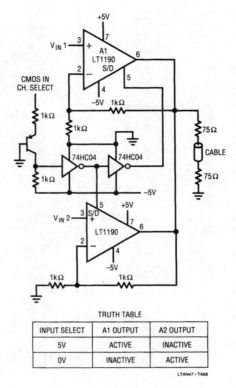

Fig. 5-3

TRUTH TABLE

INPUT SELECT	A1 OUTPUT	A2 OUTPUT
5V	ACTIVE	INACTIVE
0V	INACTIVE	ACTIVE

LTAN47 · TA68

Two-channel video amplifier
Figure 5-3 shows a simple way to multiplex two video amplifiers onto a single 75-
Ω cable. The appropriate amplifier is activated in accordance with the truth table.
Amplifier performance includes 0.02% differential-gain and 0.1° differential-phase
errors. The 75-Ω back termination looking into the cable means that the amplitude
must swing 2 Vp-p to produce 1-Vp-p at the cable output. LINEAR TECHNOLOGY,
APPLICATION NOTE 47, P. 33.

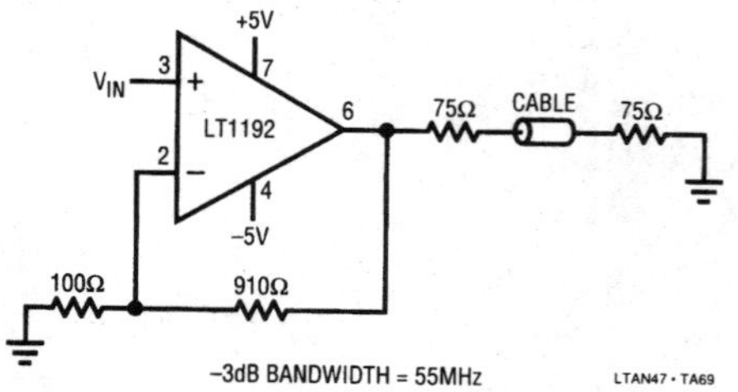

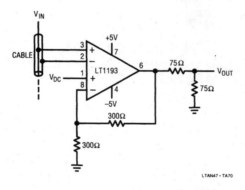

Fig. 5-4

Simple video amplifier

Figure 5-4 is a simpler version of Fig. 5-3, but with only one video channel. The values provide for a gain of 10, with a bandwidth of 55 MHz. LINEAR TECHNOLOGY, APPLICATION NOTE 47, P. 33.

Fig. 5-5

Cable-sense amplifier for loop-through connections

Figure 5-5 shows a differential amplifier used to extract signals from a distribution cable. The amplifier differential inputs reject common-mode signals. Amplifier performance includes 0.02% differential-gain and 0.1° differential-phase errors. A separate input permits dc adjustment. LINEAR TECHNOLOGY, APPLICATION NOTE 47, P. 33.

High-speed (wideband) circuits

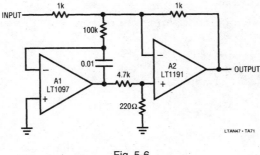

Fig. 5-6

Dc stabilization using a summing point
Figure 5-6 shows a circuit where the precision offset of a dc amplifier (LT1097) is combined with the bandwidth of a fast device (LT1191) to provide a dc stabilized, wideband amplifier. The LT1097 monitors the summing node (the two 1-kΩ resistors), compares the node to ground, and drives the LT1191 noninverting input. The resulting circuit is a unity-gain inverter with 35-µV offset, 1.5-V/°C drift, 450-V/µs slew rate, and 90-MHz bandwidth. Bias current, dominated by the LT1191, is about 500 nA. LINEAR TECHNOLOGY, APPLICATION NOTE 47, P. 33.

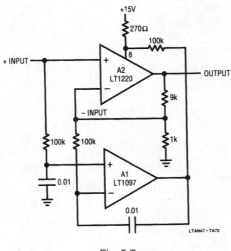

Fig. 5-7

High-speed (wideband) circuit titles and descriptions

Dc stabilization using differential sensing

Figure 5-7 is similar to Fig. 5-6, except that the sensing is done differentially, preserving access to both fast amplifier inputs. The combined characteristics of these amplifiers yields the following performance: 50-μV offset voltage, 1-μV/°C offset drift, 250-V/μs slew rate, and 45-MHz gain bandwidth. LINEAR TECHNOLOGY, APPLICATION NOTE 47, P. 34.

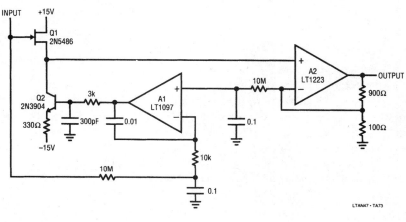

Fig. 5-8

Dc stabilization with high input impedance

Figure 5-8 shows a dc-stabilized, wideband amplifier using a FET input for high input impedance. Input capacitance is about 3 pF, the bandwidth is 100 MHz, and the gain is 10 (using the feedback values shown). A1 stabilizes the circuit by controlling the Q1 channel current through feedback. LINEAR TECHNOLOGY, APPLICATION NOTE 47, P. 34.

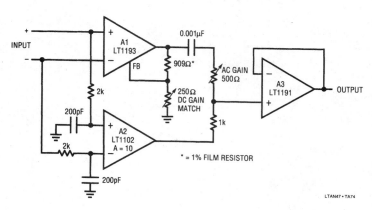

Fig. 5-9

High-speed (wideband) circuits

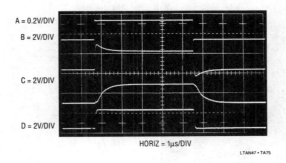

A = 0.2V/DIV

B = 2V/DIV

C = 2V/DIV

D = 2V/DIV

HORIZ = 1µs/DIV

LTAN47 • TA75

Fig. 5-10

Dc stabilization with differential input and gain of 10
Figure 5-9 shows a way to get full differential inputs with dc-stabilized operation. Figure 5-10 shows the waveforms. Trace A is one side of the differential input applied to the circuit, trace B is the A1 output (taken at the 500-Ω pot and 0.001-µF junction), trace C is the A2 output, and trace D is the A3 output. To adjust, trim the AC GAIN pot for the squarest corners, and the DC GAIN pot for a flat top, of the output signal (trace D). Circuit gain is 10, bandwidth exceeds 35 MHz, slew rate is 450 V/µs, and dc offset is about 200 µV. LINEAR TECHNOLOGY, APPLICATION NOTE 47, P. 35.

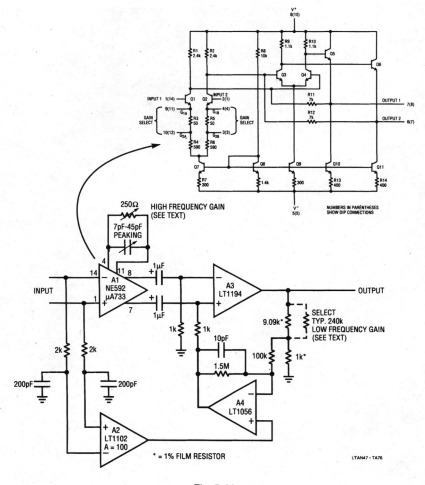

Fig. 5-11

High-speed (wideband) circuits

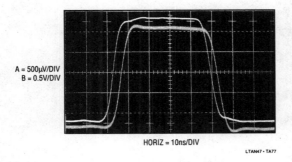

A = 500μV/DIV
B = 0.5V/DIV

HORIZ = 10ns/DIV

LTAN47 · TA77

Fig. 5-12

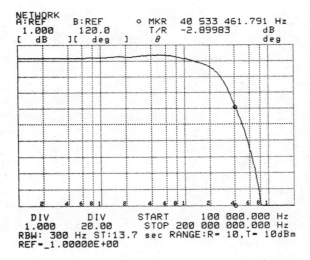

NETWORK
A:REF   B:REF   ○ MKR   40 533 461.791 Hz
 1.000   120.0   T/R   -2.89983   dB
[dB][deg]   θ   deg

 DIV   DIV   START   100 000.000 Hz
 1.000   20.00   STOP 200 000 000.000 Hz
RBW: 300 Hz ST:13.7 sec RANGE:R= 10,T= 10dBm
REF=_1.00000E+00

Fig. 5-13

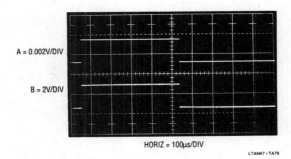

A = 0.002V/DIV

B = 2V/DIV

HORIZ = 100μs/DIV

LTAN47 · TA79

Fig. 5-14

High-speed (wideband) circuit titles and descriptions

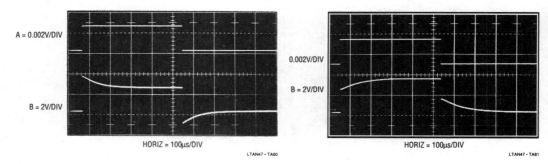

Fig. 5-15   Fig. 5-16

Dc stabilization with differential input and a gain of 1000
Figure 5-11 is similar to Fig. 5-9, but with a gain of 1000. The Fig. 5-11 circuit bandwidth is about 35 MHz, with full-power response available to 10 MHz. The rise time is 7 ns and delay is less than 7.5 ns. Input noise is about 15 μV broadband. Figure 5-12 shows the circuit response to a 60-ns, 2.5-mV pulse (trace A). The ×1000 output (trace B) has delay and rise times in the 5-ns to 7-ns range. Peaking can be trimmed with the adjustment at A1. Figure 5-13 plots the circuit gain versus frequency. Gain is flat within 0.5 dB to 20 MHz, with the −3-dB point at 38 MHz. The edge peaking shown in Fig. 5-12 shows up in Fig. 5-13 as a very slight gain increase, starting at about 1 MHz and continuing out to about 15 MHz. (This gain increase can be trimmed with the peaking adjustment, if necessary.) To use the circuit, apply a low-frequency or dc signal of known amplitude and adjust the low-frequency gain (parallel resistor) for a ×1000 output (after the output has settled). Next, adjust the high-frequency gain pot so that the output signal front and rear corners have amplitudes identical to the settled portion. Finally, trim the peaking-adjustment capacitor for best settling of the output pulse front and rear corners. Figure 5-14 shows the input (trace A) and output (trace B) waveforms with all adjustments properly set. Figure 5-15 shows the effects of too much ac gain (excessive peaking). Figure 5-16 shows the effects of too much dc gain (long trailing response, with incorrect amplitude). LINEAR TECHNOLOGY, APPLICATION NOTE 47, P. 36, 37.

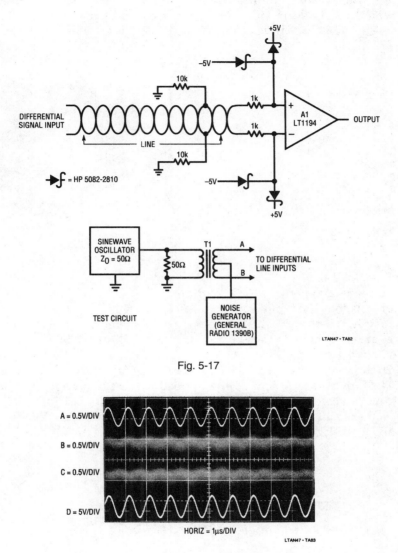

Fig. 5-17

Fig. 5-18

High-speed differential line receiver
Figure 5-17 shows a simple, fast differential line receiver using an LT1194 gain-of-ten amplifier. A test circuit is also included on Fig. 5-17. Figure 5-18 shows the test results. The sine-wave oscillator drives T1 (trace A, Fig. 5-18), producing a differential-line output at the secondary. The T1 secondary is returned to ground through a broadband noise generator, flooding the inputs with common-mode

High-speed (wideband) circuit titles and descriptions

noise (traces B and C are A1 inputs.) Trace D, the A1 ×10 version of the differential signal at the input, is clean with no visible noise or disturbances, in spite of the 100:1 noise-to-signal ratio. LINEAR TECHNOLOGY, APPLICATIONS NOTE 47, P. 38.

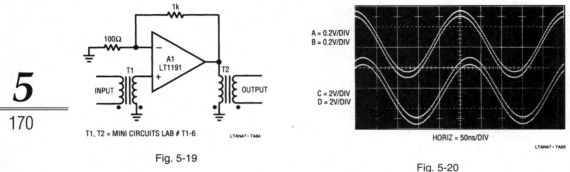

Fig. 5-19

Fig. 5-20

Transformer-coupled amplifier
Figure 5-19 shows another way to get high common-mode rejection, together with true 3-port isolation. The input, gain stage, and output are all galvanically isolated from each other. As such, this circuit is useful where large common-mode differences are encountered or where ground integrity is uncertain. With the values shown, A1 has a gain of 11. T1 feeds the A1 input, and the output is taken from T2. Figure 5-20 shows results for a 4-MHz input, with all of the polarity dots referred to ground. The input (trace A, Fig. 5-20) is applied to T1. The output of T1 (trace B) feeds A1. A1 produces gain, and the A1 output (trace C) feeds T2. The T2 output (trace D) is the circuit output. Using the specified transformers, the low-frequency cutoff is about 10 kHz. LINEAR TECHNOLOGY, APPLICATION NOTE 47, P. 39.

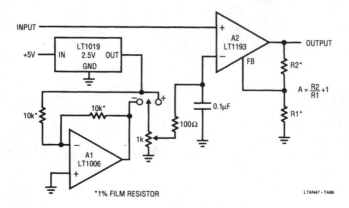

Fig. 5-21

High-speed (wideband) circuits

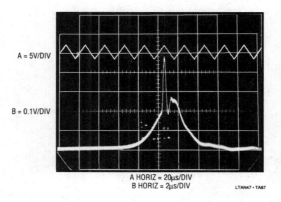

A = 5V/DIV

B = 0.1V/DIV

A HORIZ = 20μs/DIV
B HORIZ = 2μs/DIV

LTAN47 · TA87

Fig. 5-22

Fast differential comparator-amplifier with adjustable offset
Figure 5-21 shows a circuit that permits one particular portion of a signal to be amplified (or examined) with all other portions rejected. Figure 5-22 shows what happens when the output of a triangle-wave generator (trace A) is applied to the circuit. Setting the bias level just below the triangle peak permits high-gain, detailed observation of the turnaround at the peak. Switching residue in the generator output is observable in trace B. Appropriate variations in voltage-source setting will permit more of the triangle slopes to be observed, (with a loss of resolution because of scope-overload). Similarly, increasing the A2 gain allows more amplitude detail, while placing restrictions on how much of the waveform can be displayed. In effect, this circuit performs the same functions as differential plug-in units for scopes. The circuit output is accurate and settled to 0.1% about 100 ns after entering the linear region. LINEAR TECHNOLOGY, APPLICATION NOTE 47, P. 39, 40.

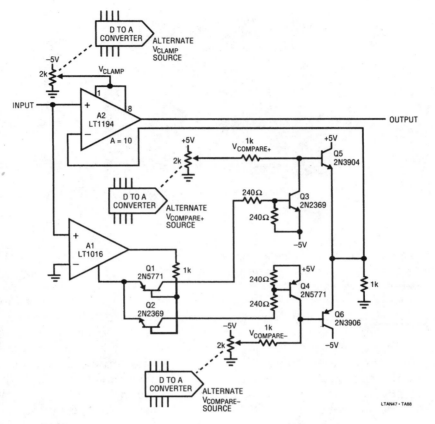

Fig. 5-23

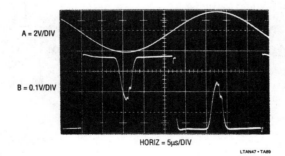

Fig. 5-24

High-speed (wideband) circuits

Differential comparator-amplifier with adjustable limiting and offset
Figure 5-23 is similar to Fig. 5-21, except that the dual channels permit observations of information between two adjustable amplitude-defined points. The setpoints are adjustable in both magnitude and sign. In the circuit of Fig. 5-23, the polarity of the offset applied to the A2 inverting input is determined by the comparator A1 output state. Figure 5-24 shows the circuit output for a sine input (trace A). The $+V_{\text{COMPARE}}$ and $-V_{\text{COMPARE}}$ voltages are set just below the sine-wave peaks, with V_{CLAMP} set to restrict amplification to the peak excursion. The circuit output (trace B) simultaneously shows the amplitude detail of both sine peaks. LINEAR TECHNOLOGY, APPLICATION NOTE 47, P. 40, 41.

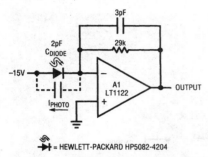

= HEWLETT-PACKARD HP5082-4204

RESPONSE DATA

LIGHT (900nM)	DIODE CURRENT	CIRCUIT OUTPUT
1mW	350µA	10.0V
100µW	35µA	1V
10µW	3.5µA	0.1V
1µW	350nA	0.01V
100nW	35nA	0.001V

LTAN47 • TA90

Fig. 5-25

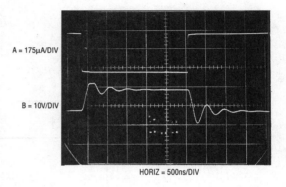

HORIZ = 500ns/DIV

Fig. 5-26

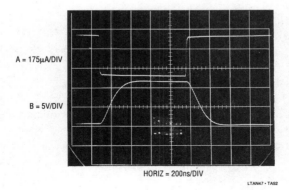

HORIZ = 200ns/DIV

LTAN47 • TA92

Fig. 5-27

High-speed (wideband) circuit titles and descriptions

Photodiode amplifier

Figure 5-25 shows a basic photodiode amplifier. The table shows various circuit outputs for different diode currents. Figure 5-26 shows circuit response to a photo input (trace A) with the 3-pF feedback capacitor removed. Notice that the output overshoots and saturates before finally ringing down to a level value. Figure 5-27 shows circuit performance with the 3-pF feedback capacitor in place. The same input pulse (trace A, Fig. 5-27) produces a cleanly damped output (trace B). However, the circuit is about 50% faster with the feedback capacitor removed. LINEAR TECHNOLOGY, APPLICATION NOTE 47, P. 41, 42.

5

174

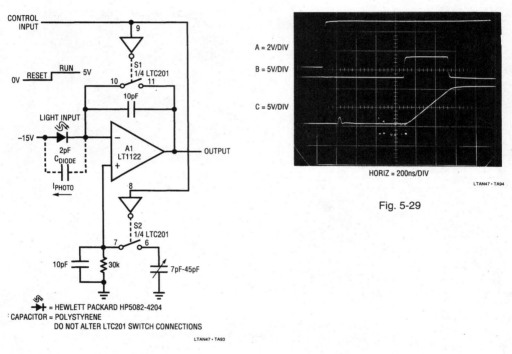

Fig. 5-28

Fig. 5-29

Fast photo integrator

Figure 5-28 shows a photodiode-amplifier circuit that was specifically designed for situations where the total energy in a light pulse (or pulses) must be measured. The circuit is a very fast integrator, with S1 used as a reset switch. S2, switched simultaneously with S1, compensated the S1 charge-injection error. With the control input (trace A, Fig. 5-29) low and no photocurrent, S1 is closed and A1 looks like a grounded follower. Under these conditions, the A1 output (trace C) remains 0 V. When the control input becomes high, A1 becomes an integrator as

High-speed (wideband) circuits

soon as S1 opens. At this point, the integrator is ready to receive and record a photo pulse. When light falls on the photodiode (trace B triggers a light pulse "seen" by the photodiode), A1 responds by integrating, and the A1 voltage (after the light event is over) is related to the total energy at the diode during the event. To adjust, cover the photodiode and set the trimmer capacitor for 0-V output from A1 (immediately after the S1/S2 switching function). LINEAR TECHNOLOGY, APPLICATION NOTE 47, P. 42.

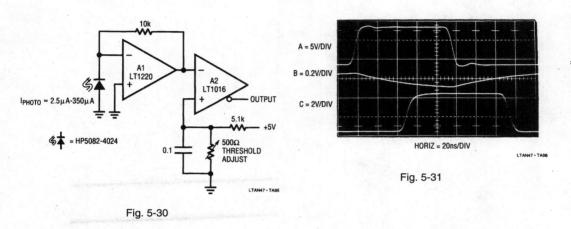

Fig. 5-30

Fig. 5-31

Fiber-optic receiver
Figure 5-30 shows a simple, high-speed fiber-optic receiver. Figure 5-31 shows the waveforms. Trace A is a pulse associated with a photo input. Trace B is the A1 response, and trace C is the A2 output. A2 compares the A1 output to a dc level that is established by the threshold-adjust setting, thus producing a logic-compatible output. LINEAR TECHNOLOGY, APPLICATION NOTE 47, P. 43.

High-speed (wideband) circuit titles and descriptions

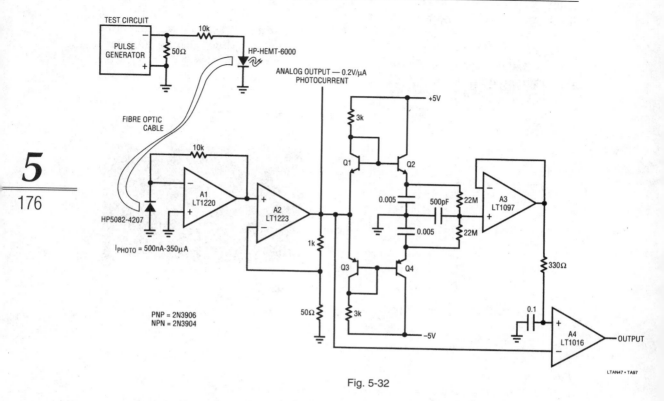

Fig. 5-32

Fig. 5-33

40-MHz fiber-optic receiver with adaptive trigger
Figure 5-32 shows a circuit that will reliably condition fiber-optic inputs up to 40
MHz with input amplitudes that vary by as much as 40 dB. The digital output
features an adaptive threshold trigger which accommodates varying signal
intensities. An analog output is also available to monitor the detector output.
Figure 5-33 shows the results of using the test circuit in Fig. 5-32. Trace A is the

High-speed (wideband) circuits

pulse-generator output. Trace B is the A2 output (analog output monitor). Trace C is the LT1016 output. These waveforms were recorded with a 5-μA photocurrent at about 20 MHz. LINEAR TECHNOLOGY, APPLICATION NOTE 47, P. 44.

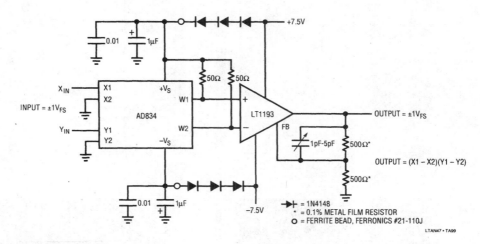

Fig. 5-34

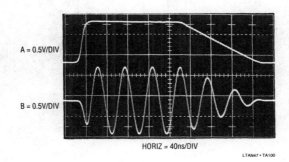

HORIZ = 40ns/DIV

Fig. 5-35

50-MHz high-accuracy analog multiplier with single-ended output
Figure 5-34 shows an analog multiplier with single-ended output. The error is within 2% over the range from dc to 50 MHz, with feedthrough below −50 dB. Figure 5-35 shows the performance when a 20-MHz sine input is multiplied by the trace A waveform. The output (trace B) is a clean instantaneous representation of the X/Y input products, with ±1 V at full scale. To trim, adjust the capacitor for minimum output square-wave peaking. LINEAR TECHNOLOGY, APPLICATION NOTE 47, P. 45.

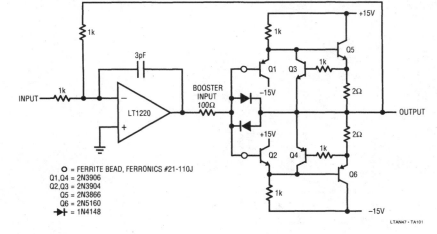

Fig. 5-36

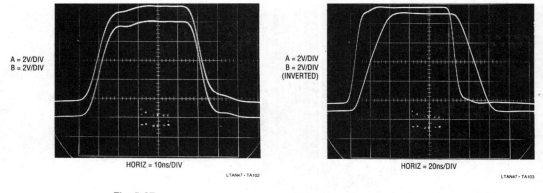

Fig. 5-37   Fig. 5-38

Power booster

Figure 5-36 shows a 200-mA power booster used with an LT1220 amplifier. Figure 5-37 shows booster performance with the LT1220 removed. The input pulse (trace A) is applied to the booster input, with the output (trace B) taken at the indicated point. The total delay is about 1 ns. Figure 5-38 shows pulse response with the LT1220 installed and a 50-Ω load. The input (trace A) produces a slew-limited output (trace B). LINEAR TECHNOLOGY, APPLICATION NOTE 47, P. 46.

High-speed (wideband) circuits

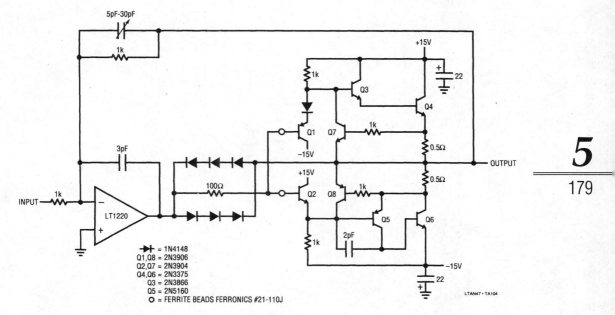

Fig. 5-39

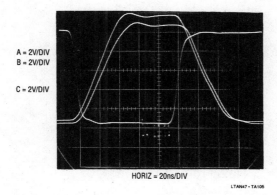

Fig. 5-40

High-power booster   Figure 5-39 is a 1-A version of the Fig. 5-36 circuit. Figure 5-40 shows the waveforms for a 10-V negative input step (trace A) with a 10-Ω load. The amplifier responds (trace B) driving the booster to close the loop. The booster output, lagging by nanoseconds (trace C), drives the load with only minor peaking (which can be minimized with the feedback capacitance trimmer). LINEAR TECHNOLOGY, APPLICATION NOTE 47, P. 47.

Op-amp and
comparator circuits

This chapter is devoted to op-amp and comparator circuits. It is assumed that you already are familiar with amplifier basics (amplification principles, bias operating points, etc.), practical considerations (heatsinks, power dissipation, and component-mounting techniques), and simplified design of amplifiers (frequency limitations). If you are not familiar with these basics, read *Lenk's Audio Handbook*, McGraw-Hill, 1991; *Lenk's RF Handbook*, McGraw-Hill, 1991; and *Simplified Design of IC Amplifiers*, Butterworth-Heinemann, 1996. The following paragraphs summarize both testing and troubleshooting of amplifier circuits. This information is included so that readers not familiar with electronic procedures can both test the circuits described here and localize problems if the circuits fail to perform as shown.

Amplifier tests

This section covers the basic tests for IC amplifiers. The section begins with a review of typical amplifier test equipment and then goes on to describe test procedures that can be applied to the circuits of this chapter. Actual circuit test results are used where practical. If the circuits pass these basic tests, use the circuits immediately. If not, use the tests as a starting point for the troubleshooting procedures that are described in this chapter. Keep in mind that all amplifier circuits need not be subjected to all tests described here. However, if an amplifier circuit produces the desired results for all of the tests, you can consider the circuit to be a successful design.

Amplifier test equipment

The tests described in this chapter can be performed using meters, oscilloscopes, generators, power supplies, assorted clips, patch cords, and so on. So, if you have a good set of test equipment that is suitable for other electronic work, you can probably get by. A possible exception is a distortion meter (especially if you are interested in audio amplifiers). Here are some points that you should consider when selecting and using test equipment.

Matching test equipment to the circuits   No matter what test instrument is involved, try to match the capabilities of the test equipment to the circuit. For example, if you are going to measure pulses, square waves, or complex waves (as you might for any IC amplifier test), a peak-to-peak meter can provide meaningful indications, but a scope is the logical instrument of choice.

Voltmeters/multimeters   In addition to making routine voltage and resistance checks, the main functions of a meter in amplifier work is to measure frequency response and trace signals from input to output. Many technicians prefer scopes for these procedures. The reason for the preference is that scopes also show distortion of the waveform during measurement or signal tracing. Other technicians prefer the simplicity of a meter, particularly in such procedures as voltage-gain and power-gain measurements.

You can sometimes get by with any ac meter (even a basic multimeter, analog or digital) for all amplifier work. However, for accurate measurements, use a wideband meter, preferably a dual-channel model. (Obviously, the meter must have a bandwidth greater than the amplifier circuit being tested!) The dual-channel feature makes it possible to monitor both channels of a stereo circuit simultaneously. This feature is particularly important for stereo frequency-response and crosstalk measurements but is of no great importance for nonstereo amplifiers.

Scopes   If you have a good scope for TV and VCR work, use that scope for all amplifier-circuit measurements. If you are considering buying a new scope, remember that a *dual-channel* instrument lets you monitor both channels of a stereo circuit (as is the case with a dual-channel voltmeter). Of greater importance, a dual-channel scope lets you monitor the input and output of an amplifier simultaneously. A scope also has the advantage over a meter in that the scope can display such common IC-amplifier conditions as distortion, hum, ripple, overshoot, and oscillation. It is an indispensable tool for measuring such characteristics as settling time, slew rate, and noise. (However, the meter is easier to read when you are measuring only gain.)

Distortion meters   If you are already in audio/stereo work, you probably have distortion meters (and know how to use them effectively). The two types of distortion measurements are: *harmonic* and *intermodulation*. No particular meter is described here. Instead, descriptions are included of how harmonic and intermodulation distortion measurements are made.

Decibel measurement basics

The *decibel* (dB) is widely used in amplifier work to express logarithmically the ratio between two power or voltage levels. For example, a typical IC op-amp data sheet lists voltage gain, power gain, and common-mode rejection ratio in dB. The decibel is one-tenth of a bel. (The bel is too large for most practical applications.)

Although there are many ways to express a ratio, the decibel is used in amplifiers for two reasons: (1) the decibel is a convenient unit to use for all types of amplifiers and (2) the decibel is related to the reaction of the human ear and is thus well suited for use with audio amplifiers.

Humans can listen to ordinary conversation comfortably and can hear thunder (which is 100,000 times louder than conversation) without damage to the ear. This capability exists because the response of the human ear to sound waves is approximately proportional to the logarithm of the sound-wave energy and is not directly proportional to the energy.

The common logarithm ($\log_{10}$) of a number is the number of times 10 must be multiplied by itself to equal that number. For example, the logarithm of 100 (that is 10×10), or (10^2) is 2. Likewise, the logarithm of 100,000 (10^5) is 5. The relationship is written:

$$\log_{10} 100,000 = 5$$

In comparing two powers, it is possible to use the bel (which is the logarithm of the ratio of the two powers). For example, in comparing the power of ordinary conversation with that of thunder, the increase in sound is equal to:

$$\log_{10} \frac{power\ of\ thunder}{power\ of\ conversation} \text{ or } \log_{10} \frac{100,000}{1}$$

Using the more convenient decibel, the increase in sound from ordinary conversation to thunder is equal to:

$$10 \log_{10} \frac{100,000}{1} \text{ or 50 decibels (or 50 dB)}$$

For convenience, the same method is used in measuring the increase in amplifier power, whether the amplifiers are used with audio frequencies or not. The increase in power of any amplifier can be expressed as:

$$gain\ (in\ dB) = 10 \log_{10} \frac{power\ output}{power\ input}$$

This relationship also can be expressed as:

$$gain\ (in\ dB) = 10 \log_{10} \frac{P_2}{P_1}$$

Usually, P_2 represents power output and P_1 represents power input. If P_2 is greater than P_1, there is a power gain, expressed in positive decibels (+dB). With P_1 greater than P_2, there is a power loss, expressed in negative decibels (−dB). Whichever is the case, the ratio of the two powers (P_1 and P_2) is taken, and the logarithm of this ratio is multiplied by 10. As a result, power ratio of 10 = 10-dB gain, power ratio of 100 = 20-dB gain, power ratio of 1000 = 30-dB gain, and so on.

Doubling power ratios

Doubling the power of an amplifier produces a power gain of +3 dB. For example, if the volume control of an amplifier is turned up so that the power rises from 4 to 8 W (watts), the gain is up +3 dB. If the power output is reduced from 4 to 2 W, the gain is down −3 dB.

If the original 4 W is increased to 8 W, the power gain is +3 dB. Increasing the power output further to 16 W produces another gain of +3 dB, with a total power gain of +6 dB. At 40 W, the power is increased 10 times (from the original 4 W), and the total power gain is +10 dB, and so on.

Adding decibels

There is another convenience in using decibels for amplifier work. When several amplifier stages are connected so that one works into another (stages connected in *cascade*, as is the usual case in IC amplifiers), the gains of each stage are multiplied. For example, if three stages each with a gain of 10, are connected in cascade, there is a total power gain of $10 \times 10 \times 10$, or 1000.

In the decibel system, the decibel gains are added. Using the example, the decibel power gain is $10 + 10 + 10$, or +30 dB. Similarly, if two amplifier stages are connected, one of which has a gain of +30 dB, and the other a loss of −10 dB, the net result is $+30 - 10 = +20$ dB.

Using decibels to compare voltages and currents

The decibel system also is used to compare the voltage input and output of an amplifier. (Decibels can be used to express current ratios. However, this is generally not practical in amplifiers.) When voltages (or currents) are involved, the decibel is a function of:

$$20 \log \frac{output\ voltage}{input\ voltage}, \ 20 \log \frac{output\ current}{input\ current}$$

The ratio of the two voltages (or currents) is taken, and the logarithm of this ratio is multiplied by 20.

Notice that, although power ratios are independent of source and load impedance values, *voltage* and *current ratios* in those equations hold true only when the source and load impedances are equal. In amplifiers where input and output impedances differ, voltage and current ratios are calculated as:

$$20 \log \frac{E_1 \sqrt{R_2}}{E_2 \sqrt{R_1}}, \ 20 \log I_1 \frac{\sqrt{R_2}}{I_2 \sqrt{R_1}}$$

where R_1 is the source or input impedance and R_2 is the load or output impedance ($E_1 \sqrt{R_2}$ and $I_1 \sqrt{R_2}$ are always higher in value than $E_2 \sqrt{R_1}$ and $I_1 \sqrt{R_1}$).

Op-amp and comparator circuits

As is true for the power relationship, if the voltage output is greater than the input, there is a decibel gain (+dB). If the output is less than the input, there is a voltage loss (−dB).

Notice that doubling the voltage produces a gain of +6 dB. Conversely, if the voltage is cut in half, there is a loss of −6 dB. To get the net effect of several voltage-amplifier stages working together, add the decibel gains (or losses) of each.

Decibels and reference levels

When an amplifier has a power gain of +20 dB, this has no numerical meaning in actual power output. Instead, it means that the power output is 100 times as great as the power input. For this reason, decibels are often used in specific reference levels.

The most common reference levels for audio amplifiers are the *VU* (volume unit) and the *dBm* (decibel meter).

When VU is used, it is assumed that the zero level is equal to 0.001 W (1 milliwatt, mW) across a 600-Ω impedance. Thus:

$$VU = 10 \log \frac{P_2}{0.001} = 10 \log \frac{P_2}{10^{-3}} = 10 \log 10^3 \, P_2$$

where P_2 is the output power.

Both the dBm and VU have the same zero level base. A dBm scale is normally found on meters when the signal to be measured is a sine wave (normally 1 kHz—kilohertz), and the VU is used for complex audio waveforms.

Frequency response

You can measure amplifier frequency response with a generator and a meter or scope. Tune the generator to various frequencies and measure the resultant output response at each frequency. Plot the results in the form of a graph or response curve. Figure 6-A shows the test connections to measure *open-loop gain* (A_{OL}) for a typical op amp (a Harris CA3450). *Open-loop gain* applies to the IC-amplifier gain without feedback. Figure 6-B shows the plot (sometimes called a *Bode* plot) or graph for the IC when frequency response is measured in the open-loop condition. (Notice that Fig. 6-B also shows the phase shift that occurs at various frequencies.)

Figure 6-C shows the test connections for the same IC when closed-loop gain is being measured (with direct feedback between pins 3 and 6 for unity gain and with resistors at pins 3 and 6 for a voltage gain, A_V, of 10). Figures 6-D and 6-E show the graphs for closed-loop gain.

The frequency at which the output begins to drop is called the *rolloff point*. The specifications for some IC amplifiers consider the rolloff point to start when the output drops 3 dB below the flat portion of the curve. In the graph of Fig. 6-E, the rolloff begins at about 10 MHz and drops 3 dB at about 20 MHz.

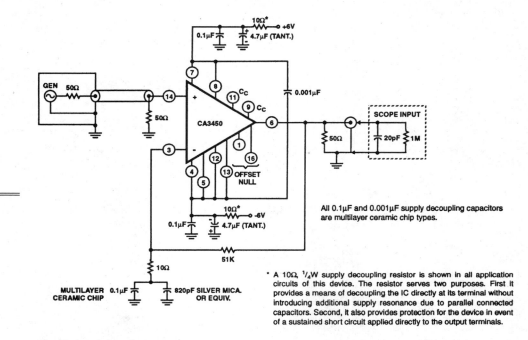

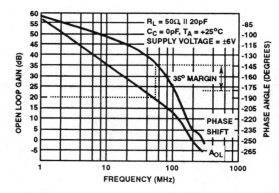

Fig. 6-A Open-loop gain test connections HARRIS SEMICONDUCTOR, LINEAR & TELECOM ICS, 1994, P. 2–217.

Fig. 6-B Bode plot for CA3450 HARRIS SEMICONDUCTOR, LINEAR & TELECO: 2–216.

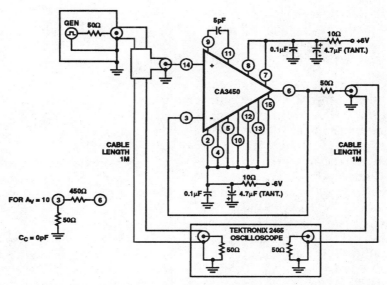

Fig. 6-C Closed-loop gain test connections Harris Semiconductor, Linear & Telecom ICs, 1994, p. 2–218.

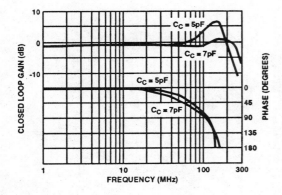

Fig. 6-D Phase/frequency plot for unity gain Harris Semiconductor, Linear & Telecom ICs, 1994, p. 2–216.

Fig. 6-E Phase/frequency plot for $A_V = 10$ Harris Semiconductor, Linear & Telecom ICs, 1994, p. 2–216.

Some IC amplifiers provide for the connection of an external compensation circuit (usually a capacitor but sometimes a capacitor-resistor combination). Such compensation circuits alter both the rolloff point and the gain-frequency relationship. No external compensation is provided in the circuit of Fig. 6-A. The circuit of Fig. 6-C shows an external compensating capacitor connected at pins 9 and 11. Figure 6-D shows the effect of different capacitor (C_c) values on gain and phase shift. Many IC amplifiers have built-in compensation, so open-loop gain-phase characteristics cannot be altered.

Amplifier tests

The procedure for measuring frequency response is to apply a constant-amplitude signal while monitoring the output. Vary the input signal frequency (but not amplitude) across the entire operating range of the amplifier. Plot the voltage output at various frequencies across the range on a graph, as follows:

1. Connect the equipment as shown in Figs. 6-A or 6-B. Keep in mind that these illustrations show the connections for a specific IC amplifier. However, the circuits have all of the elements of typical gain/frequency test circuit. The test signal is applied to the noninverting input, with the inverting input connected to the output (or possibly to ground for the open-loop test of some IC amplifiers). The output should be an amplified, noninverted version of the input. Both the input and output are terminated at some specific test value (typically 50 Ω). The input should be terminated in an impedance equal to that of the signal source, with the output terminated at the impedance of the test instrument (scope) input.

   Notice that the IC amplifier characteristics will change (sometimes drastically) with changes in output load (and external compensation). So, if the data-sheet test values are not close to those of the real load, try testing the IC with real-world values at the output, in addition to the tests with data-sheet values. (You might find that the IC will not meet your particular frequency/gain or rolloff requirements!)

   This same precaution applies to tests with and without external compensation. For example, the test circuit of Fig. 6-A shows no external compensation (no connection between pins 9 and 11). The test circuit of Fig. 6-C shows an external capacitor at pins 9/11. By comparing the graphs of Figs. 6-B, 6-D, and 6-E, notice that the gain/frequency characteristics are different with and without compensation. For example, the open-loop gain (Fig. 6-B) is about 57 dB at a frequency of 1 MHz, but drops to about 10 dB at 100 MHz. When external compensation is used, and the feedback resistors are set for a voltage gain of 10 (20 dB), the voltage gain is flat up to about 10 MHz and then drops off to 10 dB at 100 MHz.

2. Initially, set the generator frequency to the low end of the range—about 1 MHz (megahertz) for the IC amplifier. Then set the generator amplitude to the desired input level. In the absence of a realistic test input voltage, set the generator output to an arbitrary value.

   A simple method of finding a satisfactory input level is to monitor the circuit output and increase the generator amplitude at the amplifier center frequency (at 20 or 30 MHz for the IC) until the amplifier is overdriven. The amplifier is overdriven when further increases in generator output do not cause further increases in meter reading (or the output waveform peaks begin to flatten on the scope display). Set the generator output just below this point. Then, return the meter or scope to monitor the generator voltage (at the circuit input) and measure the voltage. Keep the generator at this voltage throughout the test.

   When making any voltage-output tests, be sure that you do not exceed the IC limits. For example, as shown in Fig. 6-F, the IC can deliver an output voltage

Op-amp and comparator circuits

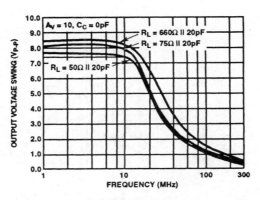

Fig. 6-F Output voltage versus frequency Harris Semiconductor, Linear & Telecom ICs, 1994, p. 2–217.

of about 5 V_{p-p} (volts, peak-to-peak) at a frequency of 20 MHz. If the closed-loop gain is set for 10, and an input signal of 0.6 V (volts) is applied at a frequency of 20 MHz, the output will be 6 V. This condition is beyond the IC capability, and will probably result in distortion of the output waveform.

3. If the circuit is provided with any operating or adjustment controls (volume, loudness, gain, treble, bass, balance, and so on), set the controls to some arbitrary point when making the initial frequency-response measurement. The response measurements can then be repeated at different control settings, if desired. Although there are no controls as such in this test circuit, the IC is provided with an offset-null feature at pins 1 and 16. If there is an unbalance in the differential input of the IC or if there is a level shift in the stages following the input, the output might not be zero with a zero input. A voltage applied at pins 1 and 16 can be adjusted to correct this condition.

   Figure 6-G shows the open-loop frequency-response test circuit for another IC amplifier (the Harris CA3094A) where an offset null is used. In this circuit, the input is shorted, and the voltage applied at pin 2 is adjusted for a zero output. Notice that the IC of Fig. 6-G is an *OTA* (operational transconductance amplifier). OTAs are covered further in this chapter.

4. Record the amplifier output voltage on the graph. Without changing the generator amplitude, increase the generator frequency by some fixed amount and record the new amplifier output voltage. The amount of frequency increase is arbitrary, and usually depends on the frequency range of the IC. In the IC, an increase of 1 MHz between measurements (up to about 10 MHz) is realistic. At frequencies higher than 10 MHz, an increase of 10 MHz between measurements should be satisfactory. Of course, small increases between measurements will show up any abnormalities (such as *peaking*). Notice that the gain peaks at frequencies between 100 and 200 MHz for the IC (Fig. 6-D).

Amplifier tests

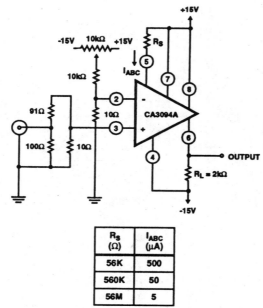

R_S (Ω)	I_{ABC} (µA)
56K	500
560K	50
56M	5

Fig. 6-G   Open-loop gain test connections for an OTA HARRIS SEMICONDUCTOR, LINEAR & TELECOM ICS, 1994, P. 2–94.

5. After the initial frequency-response measurement, check the effect of operating or adjustment controls (if any). For example, in an audio amplifier, the volume, loudness, and gain controls should have the same effect across the entire frequency range. Treble and bass controls might have some effect on all frequencies. However, a treble control should have the greatest effect at the high end, but bass controls should be most effective at the low end.

6. Remember that generator output can vary with changes in frequency (a fact that is possibly overlooked in making frequency-response tests). Monitor the generator output amplitude after each change in frequency. It is essential that the generator output amplitude remains constant over the entire frequency range of the test.

Voltage gain

The voltage gain for an amplifier is measured in the same way as frequency response. The ratio of output voltage to input voltage (at any given frequency across the entire frequency range) is the voltage gain. Because the input voltage (generator output) is held constant for a frequency-response test, a voltage-gain curve should be identical to a frequency-response curve (such as shown in Figs. 6-B, 6-D, and 6-E). Keep in mind that the voltage gain shown by Fig. 6-B depends primarily on the IC, but the gain shown in Figs. 6-D and 6-E is set by external factors (feedback resistors).

Power output and power gain

The power output of an amplifier is found by noting the output voltage across the load resistance, at any frequency across the entire frequency range. For example, in the circuit of Fig. 6-A, if the output voltage is 7 V, the power is:

$$P = \frac{E^2}{R} = \frac{7^2}{50} = 0.98 \text{ W}$$

As a practical matter, never use a wire-wound component (or any component that has reactance) for the load resistance. Reactance changes with frequency, and that causes the load to change. Use a composition resistor or potentiometer for the load.

To find the *power gain* of an amplifier, begin by finding both the input and output power. You can find the input power in the same way as the output, but you must know or calculate the input impedance. Calculating input impedance is not always practical in some circuits—especially in designs where input impedance depends on transistor gain. (The procedure for finding dynamic input impedance is described further in this chapter.) With input power known (or estimated), the power gain is the ratio of output power to input power. When the input is terminated in a resistance that is far lower than the amplifier input impedance, you can use the value of the input terminating resistance to calculate the input power.

Input sensitivity

In some IC amplifiers, an input-sensitivity specification is used in place of or in addition to power-output/gain specifications. *Input sensitivity* implies a minimum power output with a given voltage input, such as 3-W output with a 100-mV (milliwatt) input. Input sensitivity usually applies to power IC amplifiers. To find input sensitivity, simply apply the specified input and note the actual power output.

Bandwidth

Some specifications require that the IC amplifier deliver a given voltage or power output across a given frequency range. Usually, the voltage bandwidth is not the same as the power bandwidth. For example, an amplifier might produce full-power output up to 1 MHz—even though the frequency response is flat up to 10 MHz. That is, voltage (without a load) remains constant up to 10 MHz, whereas power output (with a load) remains constant up to 1 MHz. Figure 6-H shows the test connections and procedures to measure bandwidth at −3 dB points for a typical op amp (the Harris CA3020/CA3020A).

Load sensitivity

Most amplifiers, especially power amplifiers, are sensitive to changes in load. An amplifier produces maximum power gain when the output impedance is the same as the load impedance. The test circuit for load-sensitivity measurement is the same as for frequency response (Figs. 6-A and 6-G), except that the load resistance

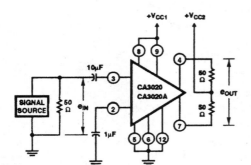

PROCEDURES:

1. Apply desired value of V_{CC1} and V_{CC2}
2. Apply 1kHz input signal and adjust for e_{IN} = 5mV (rms)
3. Record the resulting value of e_{OUT} in dB (reference value)
4. Vary input-signal frequency, keeping e_{IN} constant at 5mV, and record frequencies above and below 1kHz at which e_{OUT} decreases 3dB below reference value
5. Record bandwidth as frequency range between -3dB points

6

192

Fig. 6-H Measurement of bandwidth at –3-dB points HARRIS SEMICONDUCTOR, LINEAR & TELECOM ICS, 1994, P. 2–50.

is variable. Again, never use a wire-wound load resistance; the reactance can result in considerable error.

To find load sensitivity, measure the power output at various load-impedance and output-impedance ratings. That is, set the load resistance to various values (including a value equal to the supposed amplifier-output impedance). Record the voltage and/or power gain at each setting. Repeat the test at various frequencies. Figure 6-I shows a typical load-sensitivity response curve. Notice that if the load is twice the output impedance (as indicated by a 2:1 ratio, a normalized load impedance of 2), the output power is reduced to about 50%.

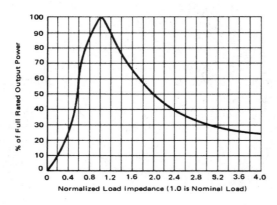

Fig. 6-I Typical load-sensitivity response curve.

Dynamic output impedance or resistance

The load-sensitivity test can be reversed to find the dynamic output impedance or resistance of an amplifier. The connections (Figs. 6-A and 6-G) and the procedures are the same, except that the load resistance is varied until *maximum power output* is found. Power is removed, the load resistance is disconnected from the circuit, and the resistance is measured with an ohmmeter. This resistance is equal to the dynamic output impedance of the amplifier (but only at that measurement frequency). The test can be repeated across the entire frequency range, as required.

Figure 6-J shows output resistance versus frequency for a typical IC amplifier (a Harris CA3450). Notice that the output resistance remains below about 10 Ω at frequencies up to about 80 MHz and then rises rapidly to more than 80 Ω as the frequency increases from 80 MHz to about 120 MHz.

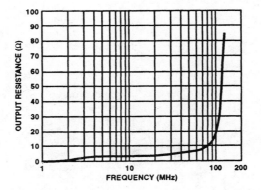

Fig. 6-J Output resistance vs. frequency Harris Semiconductor, Linear & Telecom ICs, 1994, p. 2–217.

Dynamic input impedance or resistance

Use the circuit and procedures of Fig. 6-K to find the dynamic input impedance of an amplifier. Notice that the IC shown in Fig. 6-K has two inputs to be measured. Also, notice that the accuracy of this impedance measurement (and the output impedance measurement) depends on the accuracy with which the resistance R is measured. Again, a noninductive (not wire-wound) resistance must be used for R. The impedance found by this method applies only to the frequency used during the test. Current drain, power output, efficiency, and sensitivity. Figure 6-L shows a circuit and the procedures for measuring zero-signal dc current drain, maximum-signal dc current drain, maximum power output, circuit efficiency, sensitivity, and transducer power gain. Again, the circuit of Fig. 6-L applies to a specific IC amplifier (Harris CA3020/3020A), but a similar circuit can be used for most IC amplifiers.

PROCEDURES:
Input Resistance Terminal 10 to Ground (R_{IN10})

1. Apply desired value of V_{CC1} and V_{CC2} and set S in Position 1
2. Adjust 1-kHz input for desired signal level of measurement
3. Adjust R for $e_2 = e_1/2$
4. Record resulting value of R as R_{IN10}

Input Resistance Terminal 3 to Ground (R_{IN3})

1. Apply desired value of V_{CC1} and V_{CC2} set S in Position 2
2. Adjust 1-kHz input for desired signal level of measurement
3. Adjust R for $e_2 = e_1/2$
4. Record resulting value of R as R_{IN3}

Fig. 6-K   Measurement of input resistance HARRIS SEMICONDUCTOR, LINEAR & TELECOM ICs, 1994, p. 2–51.

PROCEDURES:
Zero-Signal DC Current Drain

1. Apply desired value of V_{CC1} and V_{CC2} and reduce e_{IN} to 0V
2. Record resulting values of I_{CC1} and I_{CC2} in mA as Zero-Signal DC Current Drain

Maximum-Signal DC Current Drain, Maximum Power Output, Circuit Efficiency, Sensitivity, and Transducer Power Gain

1. Apply desired value of V_{CC1} and V_{CC2} and adjust e_{IN} to the value at which the Total Harmonic Distortion in the output of the amplifier = 10%
2. Record resulting value of I_{CC1} and I_{CC2} in mA as Maximum Signal DC Current Drain
3. Determine resulting amplifier power output in watts and record as Maximum Power Output (P_{OUT})
4. Calculate Circuit Efficiency (η) in % as follows:

$$\eta = 100 \frac{P_{OUT}}{V_{CC1}I_{CC1} + V_{CC2}I_{CC2}}$$

where P_{OUT} is in watts, V_{CC1} and V_{CC2} are in volts, and I_{CC1} and I_{CC2} are in amperes.

5. Record value of e_{IN} in mV (rms) required in Step 1 as Sensitivity (e_{IN})
6. Calculate Transducer Power Gain (G_p) in dB as follows:

$$G_p = 10\log_{10} \frac{P_{OUT}}{P_{IN}}$$

where P_{IN} (in mW) $= \dfrac{e_{IN}^2}{3000 + R_{IN(10)}}$ **

**See Figure 10 for definition of $R_{IN(10)}$

*T: PUSH-PULL OUTPUT TRANSFORMER;
LOAD RESISTANCE (R_L) SHOULD BE
SELECTED TO PROVIDE INDICATED
COLLECTOR-TO-COLLECTOR LOAD
IMPEDANCE (R_{CC})

Fig. 6-L   Measurement of current drain, power output, efficiency, and sensitivity
HARRIS SEMICONDUCTOR, LINEAR & TELECOM ICs, 1994, P. 2–50.

Op-amp and comparator circuits

Sine-wave analysis

All amplifiers are subject to distortion. That is, the output signal might not be identical to the input signal. Theoretically, the output should be identical to the input, except for the amplitude. You can check by applying a sine wave at the amplifier input (using a circuit similar to Fig. 6-A or 6-G) and monitoring both the input and output with a scope. If there is no change in the scope display, except for amplitude, there is no distortion.

In practical testing or troubleshooting, analyzing sine waves to pinpoint amplifier problems that produce distortion is a difficult job. Unless distortion is severe, it might pass unnoticed. Sine waves are best used where harmonic-distortion or intermodulation-distortion meters are combined with the scope for distortion analysis. (Distortion meters are covered further in this chapter.) If a scope is used alone for distortion analysis, square waves provide the best results. (The reverse is true for frequency-response or power measurements.)

Square-wave analysis

Distortion analysis is more effective with square waves because of the high odd-harmonic content in square waves (and because it is easier to see a deviation from a straight line with sharp corners than from a curving line). The procedure for checking distortion with square waves is essentially the same as that used with sine waves. Square waves are introduced into the amplifier input, and the output is measured with a scope (Fig. 6-M). The primary concern is deviation of the out-

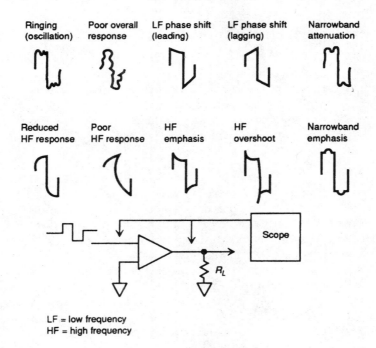

Fig. 6-M Basic square-wave distortion analysis.

put waveform from the input waveform (which also is monitored on the scope). If the scope has a dual-trace feature, the input and output can be monitored simultaneously. Also, if the scope has an invert function, the output can be inverted from the input for a better comparison of input and output.

If there is a change in the waveform, the nature of the change can sometimes reveal the cause of the distortion. Notice that the drawings of Fig. 6-M are generalized and that the same waveform can be produced by different causes. For example, poor *LF* (low-frequency) response appears to be the same as (high-frequency) emphasis.

Figure 6-N shows the waveforms that are produced by an actual circuit. Notice that the output (trace B) does a good job of following the input (trace A) at a gain of −1. That is, the output is inverted from the input, and there is no gain (unity gain). Also notice that there is some reduced high-frequency response in the output trace B, but not the exaggerated response shown in Fig. 6-M.

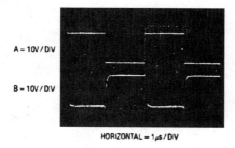

A = 10V/DIV

B = 10V/DIV

HORIZONTAL = 1μs/DIV

Fig. 6-N Amplifier response to square waves.

The third, fifth, seventh, and ninth harmonics of a clean square wave are emphasized. If an amplifier passes a given frequency and produces a clean square-wave output, it is reasonable to assume that the frequency response is good up to at least nine times the square-wave frequency.

Harmonic distortion

No matter what IC amplifier is used or how well the IC is designed, there is a possibility of odd or even harmonics being present with the fundamental signal. These harmonics combine with the fundamental and produce distortion, as is the case when any two or more signals are combined. The effects of second- and third-harmonic distortion are shown in Fig. 6-O.

Harmonic-distortion meters operate on the *fundamental-suppression principle*. A sine wave is applied to the amplifier input, and the output is measured on a scope or meter. The output is then applied through a filter that suppresses the fundamental frequency. Any output from the filter is then the result of harmonics. Figure 6-P shows typical connections and procedures for measurement of harmonic

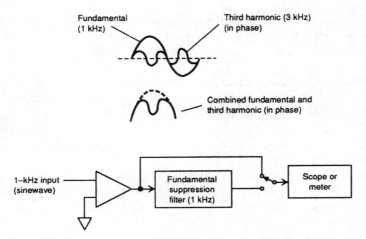

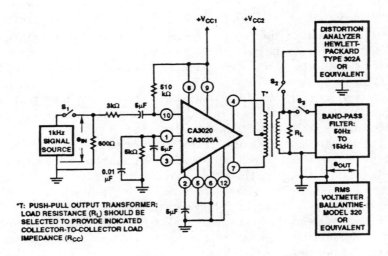

Fig. 6-O   Basic harmonic-distortion analysis.

Fig. 6-P   Measurement of THD and S/N ratio. HARRIS SEMICONDUCTORS, LINEAR & TELECOM ICS, 1994, P. 2–51.

PROCEDURES:

Signal-to-Noise Ratio

1. Close S_1 and S_3; open S_2
2. Apply desired values of V_{CC1} and V_{CC2}
3. Adjust e_{IN} for an amplifier output of 150mW and resulting value of E_{OUT} in dB as e_{OUT1} (reference value)
4. Open S_1 and record resulting value of e_{OUT} in dB as e_{OUT2}
5. Signal-to-Noise Ratio $(S/N) = 20\log_{10}\dfrac{e_{OUT1}}{e_{OUT2}}$

Total Harmonic Distortion

1. Close S1 and S2; open S3
2. Apply desired values of V_{CC1} and V_{CC2}
3. Adjust e_{IN} for desired level amplifier output power
4. Record Total Harmonic Distortion (THD) in %

Amplifier tests

distortion, where a Hewlett-Packard Type 302A, or equivalent, analyzer is used to measure the *THD* (total harmonic distortion) of a Harris CA3020/CA3020A. This circuit also is used for *S/N* (signal-to-noise) measurements, as described further in this chapter.

In some tests, particularly in audio-amplifier tests, a scope is combined with a harmonic-distortion meter to find the harmonic frequency. For example, if the input is 1 MHz and the output (after filtering) is 3 MHz, third-harmonic distortion is indicated. (Reduce the scope horizontal sweep so that you can see one input cycle. If there are three cycles at the output for the same time period as one input cycle, this indicates third-harmonic distortion.)

The percentage of harmonic distortion also is determined by this method. For example, if the output is 100 mV (millivolts) without the filter and 3 mV with the filter, this indicates a 3% harmonic distortion. Notice that total harmonic distortion varies with the power output of the amplifier. For that reason, it is generally necessary to adjust the input voltage for a given power output, as shown in Fig. 6-P. Also notice that THD depends on load.

Intermodulation distortion

When two signals of different frequencies are mixed in an amplifier, it is possible that the lower-frequency signal will modulate the amplitude of the higher-frequency signal. This modulation produces a form of distortion that is known as *IMD* (intermodulation distortion). Figure 6-Q shows the basic elements of IMD meters (a signal generator and a highpass filter). The generator portion produces a higher-frequency signal (usually 7 kHz for standard recording-industry testing) that is modulated by a low-frequency signal (usually 60 Hz).

The mixed signals are applied to the amplifier input, with the output connected through a highpass filter to a scope. The highpass filter removes the low-frequency (60 Hz) signal. The only signal that appears on the scope should be the 7 kHz. If any 60-Hz signal is present on the scope, the 60-Hz signal is being passed through as modulation on the 7-kHz signal.

Figure 6-Q also shows a simple IMD test circuit that can be made up in the shop. The highpass filter is designed to pass signals that are about 200 Hz and higher. The purpose of the fixed 40- and 10-kΩ resistors is to set the 60-Hz signal at four times the amplitude of the 7-kHz signal (assuming that both signals leave the generator at the same amplitude). Adjust the 10-kΩ potentiometer that controls the mixed 60-Hz/7-kHz signals to some level that does not overdrive the amplifier being tested.

Calculate the percentage of IMD using the equation shown in Fig. 6-Q. For example, if the maximum output (shown on the scope) is 1 V, and the minimum is 0.99 V, the percentage of IMD is about:

$$\frac{1.0 - 0.99}{1.0 + 0.99} = 0.005 \times 100 = 0.5\%$$

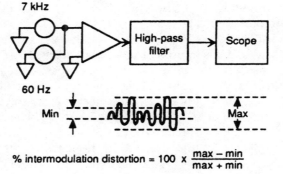

$$\% \text{ intermodulation distortion} = 100 \times \frac{\text{max} - \text{min}}{\text{max} + \text{min}}$$

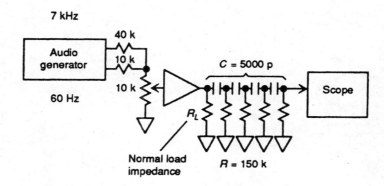

Fig. 6-Q   Basic intermodulation-distortion analysis.

Background noise

If a scope is sufficiently sensitive, it can be used to check and measure the background-noise level of an amplifier, as well as to check for the presence of hum, oscillation, and so on. The scope should be capable of measurable deflection with an input below 1 mV (and considerably less if an IC amplifier is involved).

The basic procedure consists of measuring amplifier output with the volume or gain controls (if any) at maximum but without an input signal. A meter can be used, but the scope is better because the frequency and nature of the noise (or other signals) are displayed visually. The scope gain must be increased until there is a noise or *hash* indication.

A noise indication might be caused by pickup in the leads between the amplifier and scope. If in doubt, disconnect the leads from the amplifier, but not the scope. If you suspect that 60-Hz power-line hum is present in the amplifier output (picked up from the power supply or other source), set the scope sync controls to the line position. If a stationary signal pattern appears, the signal is the result of line hum getting into the circuit. If a signal appears that is not at the line frequency, the signal can be the result of oscillation in the amplifier or stray pickup. Short the amplifier input terminals. If the same signal remains, suspect that oscillation is occurring in the amplifier circuits.

With present-day IC amplifiers, the internal or background noise is considerably less than 1 mV, and it is impossible to measure directly—even with a sensitive scope. You must use a circuit that amplifies the output of the IC under test before the output is applied to the scope. Figure 6-R is such a circuit (and is used for noise tests of an OP-77 IC). The IC under test is connected for high voltage gain, as is the following amplifier. (The total voltage gain is 50,000.) This gain makes it possible to monitor (and record) noise on a chart recorder (Fig. 6-S). Noise is measured over a 10-s (second) interval, noting the peak-to-peak value, which is about 25 nV (nanovolts) in Fig. 6-S.

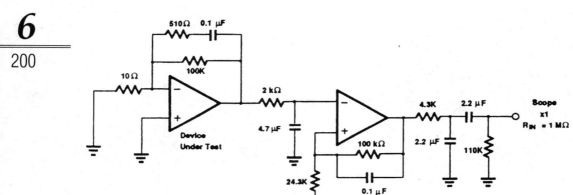

Fig. 6-R   0.1- to 10-Hz noise test circuit.

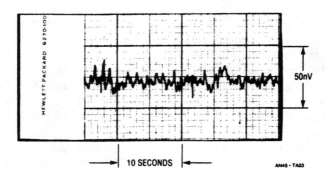

Fig. 6-S   Typical amplifier background noise as measured on a chart recorder over a 10-s interval.

Signal-to-noise ratio

Some IC amplifiers are tested for signal-to-noise (S/N) ratio instead of (or in addition to) background noise. Figure 6-P shows the circuit connections and procedures for S/N measurement of the CA3020/CA3020A. (This circuit is the same as for THD, except that the distortion analyzer is not connected when S/N is measured.)

A signal-to-noise test shows the relationship of background noise to signal amplitude, when the amplifier is operated under specific conditions. For example, in the circuit of Fig 6-P, the input signal is increased in amplitude until the output is 150 mW, and the output voltage is recorded in dB. Then the input signal is removed, but the input terminals remain connected together through resistors and capacitors, so the only output is the noise voltage within the IC. This background-noise voltage also is recorded in dB, and the S/N ratio is calculated as shown.

Slew rate (transient response)

Amplifier slew rate is the maximum rate of change in output voltage that the amplifier can produce when maintaining linear characteristics (symmetrical output without clipping). Slew rate is often listed under the heading of *transient response* in data sheets. Other transient response characteristics include *rise time, settling time, overshoot,* and possibly *error band.* All these topics are covered further in this chapter.

Slew rate is expressed by the difference in output voltage divided by difference in time, d_{V_0}/d_t. Usually, slew rate is listed in volts per microsecond. For example, if the output voltage from an amplifier can change 7 V in 1 µs (microseconds), the slew rate is 7 (which can be listed as 7 V/µs). The major effect of slew rate on circuit performance is that (all other factors being equal) a higher slew rate results in higher power output.

You can estimate the *approximate power bandwidth* of an amplifier if you know the slew rate. The equation is:

$$full\ power\ bandwidth\ in\ megahertz = \frac{(slew\ rate)}{(6.28 \times peak\ output\ voltage)}$$

For example, the slew rate for a Harris HA-2529 is listed as 150 (typical) when the peak-to-peak output voltage is ±10 V (a 10-V peak output voltage). Using the equation, the power bandwidth is:

$$\frac{150}{62.8} = 2.39\ MHz$$

The data sheet for the HA-2529 shows a full-power bandwidth of 2.1 MHz (minimum) and 2.6 MHz (typical).

A simple way to find amplifier slew rate is to measure the slope of the output waveform when a square-wave input is applied, as shown in Fig. 6-T. The input square wave must have a rise time that exceeds the slew-rate capability of the amplifier. As a result, the output does not appear as a square wave, but as an inte-

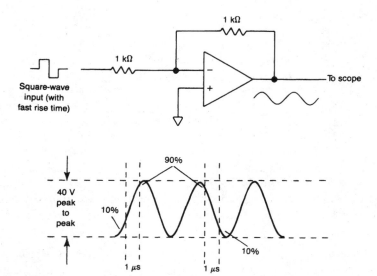

Example shows a slew rate of about 40 (40 V/μs) at unity gain

Fig. 6-T   Slew-rate measurement.

grated wave. In the example shown, the output voltage rises (and falls) about 40 V in 1 μs. Notice that slew rate is usually measured in the closed-loop condition (with negative feedback) and that slew rate increases with higher gain.

Figure 6-U shows the slew-rate and transient-response test circuits for a typical IC op amp (Harris HA-2529). Figure 6-U also includes some definitions for slew rate, settling time, rise time, overshoot, and error band, which are covered in the following section.

Rise time, settling time, and overshoot

There are many ways to measure these transient-response characteristics. Figure 6-V shows the settling time test circuit.

For this IC amplifier (HA-2529), rise time is specified with an output of 200 mV and a gain of 3. Thus, the small-signal response scope displays must be used. As shown, the rise time (measured from the 10% point to the 90% point, Fig. 6-U) is about 40 ns (nanoseconds). The data sheet specifies a typical rise time of 20 ns and a maximum of 50 ns.

Settling time is the total length of time from input-step application until the output remains within a specified error band or point around the final value. For the HA-2529, settling time is specified to 0.1% of the final value with an output of 10 V and a gain of −3. Thus, the large-signal response must be used. As shown, settling time is somewhat less than 150 ns (from the start of the input, through the overshoot, and back to where the output levels to 10 V). The data sheet specifies a typical settling time of 200 ns. From a circuit-performance standpoint, an increase in rise time, settling time, and overshoot lowers the frequency response and band-

Op-amp and comparator circuits

Test Circuits

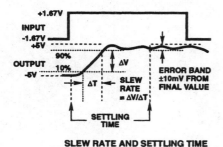

SLEW RATE AND SETTLING TIME

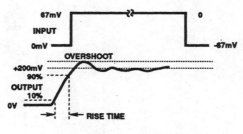

TRANSIENT RESPONSE

NOTE: Measured on both positive and negative transitions from 0V
to +200mV and 0V to -200mV at the output.

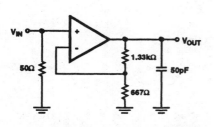

SLEW RATE AND TRANSIENT RESPONSE

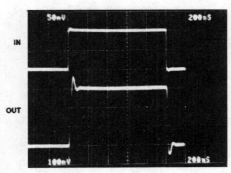

SUGGESTED V$_{OS}$ ADJUSTMENT AND COMPENSA-
TION HOOK UP

Tested offset adjustment range is IV$_{OS}$ + 1mVI minimum referred to
output. Typical ranges are +28mV to -18mV with R$_T$ = 20kΩ

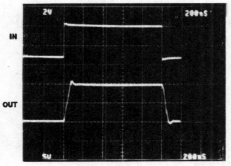

Fig. 6-U   Slew-rate and transient-response tests.

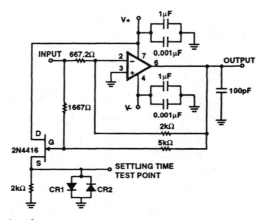

- $A_V = -3$
- Feedback and summing resistor ratios should be 0.1% matched.
- Clipping diodes CR1 and CR2 are optional. HP5082-2810 recommended.

Fig. 6-V Settling time test circuit. HARRIS SEMI-CONDUCTORS, LINEAR & TELECOM ICs, 1994, P. 2–310.

width. If the IC amplifier is used in pulse applications, rise times and settling times can distort the output pulse.

Phase shift

The phase shift between input and output of some IC amplifiers is not a significant design factor, but is a critical factor in others, particularly for op amps (and other ICs operating as op amps). This is because an op amp generally uses the principle of feeding back output signals to the input. Under ideal open-loop conditions, the output should be exactly 180° out of phase with the inverting input and in phase with the noninverting input. Any substantial deviation from this condition can cause op-amp circuit problems.

For example, assume that an op-amp circuit uses the inverting input (with the noninverting input grounded) and the circuit output is fed back to the inverting input. If the output is not shifted the full 180° (for example, if the shift is only a few degrees), the circuit might oscillate (because the output being fed back is almost in phase with the input). Even if there is no oscillation, the amplifier gain will not be stabilized, and the circuit will not operate properly.

A dual-trace scope, connected as shown in Fig. 6-W, is the ideal tool for phase measurement. For the most accurate results, the cables that connect the input and output should be of the same length and characteristics. At higher frequencies, a difference in cable length or characteristics can introduce a phase shift. For simplicity, adjust the scope controls until one cycle of the input signal occupies exactly nine screen divisions (typically 9 centimeters) horizontally. Then, find the

Op-amp and comparator circuits

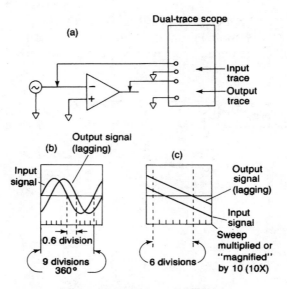

Fig. 6-W   Phase-shift measurement.

phase/division factor of the input signal. For example, if 9 centimeters represents one cycle (360°), 1 centimeter represents 40° (360/9 = 40).

With the phase/division factor established, measure the horizontal distance between corresponding points on the two waveforms (input and output signals). Then multiply the measured distance by the phase/division factor of 40°/cm (degrees per centimeter) to find the phase difference. For example, if the horizontal distance is 0.6 cm with a 40°/centimeter factor, the phase difference is:

$$0.6 \times 40° = 24°$$

If the scope has speed magnification, you can get more accurate results. For example, if the sweep rate is increased 10 times, the magnified phase factor is:

$$40°/cm \times 0.1 = 4°C$$

Figure 6-W shows the same signal with and without sweep magnification. With a 10× magnification, the horizontal distance is 6 centimeters, and the phase difference is:

$$6 \times 4 = 24°$$

Feedback measurement

Because IC-amplifier circuits can include feedback (particularly op-amp circuits), it is sometimes necessary to measure feedback voltage at a given frequency with given operating conditions. The basic feedback-measurement connections are

Amplifier tests

shown in Fig. 6-X. Although it is possible to measure the feedback voltage as shown in Fig. 6-X(A), a more accurate measurement is made when the feedback lead is terminated in the normal operating impedance, as shown in Fig. 6-X(B).

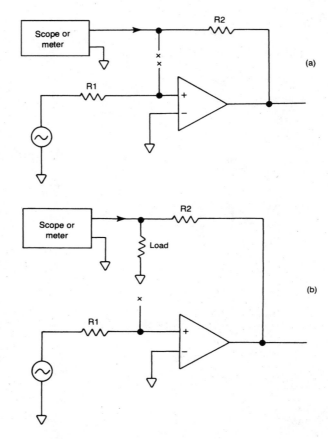

Fig. 6-X   Feedback measurement.

If an input resistance is used in the circuit, and this resistance is considerably lower than the IC input resistance, use the circuit-resistance value. If in doubt, measure the input impedance of the IC and terminate the feedback lead in that value (to measure open-loop feedback voltage). Remember that open-loop voltage gain must be substantially higher than the closed-loop voltage gain for most op-amp circuits to perform properly.

Input-bias current

IC-amplifier input-bias current is the average value of the two input-bias currents of the differential-input stage. In circuit performance, the significance of input-bias current is that the resultant voltage drops across input resistors (such as the resis-

Op-amp and comparator circuits

tors at pin 3 of the IC in Fig. 6-G) restrict the input common-mode voltage range at higher impedance levels. The input-bias current produces a voltage drop across the input resistors. This voltage drop must be overcome by the input signal (which can be a problem if the input signal is low and the input resistors are large).

Input-bias current can be measured using the circuit of Fig. 6-Y. Any resistance value for R1 and R2 can be used, provided that the value produces a measurable voltage drop and that the resistance values are equal. A value of 1 kΩ (with a tolerance of 1% or better) for both R1 and R2 is realistic for typical op amps.

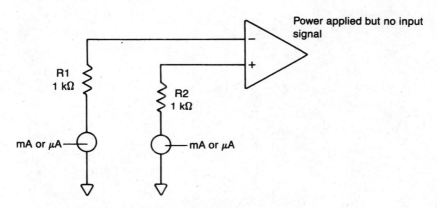

Fig. 6-Y Input-bias current measurement.

If it is not practical to connect a meter in series with both inputs (as shown), measure the voltage drop across R1 and R2, and calculate the input-bias current. For example, if the voltage is 3 mV across 1-kΩ resistors, the input-bias current is 3 μA (microamperes). Try interchanging R1 and R2 to see if any difference is the result of a difference in resistor values. In theory, the input-bias current should be the same for both inputs. In the real world, the bias currents should be almost equal. Any great difference in input bias is the result of unbalance in the input differential amplifier of the IC, and it can seriously affect circuit operation (and it usually indicates that an IC is defective).

Input-offset voltage and current

Input-offset voltage is the voltage that must be applied at the input terminals to get zero output voltage, whereas input-offset current is the difference in input-bias current at the amplifier input. Offset voltage and current are usually referred back to the input because the output voltages depend on feedback.

From a circuit performance standpoint, the effect of input-offset is that the input signal must overcome the offset before an output is produced. Also, with no input, there is a constant shift in output level. For example, if an IC amplifier has a 1-mV input-offset voltage, and a 1-mV signal is applied, there is no output. If the signal is increased to 2 mV, the amplifier produces only the peaks in excess of 1 mV.

Input-offset voltage and current can be measured using the circuit of Fig. 6-Z. As shown, the output is alternately measured with R3 shorted and with R3 in the circuit. The two output voltages are recorded as E_1 (R1 closed, R3 shorted) and E_2 (S1 open, R3 in the circuit).

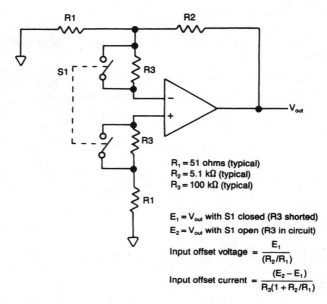

$R_1 = 51$ ohms (typical)
$R_2 = 5.1$ kΩ (typical)
$R_3 = 100$ kΩ (typical)

$E_1 = V_{out}$ with S1 closed (R3 shorted)
$E_2 = V_{out}$ with S1 open (R3 in circuit)

Input offset voltage $= \dfrac{E_1}{(R_2/R_1)}$

Input offset current $= \dfrac{(E_2 - E_1)}{R_3(1 + R_2/R_1)}$

Fig. 6-Z Input-offset voltage and current measurement.

With the two output voltages recorded, the input-offset voltage and current can be calculated using the equations of Fig. 6-Z. For example, assume that R1, R2, and R3 are at the values shown, that E_1 is 83 mV, and that E_2 is 363 mV:

$$\textit{Input-offset voltage} = \frac{83 \text{ mV}}{100} = 0.83 \text{ mV}$$

$$\textit{Input-offset current} = \frac{280 \text{ mV}}{100 \text{ k}\Omega \ (1 + 100)} = 27.7 \text{ nA}$$

Common-mode rejection

There are many definitions for *CMR* (common-mode rejection), which is sometimes listed as *CMRR* or *CM*$_{rej}$). No matter what definition is used, the first step to measure CMR is to find the open-loop gain of the IC at the desired operating frequency. Then connect the IC in the common-mode test circuit of Fig. 6-AA. In-

$$\frac{V_{out}\,(1\ mV)}{\text{open-loop gain}} = \text{equivalent differential input signal}$$

$$\frac{\text{Common-mode}}{\text{rejection}} = \frac{V_{in}}{\text{equivalent differential input signal}}$$

Fig. 6-AA   Common-mode rejection measurement.

crease the common-mode voltage (at the same frequency used for the open-loop gain test) until you get a measurable output. Do not exceed the maximum input common-mode voltage specified in the data sheet. If no such value is available, do not exceed the normal input voltage of the IC.

To simplify the calculation, increase the input voltage until the output is at some exact value, such as the 1 mV shown. Divide this value by the open-loop gain to find the equivalent differential input signal. For example, with an open-loop gain of 100 and an output of 1 mV, the equivalent differential is:

$$\frac{0.001}{100} = 0.00001$$

Now measure the input voltage that produced the 1-mV output and divide the input by the equivalent differential to find the common-mode rejection ratio. In the example, simply find the input voltage that produces the 1-mV output and move the decimal point over five places. For example, if the output is 1 mV with a 10-V input and a gain of 100, the ratio is 0.0001. This ratio can be converted to decibels (a voltage ratio of 80 dB).

Power-supply sensitivity (or PSS)

PSS (power-supply sensitivity) is the ratio of change in input-offset voltage to the change in power-supply voltage that produces the change. On some data sheets, the term is expressed in millivolts or microvolts per volt (mV/V or μV/V), which represents the change of input-offset voltage (in microvolts or millivolts) to a change (in volts) of the power supply. In other data sheets, the term *PSRR* (power-supply rejection ratio) is used instead, and it is given in decibels.

No matter what it is called, the characteristic can be measured using the circuit of Fig. 6-Z (the same test circuit as for input-offset voltage). The procedure is the same as for measurement of input-offset voltage, except that the supply voltage is changed (in 1-V steps). The amount of change in input-offset voltage for a 1-V power-supply change is the PSRR. The ratio of change can be converted to decibels if required. The circuit of Fig. 6-Z also can be used when the amplifier is operated from two power supplies. One supply voltage is changed (in 1-V steps) while the other supply voltage is held constant.

IC amplifier types

This chapter includes circuits for a number of different IC-amplifier types. Following is a summary of these types.

Operational amplifiers (op amps)

The designation *op amp* was originally used for a series of high-performance direct-coupled amplifiers in analog computers. These amplifiers performed mathematical operations (summing, scaling, subtraction, integration, and so on). Today, the availability of inexpensive IC amplifiers has made the packaged op amp useful as a replacement for any amplifier.

Figure 6-BB shows some classic op-amp functions. As shown by the equation for the *inverting amplifier,* the output voltage V_{OUT} is equal to the input voltage (V_{IN}) multiplied by the ratio R_2/R_1. If R_1 and R_2 are the same value, there is no gain (*unity gain* or a gain of 1). If R_2 (the feedback resistance) is 100 kΩ and R_1 (the input resistance is 10 kΩ), the voltage gain is 10. Most op amps are operated in this closed-loop configuration, with feedback. The main purpose of the feedback is to stabilize gain at some fixed values. With the inverting amplifier, the output voltage is inverted from the input voltage. This inversion happens because the input is applied to the inverting input (−) of the op amp at pin 2.

The *noninverting amplifier* is used when the input and output voltages must be in phase (input applied to the noninverting or + input at pin 3). With noninverting, the V_{OUT} is equal to V_{IN}, multiplied by the ratio of $(R_1 + R_2)/R_1$.

In the *difference amplifier,* V_{OUT} depends on the difference between the two input voltages ($V_2 - V_1$), multiplied by the ratio of the resistances. For the *inverting summing amplifier,* V_{OUT} is the sum of voltages V_1, V_2, and V_3. For example, if all three input voltages at 5 V, the output is 15 V (provided that R1 through R4 are all the same value).

The circuits of Fig. 6-BB require multiple power supplies (typically ±5 V, ±10 V, and ±15 V). This requirement is typical for most op-amp circuits and is one of the limitations of most op amps. (However, it is possible to operate some op amps with a single supply.)

Notice that in the inverting circuits, the noninverting input of the op amp is returned to ground, possibly through a resistor. This grounded input is typical for op amps operated as single-input circuits (even though op amps have a differential input). As a guideline, when a grounding resistor is used, the resistor value is equal to the parallel resistance of the input and feedback resistors (as shown by the equations for the inverting summing amplifier).

Notice that there is no external compensation for many of the op-amp circuits in this chapter. Early op amps often required external compensation circuits (usually capacitors or resistors, or both) to provide a given frequency-response charac-

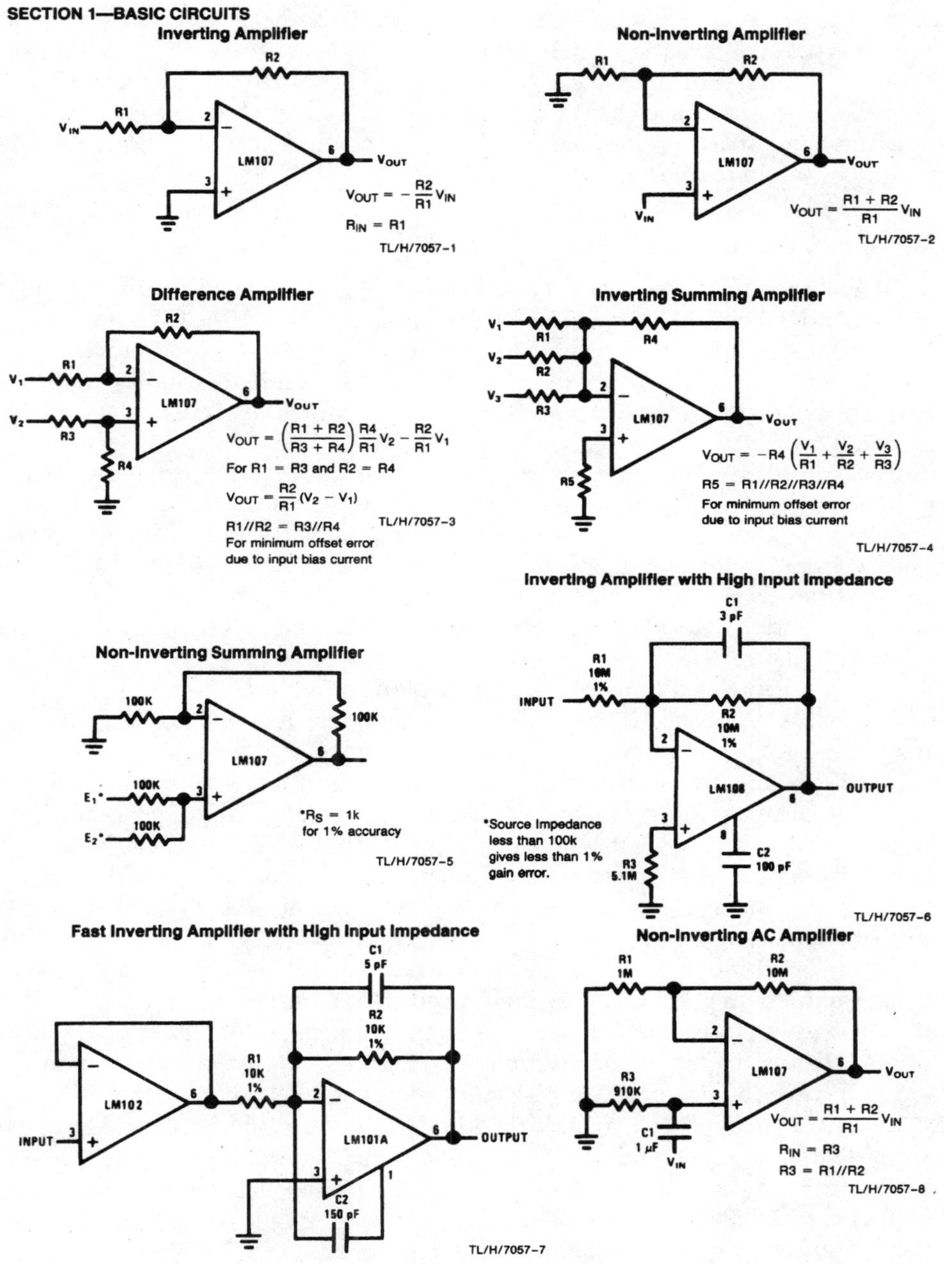
SECTION 1—BASIC CIRCUITS
Inverting Amplifier
R2
R1
VIN
LM107
VOUT
$V_{OUT} = -\frac{R2}{R1} V_{IN}$
$R_{IN} = R1$
TL/H/7057-1
Non-Inverting Amplifier
R1
R2
LM107
VOUT
VIN
$V_{OUT} = \frac{R1 + R2}{R1} V_{IN}$
TL/H/7057-2
Difference Amplifier
R2
R1
V1
V2
R3
R4
LM107
VOUT
$V_{OUT} = \left(\frac{R1 + R2}{R3 + R4}\right) \frac{R4}{R1} V_2 - \frac{R2}{R1} V_1$
For R1 = R3 and R2 = R4
$V_{OUT} = \frac{R2}{R1} (V_2 - V_1)$
R1//R2 = R3//R4
For minimum offset error
due to input bias current
TL/H/7057-3
Inverting Summing Amplifier
V1
V2
V3
R1
R2
R3
R4
R5
LM107
VOUT
$V_{OUT} = -R4 \left(\frac{V_1}{R1} + \frac{V_2}{R2} + \frac{V_3}{R3}\right)$
R5 = R1//R2//R3//R4
For minimum offset error
due to input bias current
TL/H/7057-4
Inverting Amplifier with High Input Impedance
C1
3 pF
R1
10M
1%
INPUT
R2
10M
1%
LM108
OUTPUT
*Source Impedance
less than 100k
gives less than 1%
gain error.
R3
5.1M
C2
100 pF
TL/H/7057-6
Non-Inverting Summing Amplifier
100K
100K
LM107
E1*
100K
E2*
100K
*RS = 1k
for 1% accuracy
TL/H/7057-5
Fast Inverting Amplifier with High Input Impedance
C1
5 pF
R2
10K
1%
R1
10K
1%
LM102
INPUT
LM101A
OUTPUT
C2
150 pF
TL/H/7057-7
Non-Inverting AC Amplifier
R1
1M
R2
10M
LM107
VOUT
R3
910K
C1
1 µF
VIN
$V_{OUT} = \frac{R1 + R2}{R1} V_{IN}$
$R_{IN} = R3$
R3 = R1//R2
TL/H/7057-8

Fig. 6-BB Classic op-amp functions. NATIONAL SEMICONDUCTOR, LINEAR APPLICATIONS HANDBOOK, 1994, P. 70.

teristic. Many present-day op amps have internal compensation and do not require external compensation. There are exceptions. For example, the inverting amplifier with high input impedance circuit of Fig. 6-BB requires an external compensating capacitor (pin 8 to ground). From a troubleshooting standpoint, if an op-amp circuit is working, but the frequency response is not as required, look for compensation problems.

OTA (operational transconductance amplifier)

An *OTA* (sometimes called a *programmable amplifier*) is similar to an op amp. However, OTAs and op amps are not necessarily interchangeable. The OTA not only includes the usual differential inputs of an op amp, but also has an additional control input in the form of an amplifier bias current (I_{ABC}). (Figure 6-G shows an IC amplifier controlled by an I_{ABC} current at pin 6. The amount of I_{ABC} current is set by resistor Rs.)

The control input increases the flexibility of the OTA for use in a wide range of applications. For example, if low power consumption, low input bias, and low offset current, or high input impedance are desired, then select low I_{ABC}. On the other hand, if operation into a moderate load impedance is the main consideration, then use higher levels of I_{ABC}.

The second major difference between an op amp and an OTA is that the OTA output impedance is extremely high (most op amps have very low output impedance). Because of this difference, the output signal of an OTA is best described in current that is proportional to the difference between the voltages at the two inputs (inverting and noninverting).

The OTA transfer characteristics (or input/output relationship) are best defined in *transconductance* rather than voltage gain. Transconductance (usually listed as g_m or g_{21}) is the ratio between the difference in current out (I_{OUT}) for a given difference in voltage input (E_{IN}). Except for the high output impedance and the definition of input/output relationships, OTA characteristics are similar to those of a typical op amp.

Figure 6-CC shows a basic OTA circuit, complete with external circuit components. The following summarizes the procedure for finding the I_{ABC} required to produce a given transconductance. Assume that an open-loop gain of 100 is required. Open-loop gain is related directly to load resistance R_L and transconductance. However, the actual load resistance is the parallel combination of R_L and R_F—about 18 kΩ ($R_L \times R_F)/R_L + R_F$). With an open-loop gain of 100 and an actual load of 18 kΩ, the g_m should be 100/18,000, or about 5.5 millimho (mmho).

The transconductance is set by I_{ABC}. With a data-sheet curve similar to that of Fig. 6-DD, select an I_{ABC} from the minimum-value curve to assure that the OTA provides sufficient gain. As shown in Fig. 6-DD, for a G_m of 5.5 mmho, the required I_{ABC} is approximately 20 μA.

Chopper-stabilized op amps

Chopper stabilization is often used where stability is essential over time and with variations in temperature and supply voltage. Figures 6-EE and 6-FF show the in-

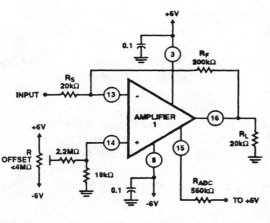

Fig. 6-CC   Basic OTA circuit. HARRIS SEMICON-DUCTORS, LINEAR & TELECOM ICs, 1994, P. 2–58.

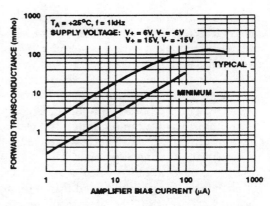

Fig. 6-DD Forward transconductance curve for OTA. HARRIS SEMICONDUCTORS, LINEAR & TELECOM ICs, 1994, P. 2–56.

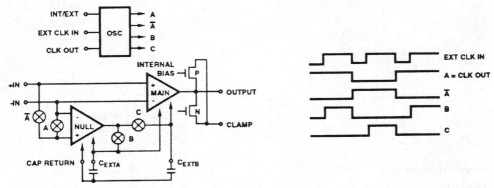

Fig. 6-EE   Internal functions of chopper-stabilized amplifier. HARRIS SEMICONDUCTOR, LINEAR & TELECOM ICs, 1994, P. 2–695.

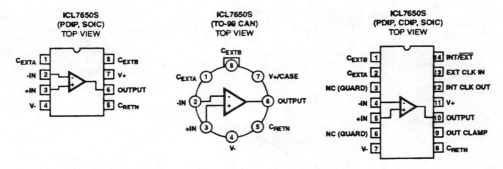

Fig. 6-FF   Pinouts of chopper-stabilized amplifier. HARRIS SEMICONDUCTOR, LINEAR & TELECOM ICs, 1994, P. 2–694.

IC amplifier types

ternal functions and pinouts, respectively, of a chopper-stabilized op amp (the Harris ICL7650S), which is a direct replacement for the industry standard ICL7650, but with improved input-offset voltage, lower input-offset temperature coefficient, reduced input-bias current, and wider common-mode voltage range.

As shown in Fig. 6-EE, there are two amplifiers: the main amplifier and the nulling amplifier. Both amplifiers have offset-null capacity. The main amplifier is connected continuously from the input to the output. The nulling amplifier, under control of the chopping oscillator and clock circuit, alternately nulls itself and the main amplifier. The two external capacitors (C_{EXTA} and C_{EXTB}) provide the required storage of the nulling potentials. The clock oscillator, and all other control circuits are self-contained. However, the 14-lead version (Fig. 6-FF) includes a provision for an external clock, if required. All of the internal functions are user-transparent, eliminating common chopper-amplifier problems (intermodulation, spikes, and overrange lock-up).

As shown in Fig. 6-GG, the chopper-stabilized IC uses the same basic connections for an inverting amplifier, noninverting amplifier, and voltage follower as a conventional op amp. However, the null/storage capacitors must be connected to the C_{EXTA} and C_{EXTB} pins, with a common connection to the C_{RETN} pin. This connection should be made directly by either a separate wire, or PC trace to avoid injection load-current IR drops into the capacitor circuits. The outside foil by each capacitor (where available) should be connected to the C_{RETN} pin.

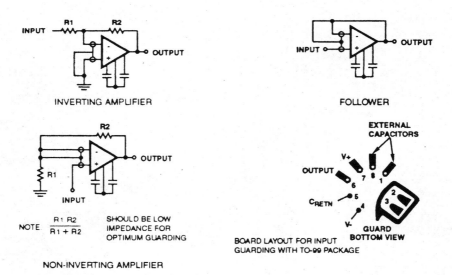

Fig. 6-GG Connections of chopper-stabilized amplifier. Harris Semiconductor, Linear & Telecom ICs, 1994, p. 2–703.

Op-amp and comparator circuits

To take full advantage of a chopper-stabilized op amp, the inputs should be guarded. Input guarding of the 8-pin TO-99 package is done with a 10-lead-pin circuit as shown in Fig. 6-GG. With this configuration, the holds adjacent to the inputs are empty when the IC is inserted into the board. The pin configuration of the 14-pin PDIP package (Fig. 6-FF) is designed to facilitate guarding, in that the pins adjacent to the inputs are not used.

Much of the information presented here is basic, and the techniques covered are of the most benefit to those readers who are unfamiliar with electronic troubleshooting. The techniques serve as a basis for understanding the step-by-step example of amplifier troubleshooting.

Signal tracing

The basic troubleshooting approach for an amplifier involves signal tracing, such as shown in Fig. 6-HH. The input and output waveforms of each stage are monitored on a scope or meter. Any stage that shows an abnormal waveform (in amplitude, waveshape, and so on) or the absence of an output (with a known-good input signal) points to a defect in the stage. Voltage and/or resistance measurements on all elements in the transistor (or IC) are then used to pinpoint the problem.

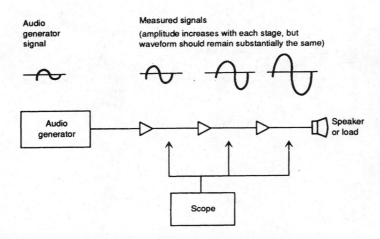

Fig. 6-HH   Basic amplifier-circuit signal tracing.

A scope is the most logical instrument to use for checking amplifier circuits (both complete amplifier systems or a single amplifier stage). The scope can duplicate every function of a meter in troubleshooting, and the scope offers the advantage of a visual display. Such a display can reveal common amplifier conditions (hum, distortion, noise, ripple, and oscillation).

When troubleshooting amplifier circuits with signal tracing, use a scope in much the same manner as a meter. Introduce a signal at the input with a generator (Fig. 6-HH). Measure the amplitude and waveform of the input signal on the scope. The input can be sine- or square-wave signals. Move the scope probe to the input and output of each stage, in turn, until the final output is reached. The gain of each stage is measured as voltage on the scope. Also, it is possible to observe any change in waveform from that which is applied to the input. Stage gain and distortion (if any) are established quickly with a scope.

Measuring gain in discrete stages

Take care when measuring the gain of discrete amplifier stages (especially in a circuit where there is feedback). For example, in Fig. 6-II(a), if you measure the signal at the base of Q1, the base-to-ground voltage is not the same as the input voltage. To get the correct value of gain, connect the low side of the measuring device (meter or scope) to the emitter and the other lead (high side) to the base. In effect, measure the signal that appears across the base-emitter junctions. This measurement includes the effect of the feedback signal.

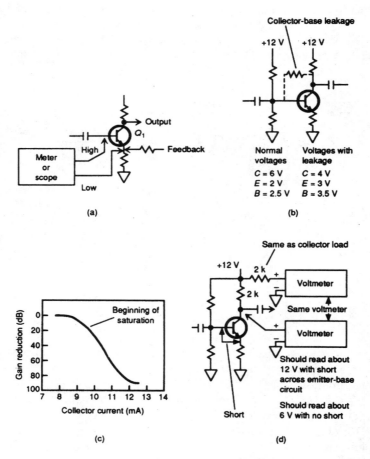

Fig. 6-II Basic discrete-circuit amplifier troubleshooting techniques.

Op-amp and comparator circuits

As a general safety precaution, never connect the ground lead of a meter or scope to the transistor base unless the lead connects back to an insulated inner chassis or board on the meter or scope. Large ground-loop currents (between the measuring device and the circuit being checked) can flow through the base-emitter junction and possibly burn out the transistor. This problem can usually be eliminated by an isolation transformer.

Low-gain problems

Low gain in a feedback amplifier can result in distortion. If gain is normal in a feedback amplifier, some distortion can be overcome. With low gain, the feedback might not be able to bring the distortion within limits. Of course, low gain by itself is sufficient cause to troubleshoot a circuit (with or without feedback).

Most feedback amplifiers have a very high open-loop gain that is set to some specific value by the ratio of resistor values (feedback-resistor value to input load-resistor value, as covered previously). If the closed-loop gain is low in an experimental circuit, this usually indicates that the resistance values are incorrect. In an existing amplifier, the problem is usually where the open-loop gain is below the point where the resistors determine gain. When troubleshooting such a situation, if waveforms indicate low gain and transistor (or IC) voltages appear normal, try replacing the transistors (or the IC).

Do not overlook the possibility that the emitter-bypass capacitors (if any) might be open or leaking. If the capacitors are leaking (acting as a resistance in parallel with the emitter resistor), there is considerable negative feedback and little gain. Of course, a completely shorted emitter-bypass capacitor produces an abnormal voltage indication on the transmitter emitter (typically 0 V or ground).

Distortion problems in discrete-amplifier stages

Distortion can be caused by improper bias, overdriving (too much gain), or underdriving (too little gain, preventing the feedback signal from countering the distortion). One problem that is often overlooked in a discrete feedback-amplifier stage with a pattern of distortion is overdriving that results from transistor leakage. Generally, it is assumed that the collector-base leakage of a transistor reduces gain because the leakage is in opposition to signal-current flow. Although this is true for a single stage, it might not be true when more than one feedback stage is involved.

Whenever there is collector-base leakage, the base assumes a voltage nearer to that of the collector (nearer than is the case without leakage). This situation increases both transistor forward bias and transistor current flow. An increase in the transistor current causes a reduction in input resistance (which might or might not cause a gain reduction, depending on where the transistor is located in the circuit). If the feedback amplifier is direct coupled, the effects of feedback are increased. This is because the operating point (set by the base bias) of the following stage is changed, which could possibly result in distortion.

Effects of leakage on discrete circuit performance

When there is considerable leakage in a discrete circuit, the gain is reduced to 0, and/or the signal waveforms are drastically distorted. Such a condition also produces abnormal waveforms and transistor voltage. These indications make troubleshooting relatively easy. The troubleshooting problem becomes very difficult when there is just enough leakage to reduce circuit gain, but not enough to distort the waveform seriously (or to produce transistor voltages that are way off).

Collector-base leakage

Collector-base leakage is the most common form of transistor leakage, and it produces a classic condition of low gain (in a single stage). When there is any collector-base leakage, the transistor is forward-biased or the forward bias is increased; see Fig. 6-II(b). Collector-base leakage has the same effect as a resistance between the collector and base. The base assumes the same polarity as the collector (although at a lower value), and the transistor is forward biased. if leakage is sufficient, the forward bias can be enough to drive the transistor into or near saturation. When a transistor is operated at or near the saturation point, the gain is reduced (for a single stage), as shown in Fig. 6-II(c).

Checking transistor leakage in-circuit

If the normal operating voltages are not known, as is the case with all experimental circuits, defective transistors can appear to be good because all of the voltage relationships are normal. The collector-base junction is reverse-biased (collector more positive than base for an npn) and the emitter-base junction is forward-biased (emitter less positive than base for an npn).

A simple way to check transistor leakage is shown in Fig. 6-II(d). Measure the collector voltage to ground. Then short the base to the emitter and remeasure the collector voltage. If the transistor is not leaking, the base-emitter short turns the transmitter off and the collector voltage rises to the same value as the supply. If there is any leakage, a current path remains (through the emitter resistor, base-emitter short, collector-base leakage path, and collector resistor). Some voltage drop occurs across the collector resistor, and the collector voltage is at some value lower than the supply.

Most meters draw current, and the current passes through the collector resistor when you measure; see Fig. 6-II(d). The current through the resistor can lead to some confusion, particularly if the meter draws heavy current (has a low ohms-per-volt rating). To eliminate any doubt, connect the meter to the supply through a resistor with the same value as the collector resistor. The voltage drop, if any, should be the same as when the transistor collector is measured to ground. If the drop is much different (lower) when the collector is measured, the transistor is leaking.

For example, assume that the collector measures 4 V with respect to ground; see Fig. 6-II(d). This voltage means that there is an 5-V drop across the collector resistor and a collector current of 4 mA ($8/2000 = 0.004$). Normally, the collector is operated at about one-half the supply voltage (at 6 V, in this example). Notice that simply because the collector is at 4 V, instead of 6 V, it does not mean that the circuit is faulty. Some circuits are designed that way.

In any event, the transistor should be checked for leakage with the emitter-base short test shown in Fig. 6-11(d). Now assume that the collector voltage rises to 10 V when the base and emitter are shorted (within 2 V of the 12-V supply). This indicates that the transistor is cutting off, but there is still some current flow through the resistor, about 1 mA (2/2000 = 0.001).

A current flow of 1 mA is high for most present-day meters. To confirm a leaking transistor, connect the same meter through a 2-kΩ resistor (same as the collector-load resistor) to the 12-V supply (preferably at the same point where the collector resistor connects to the power supply). Now, assume that the indication is 11.7 V through the external resistor. This voltage shows that there is some transistor leakage.

The amount of transistor leakage can be estimated as follows: 11.7 − 10 = 1.7-V difference, and 1.7/2000 = 0.00085 = 0.85 mA. However, from a practical troubleshooting standpoint, the presence of any current flow with the transistor supposedly cut off is sufficient cause to replace the transistor.

Example of amplifier-circuit troubleshooting

This step-by-step troubleshooting problem involves locating the defective part (or improperly connected wiring) in a combination discrete-IC audio amplifier. Figure 6-JJ shows the schematic diagram. This circuit was chosen as an example because it combines both IC and discrete components. The CA3094B is a programmable amplifier (where gain is set by the resistor at pin 5) that is similar to an OTA.

No matter what the trouble symptom, the actual value can eventually be traced to one or more of the circuit parts (transistors, ICs, diodes, capacitors, etc.), unless you have wired the parts incorrectly! Even then, the following waveform, voltage, and resistance checks will indicate which branch within the circuit is at fault.

If you were servicing this circuit in existing equipment, the first step would be to study the literature and test the circuit to confirm the trouble. In this example, the only "literature" is Fig. 6-JJ. The circuit description claims an output of 12-W into an 8-Ω load. Although the load is shown as R_L, you can assume that the circuit will be used with an 8-Ω speaker. There are no test points or waveforms, the voltage information is incomplete, and there is no resistance-to-ground information. However, with this fragmentary data, you can test the circuit, monitor the signals at various points in the circuit, and localize trouble using the test results.

The first step is to apply a signal at the input and monitor the output. The input can be applied at C1 (as shown). The output is measured at R_L, or at an 8-Ω speaker connected in place of R_L. Use the resistor or the speaker, but never operate the circuit without a load. Components Q2 and Q3, and possibly Q1, can be destroyed if they are operated without a load.

The output signal (at the junction of Q2/Q3 emitters and/or the speaker) must be about 10 V to produce 12 W across an 8-Ω load ($9.8^2/8$ = 12 W). If the circuit has a 40-dB voltage gain (as claimed), 0.1 V (100 mV) at the input should be sufficient to fully drive the speaker, depending on the setting of R1.

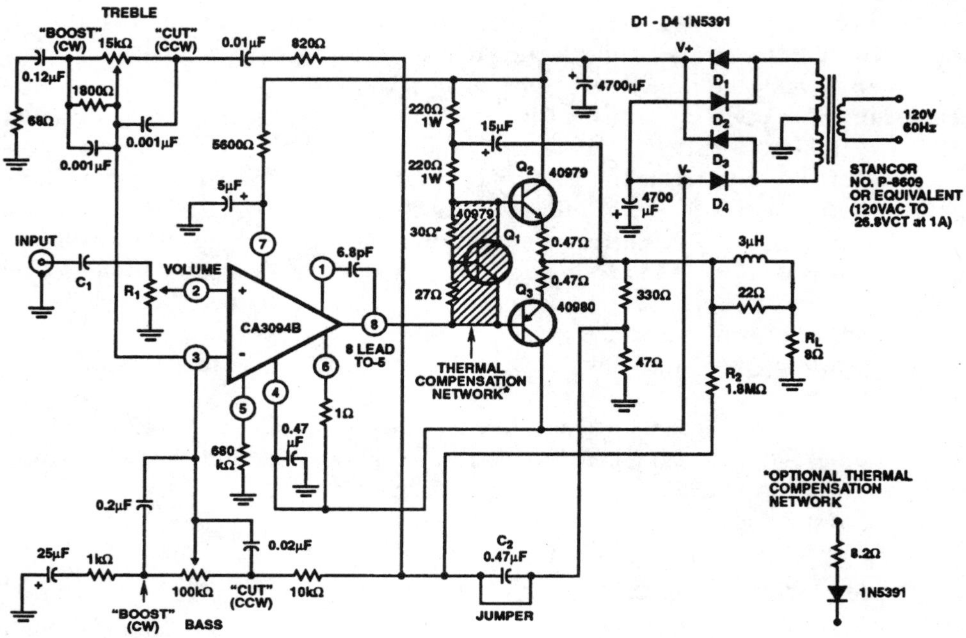

TREBLE
"BOOST" (CW) 15kΩ "CUT" (CCW) 0.01µF 820Ω
0.12µF
1800Ω
68Ω
0.001µF
0.001µF
0.01µF 5600Ω
5µF +

D1 - D4 1N5391
V+
+ 4700µF
220Ω 1W
220Ω 1W
15µF +
Q2 40979
30Ω 40979
Q1 0.47Ω
27Ω 0.47Ω
Q3 40980
V-
+ 4700 µF
D1
D2
D3
D4
120V 60Hz
STANCOR NO. P-8609 OR EQUIVALENT (120VAC TO 26.8VCT at 1A)

INPUT
C1
R1
VOLUME
2
+
7
CA3094B
3
-
1
6.8pF
8
8 LEAD TO-5
6
5
4
0.47 µF
1Ω
680 kΩ
THERMAL COMPENSATION NETWORK*
330Ω
47Ω
3µH
22Ω
R2 1.9MΩ
RL 8Ω

0.2µF
25µF 1kΩ
0.02µF
100kΩ "CUT" (CCW) 10kΩ
"BOOST" (CW) BASS
C2 0.47µF
JUMPER

*OPTIONAL THERMAL COMPENSATION NETWORK
8.2Ω
1N5391

TYPICAL PERFORMANCE DATA
For 12W Audio Amplifier Circuit

Power Output (8Ω load, Tone Control set at "Flat")
Music (at 5% THD, regulated supply) 15W
Continuous (at 0.2% IMD, 60Hz and 2kHz
mixed in a 4:1 ratio, unregulated supply)
See Figure 8 in AN6048............................ 12W
Total Harmonic Distortion
At 1W, unregulated supply.......................... 0.05%
At 12W, unregulated supply......................... 0.57%
Voltage Gain 40dB
Hum and Noise (below continuous power output).......... 83dB
Input Resistance 250kΩ
Tone Control Range................... See Figure 9 in AN6048

NOTES:
1. For standard input: Short C2; R1 = 250kΩ, C1 = 0.047µF; remove R2
2. For ceramic cartridge input: C1 = 0.0047µF, R1 = 2.5MΩ, remove jumper from C2; leave R2

Fig. 6-JJ Typical discrete-IC amplifier circuit and performance data. HARRIS SEMICONDUCTOR, LINEAR & TELECOM ICS, 1994, P. 2–100.

Connect an audio generator to the input and set the generator to produce 0.1 V at a frequency of 1 kHz. Set both the bass and treble controls to midrange and adjust R1 until you get a good tone on the speaker and/or a readable signal at the Q2/Q3-emitter junction. Adjust R1 for a 10-V signal at the speakers or emitter junctions. The tone will probably burst your eardrums at this point! Adjust R1 until the tone is reasonable and vary both the bass and treble controls. Both tone controls should have some effect on the tone, but the bass control should have the most control. Change the generator frequency to 10 kHz and repeat the tone-control test. Now the treble control should have the most effect.

If the circuit operates as described thus far, it is reasonable to assume that the circuit is good. Quit while you're ahead! If you have access to distortion meters, check distortion against the performance data on Fig. 6-JJ, both THD and IMD. Also check the actual voltage input (at pin 2 of the IC) when a 1-V signal appears at the output and determine the true amplifier voltage gain (which should be 40 dB).

If the circuit does not operate as described, set R1 and the bass/treble controls to the midrange and monitor the signal voltages at pins 2 and 8 of the IC. You can monitor all test points with an ac voltmeter, dc voltmeter with a rectifier probe, or with a scope. The scope is preferred because any really abnormal distortion at the test points appears on the scope display (as does the voltage).

If there is a signal at pin 2 of the IC, but not at pin 8 (or if the signal at pin 8 shows little gain over the pin-2 signal), the problem is at the IC portion of the circuit. Check all voltages at the IC. You do not know the exact values, but here are some hints.

The transformer has a 26.8-V center-tapped secondary, so $V+$ and $V-$ should be about 12 to 15 V (and should be substantially the same). In any event, pins 4 and 6 of the IC should be about -12 V, and pin 7 should be about ±12 V (although pin 7 will probably be lower than pins 4 and 6 because of the 5600-Ω resistor at pin 7).

If the IC voltages appear to be good, but the gain at pin 8 is low, it is possible that the IC is bad, that the resistor at pin 5 is not of the correct value (this resistor determines IC gain, as mentioned for OTAs and programmable amplifiers), or that there is too much feedback at pin 3 (from the output through C2 and the tone controls). Remember that most of the voltage gain for this amplifier is in the IC portion of the circuit. Q2 and Q3 provide power output.

If there is a good signal at pin 8 of the IC but not at the output (speaker or R_L), the problem is at the discrete portion of the circuit (Q1, Q2, and Q3). Check the collector voltages of Q2 and Q3. These voltages should be about 12 V and should be substantially the same, except of opposite polarity. Also, the Q2/Q3 voltages should be substantially the same as at pins 4, 6, and 7 of the IC.

The waveform (signal) and voltage checks that are described here should be sufficient to locate any major defect in the circuit (including an improperly wired experimental circuit!). Of course, if the circuit operates, but performance is not as claimed, it is possible that the problem is one of poor physical layout, wrong component values, and so on. Also, the basic techniques described here can be applied to the other circuits of this chapter.

Comparator operation

Comparator ICs are essentially high-gain op amps designed for open-loop operation. Typically, a comparator produces an output when the input goes above or below a certain level or if it crosses zero. For example, the LM111 comparator IC shown in Figs. 6-KK and 6-LL produces a logic-1 output at pin 7 with a positive signal between the two inputs. A logic-0 output is produced with a negative input.

Fig. 6-KK   Level detector for photodiode.

Parameter	Limits			Units
	Min	Typ	Max	
Input Offset Voltage		0.7	3	mV
Input Offset Current		4	10	nA
Input Bias Current		60	100	nA
Voltage Gain		100		V/mV
Response Time		200		ns
Common Mode Range	0.3		3.8	V
Output Voltage Swing			50	V
Output Current			50	mA
Fan Out (DTL/TTL)	8			
Supply Current		3	5	mA

Fig. 6-LL Characteristics of an LM111 comparator.

Threshold or level detection is accomplished by putting a reference voltage on one input and the signal to be compared on the other input. The output then changes states when the signal input goes above or below the reference-input level. An op amp can be used as a comparator, except that op-amp response time is typically in the tens of microseconds (often too slow for many applications). Figure 6-LL shows the characteristics of a classic comparator IC (the LM111 operated with a 5-V supply at 25°C).

IC comparator application tips

One of the problems with any IC comparator is the tendency to oscillate. Here are some tips to minimize oscillation and other comparator problems.

Keep output and balance leads (such as pins 5, 6, and 7 of Fig. 6-MM) apart, if possible, to avoid stray coupling between the output and balance. If the balance is not used (Fig. 6-JJ), tie the balance pins together. When balance is required, try connecting a 0.1-μF capacitor between the balance leads to minimize oscillation. Normally, bypass capacitors are not required for IC comparators. If it is required to eliminate large voltage spikes into the supplies when the comparator changes states, keep the leads as short as possible between the IC and bypass capacitors.

When source resistances between 1 and 10 kΩ are used, the impedance (both capacitive and resistive) on both inputs should be equal. The equal impedance tends to reject the feedback signal (and oscillation). Use positive feedback to increase hysteresis, as shown in Figs. 6-NN and 6-OO.

When driving the inputs from a low-impedance source, use a limiting resistor in series with the input lead. The resistor limits peak current and is especially important when the inputs go where they can accidentally be connected to a high-voltage source. Low-impedance sources do not cause a problem unless the output voltage exceeds the negative supply. However, because the supplies go to zero when turned off, isolation might be needed.

Op-amp and comparator circuits

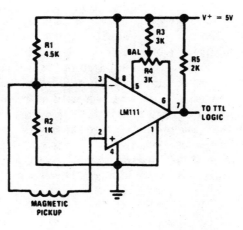

Fig. 6-MM Zero-crossing detector for magnetic transducer.

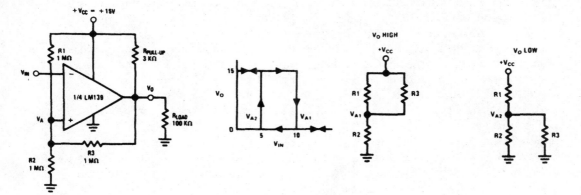

Fig. 6-NN Inverting comparator with hysteresis.

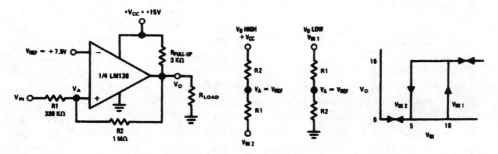

Fig. 6-OO Noninverting comparator with hysteresis.

Large capacitors at the input (greater than 0.1 μF) should be treated as a low source impedance and should be isolated with a resistor. Such capacitors can hold a charge larger than the supply when the supply is abruptly shut off.

Avoid reversing the supplies on comparators (or any IC). Typically, reverse voltage in excess of 1 V can melt the aluminum interconnections if the current is high. Use a clamp diode with adequate peak-current rating across the supply bus.

Do not operate an IC comparator with the ground terminal at a voltage that exceeds either supply. Also, the output voltage (such as the 50 V in Fig. 6-LL) generally applies to the potential between the output and the V-terminal. Therefore, if the comparator is operated from a negative supply (Fig. 6-KK), the maximum output voltage must be reduced by an amount equal to (or less than) the V-voltage.

Comparator tests

The test for any comparator circuit is to change the input and check that the output changes. For example, in the circuit of Fig. 6-KK, the output should change states when the diode D1 current reaches 1 μA. Check by covering D1, measuring the voltage at pin 7, and then exposing D1 to light. Typically, the output should switch between 0 V and near −10 V.

In the circuits of Figs. 6-NN and 6-OO, monitor the output across R_{LOAD} while varying the input voltage above and below the V_{REF} point (7.5 V using the values shown). Typically, the output switches between 0 and near 15 V. For the circuit of Fig. 6-NN, the output should go to 15 V when the input is at 5 V, and it should drop to 0 V when the input is increased to 10 V. The opposite should occur for the circuit of Fig. 6-OO.

In the circuit of Fig. 6-PP, apply voltage to V_{IN} and check that the lamp L1 turns on and off. If V_{IN} is greater than V_A, or less than V_B, the lamp should be off. Lamp L1 goes on only when V_{IN} is less than V_A but greater than V_B. The voltage range where L1 is on depends on the values of R_1, R_2, and R_3.

In the circuit of Fig. 6-QQ, apply a sine-wave signal at the input and check that the output produces a square wave of the same frequency. The square-wave amplitude depends on the value of the pull-up resistance in relation to R5/R6.

Comparator troubleshooting

The first step in troubleshooting comparator circuits is to check that the desired output-voltage change is produced for a given change at the input. If not, try correcting the problem with adjustment (as described for the circuit).

If the problem cannot be corrected by adjustment (or there is no adjustment), trace signals using a meter or scope from input (typically, a voltage level) to output (typically, a rapid voltage-level change, or possibly a square-wave/pulse-signal output). Follow this with voltage and/or point-to-point resistance measurement. The following are some typical examples.

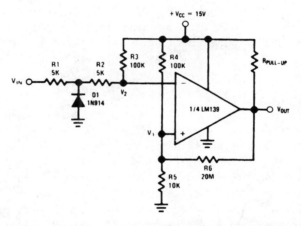

Fig. 6-PP Limit comparator with lamp driver.

Fig. 6-QQ Zero-crossing detector for squaring a sine wave.

In the circuit of Fig. 6-KK, if the output does not change states when D1 is alternately exposed to light and dark, check for any change at pin 2 of the LM111, with D1 covered and uncovered. Although the voltage change will be small (D1 produces about 1 μA of current), the change should be measurable. If there is no change at pin 2 of the LM111, suspect D. If there is a measurable change at pin 2, but not at pin 7, suspect the LM111.

Comparator troubleshooting

In the circuits of Figs. 6-NN and 6-OO, check that the output switches between zero and near 15 V when the input is varied between 5 and 10 V. Of course, the circuits produce opposite results or outputs for the same input, as shown by the hysteresis graphs (Fig. 6-NN is inverting, and Fig. 6-OO is noninverting). If there is no change in output for a given input change, suspect problems with the LM139 (unless, of course, the circuit is not properly wired, in which case you shall have no pie!).

In the circuit of Fig. 6-PP, if lamp L1 stays on or off when V_{IN} is varied above and below the limits of V_A and V_B, check for a change at the base of Q1 when V_{IN} is varied. If there is a change at Q1, but the lamp does not respond, suspect Q1.

For example, if the base of Q1 goes negative, the lamp should turn off, and vice versa. Of course, if the lamp is always off, the problem could be the lamp (which you should have checked first!). Remember that the point at which the lamp turns on is set by the values of R1, R2, and R3. Assuming a V_{CC} of 10 V, and that R1, R2, and R3 are all the same value, L1 should be off if V_{IN} is greater than 6.6 V, or less than 3.3 V. Lamp L1 stays on when V_{IN} is greater than 3.3 V, but less than 6.6 V.

In the circuit of Fig. 6-QQ, if there is no square-wave output for a sine-wave input, suspect the LM139 (again assuming good wiring and proper resistor values). Of course, if D1 is leaking badly (shorted), the input signal might be prevented from reaching the LM139. A level of about −700 mV at the junction of R1 and R2 indicates that D1 is probably good.

Remember that the zero reference is set by the values of R4 and R5. With $R_1 + R_2$ equal to R_5, V_1 should equal V_2 when V_{IN} is zero, and the square-wave output should switch states each time the sine-wave input crosses zero. With the values shown, the no-signal voltage at both inputs of the LM139 is about 1.5 V.

Comparator response-time problems

One difficult troubleshooting problem for comparators is when the circuit operates, but the response time is not correct (too slow). It is difficult to tell if the problem is one of circuit components or the comparator IC. The test connections and corresponding response-time graphs shown in Fig. 6-RR can help pinpoint the problem. Notice that there are no external components (except for the 5.1-kΩ load). The output is monitored on a scope, and a pulse generator output is applied at the input. (The pulse generator must have a rise time faster than the anticipated comparator response time.) The graphs of Fig. 6-RR show the response to both positive and negative pulses. Notice that response time increases for lower input voltages, so the tests should be made with the same pulse voltage as is used with the circuit. If the response time of the comparator IC is well within the required tolerance (when tested without external components) the problem is localized. However, if response time for the IC is too slow, use a faster IC such as the RM4805 comparator shown in Fig. 6-SS.

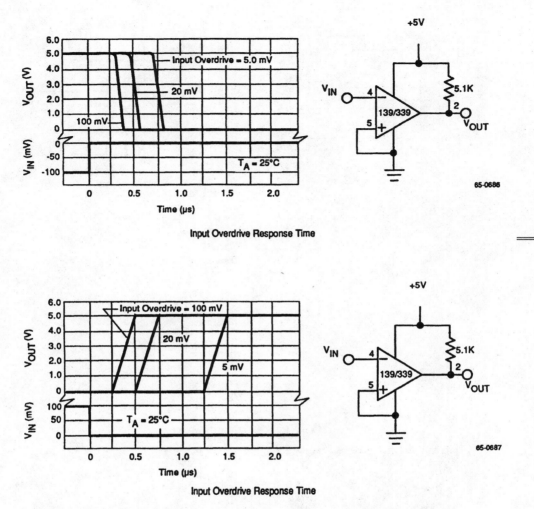

Fig. 6-RR   Response-time test circuit and graphs (low speed).

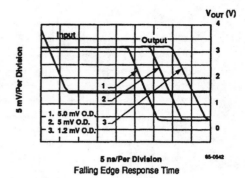

Rising Edge Response Time

Falling Edge Response Time

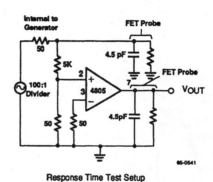

Response Time Test Setup

Response to 25 MHz Sine Wave

Response to 50 MHz Sine Wave

Fig. 6-SS Response-time test circuit and graphs (high speed).

Op-amp/comparator circuit titles and descriptions

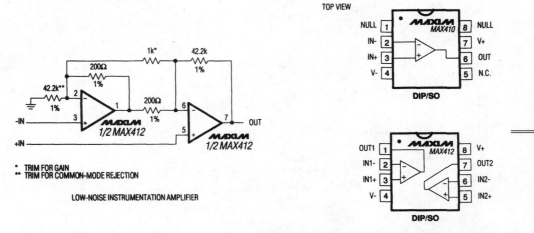

* TRIM FOR GAIN
** TRIM FOR COMMON-MODE REJECTION

LOW-NOISE INSTRUMENTATION AMPLIFIER

Pin Configurations continued on last page.

Fig. 6-1

Low-noise instrumentation amplifier
Figure 6-1 shows a low-noise instrumentation amplifier using both sections of a MAX412. Input voltage-noise density is less than 2.4 nV/$\sqrt{\text{Hz}}$ at 1 kHz. The output voltage swing is 7.3 Vp-p into 2 kΩ from ±5-V supplies. Supply current is 2.5 mA per amplifier, unity-gain bandwidth is 28 MHz, slew rate is 4.5 V/µs, maximum offset voltage is 250 µV, and minimum voltage gain is 115 dB. MAXIM HIGH-RELIABILITY DATA BOOK, 1993, P. 3-9.

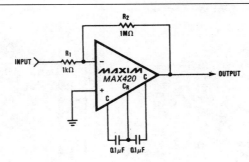

Inverting Amplifier

Fig. 6-2

Op-amp/comparator circuit titles and descriptions

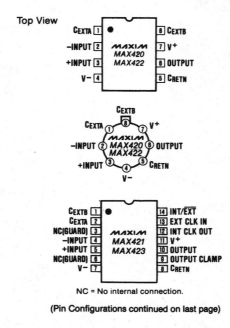

NC = No internal connection.

(Pin Configurations continued on last page)

Fig. 6-2 Continued

Inverting amplifier

Figure 6-2 shows a chopper-stabilized MAX420 connected as an inverting amplifier. The offset voltage is 5 µV maximum; the input voltage range is +12 V to −15 V with ±15-V supplies; the input noise is 0.3 µVp-p from dc to 1 Hz; the gain, CMRR, and PSRR are all 120 dB; the maximum supply current is 0.5 mA; and the input bias current is 30 pA. MAXIM HIGH-RELIABILITY DATA BOOK, 1993, P. 3-11.

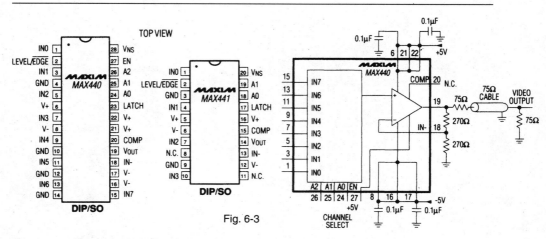

Fig. 6-3

Op-amp and comparator circuits

Multiplexed cable driver (8 channel)
Figure 6-3 shows a multiplexer/amplifier MAX440 connected to provide 8 channels of video signal to a single cable. Unity-gain bandwidth is 160 MHz, 6-dB-gain bandwidth is 110 MHz, channel-switch time is 15 ns, slew rate is 370 V/μs, and on/off input capacitance is 4 pF. Maxim New Releases Data Book, 1993, p. 3-15.

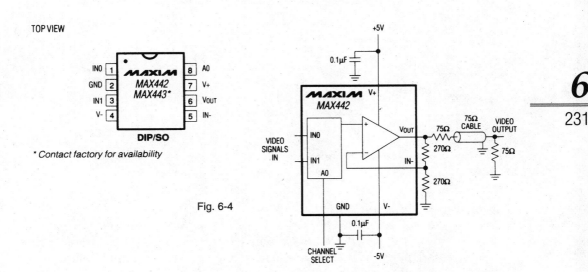

Fig. 6-4

* Contact factory for availability

Multiplexed cable driver (2 channel)
Figure 6-4 shows a multiplexer/amplifier MAX442 connected to provide 2 channels of video signal to a single cable. Unity-gain bandwidth is 140 MHz, channel-switch time is 36 ns, and slew rate is 250 V/μs. Maxim New Releases Data Book, 1993, p. 3-17.

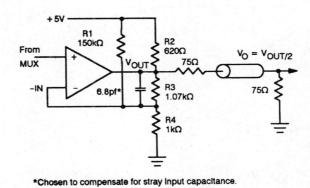

*Chosen to compensate for stray input capacitance.

Fig. 6-5

Op-amp/comparator circuit titles and descriptions

Figure 6-5 shows how the circuits of Figs. 6-3 and 6-4 can be connected to provide minimum phase distortion. MAXIM NEW RELEASES DATA BOOK, 1993, P. 3-20.

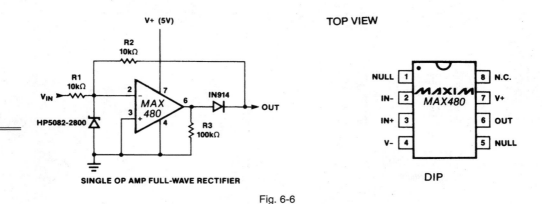

Fig. 6-6

Single op-amp full-wave rectifier
Figure 6-6 shows a MAX480 connected to form a full-wave rectifier. Maximum offset-voltage drift is 1.5 μV/°C, maximum supply current is 20 μA, minimum output drive is 5 mA, maximum input-offset voltage is 70 μV, maximum input bias current is 3 nA, and minimum open-loop gain is 700 V/mV. MAXIM NEW RELEASES DATA BOOK, 1993, P. 3-23.

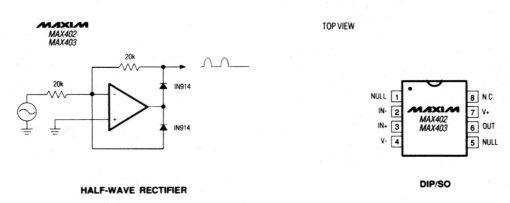

Fig. 6-7

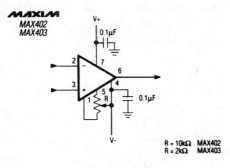

R = 10kΩ   MAX402
R = 2kΩ   MAX403

Fig. 6-8

6

233

Half-wave rectifier
Figure 6-7 shows a MAX402/03 connected to form a half-wave rectifier. The MAX402 has a 5-V/μs slew rate and 1.4-MHz bandwidth with 75 μA for supply current. The MAX403 has a 25-V/μs slew rate and 7-MHz bandwidth, with 375 μA for supply current. Both op amps are unity-gain stable and operate from ±3 V to ±5 V, or a single supply from +6 V to +10 V. Figure 6-8 shows a null circuit. MAXIM NEW RELEASES DATA BOOK, 1992, P. 3-15, 3-21.

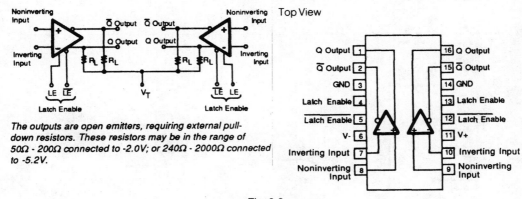

The outputs are open emitters, requiring external pull-down resistors. These resistors may be in the range of 50Ω - 200Ω connected to -2.0V; or 240Ω - 2000Ω connected to -5.2V.

Fig. 6-9

Ultra-fast ECL output comparator with latch enable
Figure 6-9 shows a MAX9687 connected as a fast (600 MHz) output comparator under latch control. The latch-enable inputs can be driven from a standard ECL gate. When LE is high and $\overline{\text{LE}}$ is low, the comparator function is normal. When LE is low and $\overline{\text{LE}}$ is high, the comparator outputs are locked in the logic states

Op-amp/comparator circuit titles and descriptions

determined by the input conditions at the time of the latch transition. If the latch-enable function is not used, connect LE to ground and leave $\overline{\text{LE}}$ open. The propagation delay is 1.4 ns, the latch setup time is 0.5 ns, the latch-enable pulse width is 2.0 ns, and the power supplies are +5 V and −5.2 V. MAXIM NEW RELEASES DATA BOOK, 1993, P. 3-39.

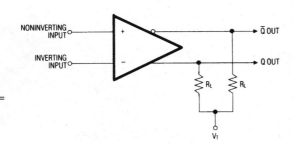
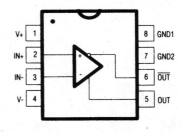

THE OUTPUTS ARE OPEN EMITTERS, REQUIRING EXTERNAL PULLDOWN RESISTORS. THESE RESISTORS MAY BE WITHIN 50Ω - 200Ω CONNECTED TO -2.0V, OR 240Ω - 2KΩ CONNECTED TO -5.2V.

Fig. 6-10

Ultra-fast ECL output comparator
Figure 6-10 is similar to Fig. 6-9, but without the latch-enable function. Propagation delay is 1.3 ns. MAXIM NEW RELEASES DATA BOOK, 1993, P. 3-41.

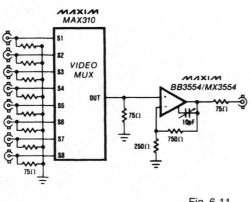
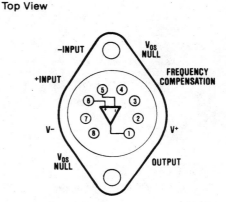

Top View

Fig. 6-11

Note:
There is no internal case connection

Op-amp and comparator circuits

Wideband fast-settling op amp with multplex input
Figure 6-11 shows a BB3554/MX3554 used with a MAX310 to provide eight channels of video signal to a single cable. The op amp slews at 1000 V/µs and provides up to ±100 mA output with ±10-V supplies. MAXIM NEW RELEASES DATA BOOK, 1993, P. 3-43.

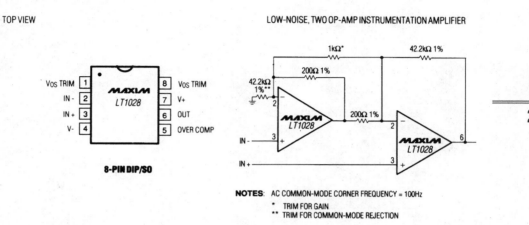

Fig. 6-12

High-speed low-noise instrumentation amplifier
Figure 6-12 is similar to Fig. 6-1, except for the op-amp characteristics. For the LT1028, the input voltage-noise density is 1.1 nV/√Hz maximum at 1 kHz, 0.85 nV/√Hz typical at 1 kHz, 1.0 nV/√Hz typical at 10 Hz, and 35 nVp-p typical at 0.1 Hz to 10 Hz. Minimum gain-bandwidth is 50 MHz, minimum slew rate is 10 V/µs, maximum offset voltage is 40 µV, and maximum offset drift is 0.8 µV/°C. Minimum voltage gain is 7 million. MAXIM NEW RELEASES DATA BOOK, 1993, P. 3-59.

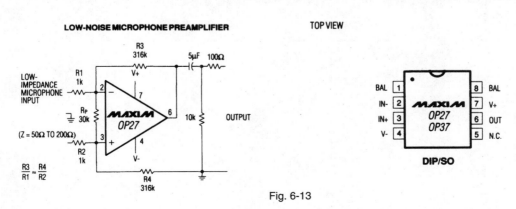

Fig. 6-13

Op-amp/comparator circuit titles and descriptions

Low-noise microphone preamplifier
Figure 6-13 shows an OP27 connected as a low-noise microphone preamp. The input voltage noise is 3 nV/$\sqrt{\text{Hz}}$ at 1 kHz. The OP27 has an 8-MHz gain-bandwidth product and a 2.8-V/μs slew rate. The OP37 has a 63-MHz gain-bandwidth with a 17-V/μs slew rate. Input-offset voltage is 10 μV, drift is 0.2 μV/°C, and output swing is ±10 V into 600 Ω. MAXIM NEW RELEASES DATA BOOK, 1993, P. 3-67.

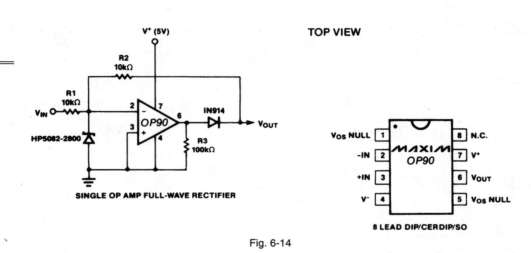

Fig. 6-14

Op-amp full-wave rectifier
Figure 6-14 shows an OP90 connected to form a full-wave rectifier. This is similar to the circuit of Fig. 6-6, except that the OP90 has a maximum input-offset voltage of 150-μV (higher than the maximum of 70 μV for the MAX480). Similarly, the input bias current, input offset current, and drift specifications of the OP90 are higher than the MAX480. MAXIM NEW RELEASES DATA BOOK, 1993, P. 3-69.

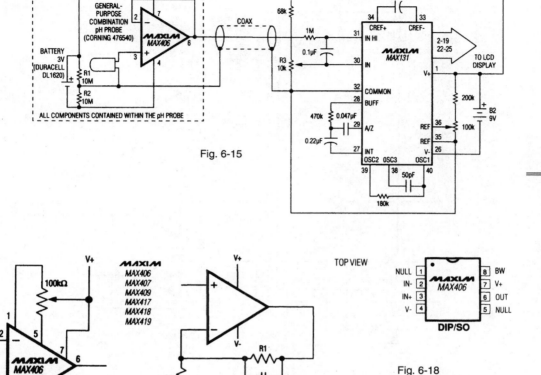

Fig. 6-15

Fig. 6-16

Fig. 6-17

Fig. 6-18

Buffered pH probe

Figure 6-15 shows a MAX406 combined with a MAX131 ADC to form a buffered pH probe. This circuit eliminates expensive low-leakage cables that often connect pH probes to meters. The MAX406 and a lithium battery are included in the probe housing. A conventional low-cost cable carries the buffered pH signal to the MAX131. In most cases, the 3-V battery life exceeds the functional life of the probe. Figures 6-16, 6-17, and 6-18 show the offset-voltage adjustment, feedback compensation, and pin configurations, respectively for the MAX406. Notice that the feedback compensation (Fig. 6-17) is not required in all cases. MAXIM NEW RELEASES DATA BOOK, 1993, P. 3-5, 3-28, 3-29.

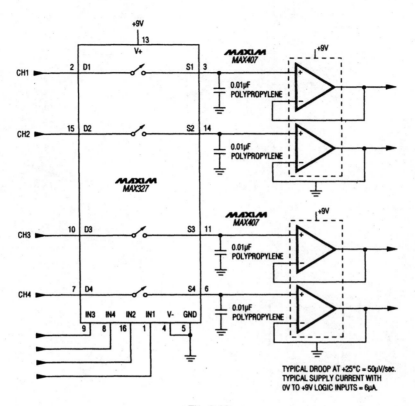

Fig. 6-19

Four-channel sample-and-hold (S/H)
Figure 6-19 shows two MAX407s combined with a MAX327 to form a four-channel S/H circuit. See Fig. 6-18 for MAX407 pin configurations. MAXIM NEW RELEASES DATA BOOK, 1993, P. 3-30.

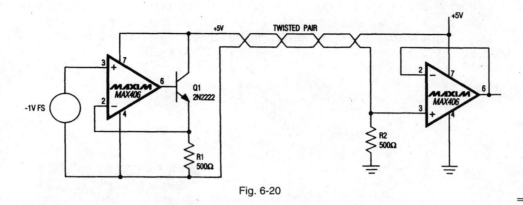

Fig. 6-20

Remotely powered sensor amplifier

Figure 6-20 shows a simple two-wire current transmitter that uses no power at the transmitting end, except from the transmitted signal itself. At the transmitter, a 0-V to 1-V input drives both a MAX406 and an NPN transistor connected as a voltage-controlled current sink. The 0-mA to 2-mA output is sent through a twisted pair to the receiver and it develops a voltage across receiver sense resistor R2. The resulting sense voltage is buffered by another MAX406, producing a 0-V to 1-V ground-referenced output signal. R1 and R2 should be well matched. The MAX406 supply current is added to the 0-mA to 2-mA signal, resulting in a 500-μV offset at the output. This offset, in addition to the MAX406 input offset, varies with temperature. MAXIM NEW RELEASES DATA BOOK, 1994, P. 3-29.

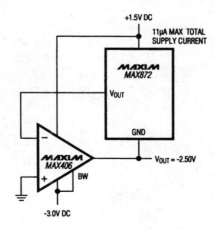

Fig. 6-21

Negative reference (−2.5 V)
Figure 6-21 shows a MAX406 combined with a low-dropout MAX872 to form a precise −2.5-V reference that requires no external components. Typically, the MAX872 requires two external resistors. Maximum current drain is 11 μA. There is no degradation of voltage because of load regulation and no compensation is needed for load capacitance. The supplies need not have precise regulation. The positive supply can be as low as 1.1 V with the negative supply as low as 2.7 V. MAXIM NEW RELEASES DATA BOOK, 1994, P. 3-31.

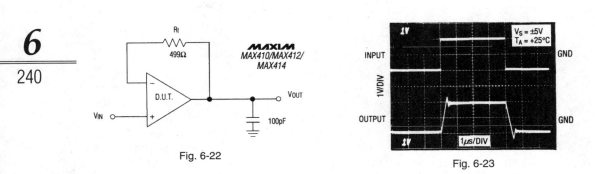

Fig. 6-22

Fig. 6-23

Voltage follower with capacitive load (100 pF)
Figure 6-22 shows a basic voltage-follower circuit with a capacitive load. Figure 6-23 shows the waveforms (see Fig. 6-1 for pin configurations). This circuit is suitable for capacitive loads up to about 100 pF. MAXIM NEW RELEASES DATA BOOK, 1994, P. 3-41.

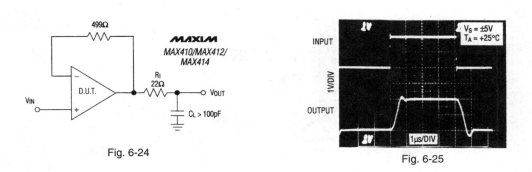

Fig. 6-24

Fig. 6-25

Op-amp and comparator circuits

Voltage follower with capacitive load (above 100 pF)
Figure 6-24 shows a voltage-follower circuit with a capacitive load above 100 pF.
Figure 6-25 shows the waveforms. (see Fig. 6-1 for pin configurations). Resistor R1
isolates the load capacitance from the amplifier output to prevent oscillation.
MAXIM NEW RELEASES DATA BOOK, 1994, P. 3-41.

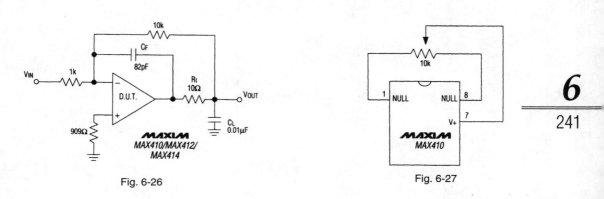

Fig. 6-26   Fig. 6-27

Voltage follower with increased accuracy (0.01 μF)
Figure 6-26 shows a voltage follower with increased accuracy (see Fig. 6-1 for pin
configurations). The feedback around isolation resistor R1 increases the accuracy
and the load capability. To drive capacitive loads greater than 0.01 μF, increase the
value of C_F. Figure 6-27 shows a null offset circuit suitable for the circuits of Figs.
6-22, 6-24, and 6-26. MAXIM NEW RELEASES DATA BOOK, 1994, P. 3-42.

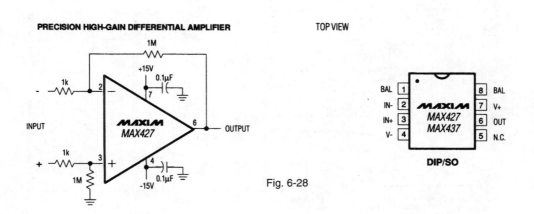

Fig. 6-28

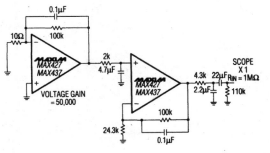

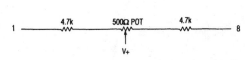

NOTE: ALL CAPACITOR VALUES ARE FOR NON-POLARIZED CAPACITORS ONLY.

Fig. 6-29

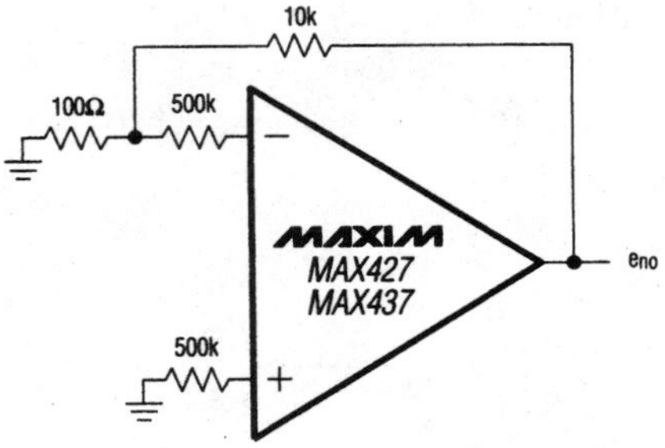

$$in = \frac{\left[e^2 - (130nV)^2\right]^{1/2}}{1M\Omega \times 100}$$

Fig. 6-30

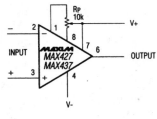

Fig. 6-31

1 —— 4.7k —— 500Ω POT —— 4.7k —— 8

V+

Fig. 6-32

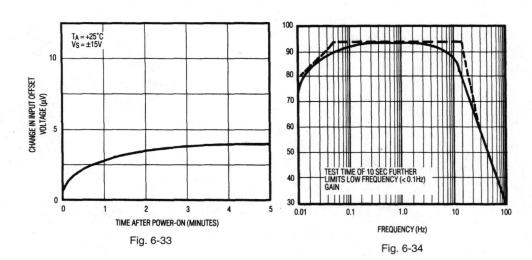

Fig. 6-33

Fig. 6-34

Op-amp and comparator circuits

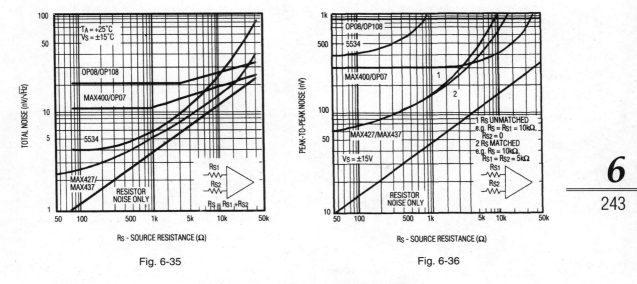

Fig. 6-35   Fig. 6-36

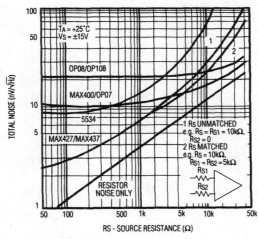

Fig. 6-37

Precision high-gain differential amplifier (low noise)

Figure 6-28 shows a MAX427 connected for differential operation. Wideband noise is 2.5 nV/$\sqrt{\text{Hz}}$, offset is less than 15 μV (5-μV typical), and drift is less than 0.8 μV/°C (0.1 μV/°C typical). The voltage gain is 20 million when driving a 2-kΩ load to ±12 V, and 12 million with a 600-Ω load to ±10 V. The MAX427 is unity-gain stable, with an 8-MHz gain-bandwidth, and a 2.5-V/μs slew rate. The decompensated MAX437 has a 60-MHz gain-bandwidth, a 15-V/μs slew rate, and is stable for closed-loop gains of five or greater. Both ICs can be operated from ±5-V supplies. Figures 6-29 and 6-30 show voltage-noise and current-noise test circuits, respectively. Figures 6-31 and 6-32 show offset-voltage adjustment circuits. Figure 6-33 shows typical offset-voltage characteristics. Figures 6-34 through 6-37 show noise characteristics. MAXIM NEW RELEASES DATA BOOK, 1994, P. 3-45.

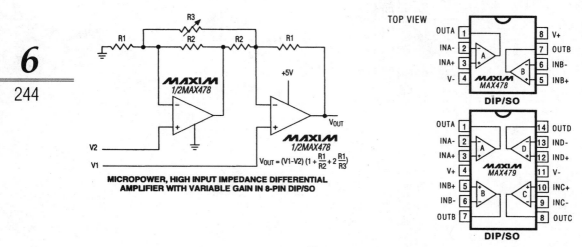

$$V_{OUT} = (V1-V2) \left(1 + \frac{R1}{R2} + 2\frac{R1}{R3}\right)$$

MICROPOWER, HIGH INPUT IMPEDANCE DIFFERENTIAL
AMPLIFIER WITH VARIABLE GAIN IN 8-PIN DIP/SO

Fig. 6-38

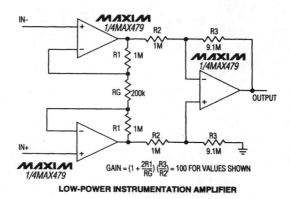

$$\text{GAIN} = \left(1 + \frac{2R1}{RG}\right)\left(\frac{R3}{R2}\right) = 100 \text{ FOR VALUES SHOWN}$$

LOW-POWER INSTRUMENTATION AMPLIFIER

Fig. 6-39

Single-supply differential amplifiers
Figure 6-38 shows a MAX478 connected as a differential amplifier with variable gain. Figure 6-39 shows the MAX479 connected as a differential instrumentation amplifier. Both circuits can operate from a single supply, either 3 V or 5 V.

Maximum supply current is 17 μA per op amp, the maximum offset voltage is 70 μV, maximum offset-voltage drift is 2.2 μV/°C (0.5 μV/°C typical) and maximum input-offset current is 250 pA. MAXIM NEW RELEASES DATA BOOK, 1994, P. 3-59, 3-69.

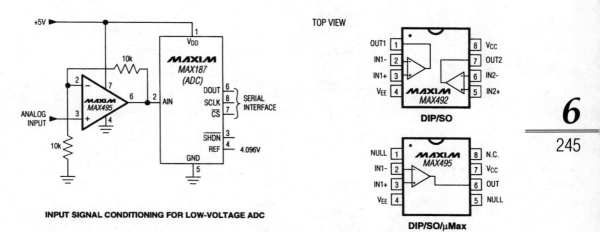

INPUT SIGNAL CONDITIONING FOR LOW-VOLTAGE ADC

Fig. 6-40

TOP VIEW

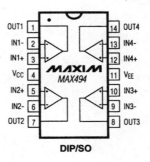

DIP/SO

Fig. 6-41

Input signal conditioner for low-voltage ADC
Figure 6-40 shows a MAX495 connected as an input signal conditioner for a MAX187 ADC. The MAX495 can operate from either a single supply (+2.7 V to +6 V) or split supplies (±1.35 V to ±3 V). Each op amp requires less than 150 μA supply current, but it can drive a 1-kΩ load. The input-referred voltage noise is 25 nV/√Hz, the offset is 200 μV, and the gain-bandwidth is 500 kHz (see Fig. 6-41 for additional pin configurations). MAXIM NEW RELEASES DATA BOOK, 1995, P. 3-23, 3-37.

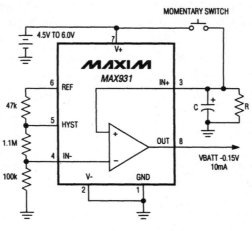

Fig. 6-42

Automatic power-off switch

Figure 6-42 shows a MAX931 comparator connected as a timed, automatic power-off circuit for a 40-mA supply. The comparator output is the supply output. With a 10-mA load, the circuit provides a voltage of ($V_{BATT} - 0.12$ V), but it draws only 3.5 μA of quiescent current. Using the values shown, the three-resistor voltage divider programs the maximum ±50 mV of hysteresis and sets the IN− voltage at 100 mV. This gives an IN+ trip threshold of about 50 mV for the IN+ falling. The RC time constant determines the maximum power-on time of the OUT pin (8) before power-down occurs. This period (in seconds) can be approximated by: R × C × 4.6. For example: 2 M × 10 μF × 4.6 = 92 (seconds). MAXIM NEW RELEASES DATA BOOK, 1995, P. 3-57.

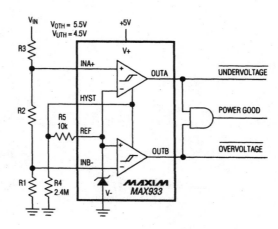

Fig. 6-43

Op-amp and comparator circuits

Window detector

Figure 6-43 shows a MAX933 comparator connected as a window detector (see Fig. 6-42 for pin connections). For the voltage thresholds shown (5.5 V upper and 4.5 V lower), use 249 kΩ, 61.9 kΩ, and 1 MΩ for R1, R2, and R3 respectively. Verify these resistor values using the following equations:

$$V_{OTH} = (V_{REF} + V_H) \times \frac{(R_1 + R_2 + R_3)}{R_1} = 5.474 \text{ V}$$

$$V_{UTH} = (V_{REF} + V_H) \times \frac{(R_1 + R_2 + R_3)}{(R_1 + R_2)} = 4.484 \text{ V}$$

where hysteresis voltage

$$V_H = V_{REF} \times \frac{R5}{R4}$$

and

$$V_{REF} = 1.182 \text{ V}$$

Other threshold values can be selected by choice of the R_1, R_2, and R_3, values, using the same equations. MAXIM NEW RELEASES DATA BOOK, 1995, P. 3-57.

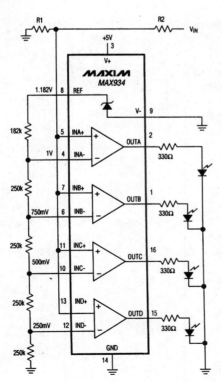

Fig. 6-44

Op-amp/comparator circuit titles and descriptions

Bar-graph level gauge

Figure 6-44 shows a MAX934 connected as a simple, four-stage level detector with LED readouts. The full-scale threshold (all LEDs on) is given by $V_{IN} = (R_1 + R_2)/R_1$ volts. The other thresholds are at ¾ full-scale, ½ full-scale, and ¼ full-scale. The output resistors limit the current into the LEDs. MAXIM NEW RELEASES DATA BOOK, 1995, P. 3-58.

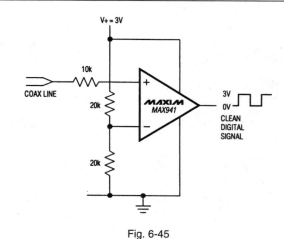

Fig. 6-45

TOP VIEW

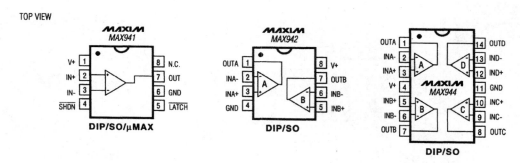

Fig. 6-46

Simple line transceiver

Figure 6-45 shows a MAX941 comparator connected as a simple line transceiver. The output is a clean square-wave signal at the input frequency. The output amplitude is equal to $V+$. See Fig. 6-46 for pin configurations. MAXIM NEW RELEASES DATA BOOK, 1995, P. 3-61, 3-69.

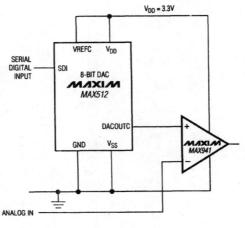

Fig. 6-47

Digitally controlled threshold detector
Figure 6-47 shows a MAX941 comparator combined with a MAX512 DAC to form a digitally controlled threshold detector. The analog signal to be compared is applied directly to the inverting input of the MAX941. The threshold point is set by the DAC output applied to the noninverting input. In turn, the DAC output is set by the serial digital input. As a result, the threshold can be set in eight discrete steps (depending on the serial input). MAXIM NEW RELEASES DATA BOOK, 1995, P. 3-69.

=7=

Power-supply circuits

The text in this chapter assumes that you are already familiar with power-supply and regulator basics (such as operation of rectifier diodes, switch-mode regulators and supplies, converters, inverters, etc.) and basic power-supply testing/troubleshooting. If not, read the author's *Simplified Design of Switching Power Supplies* and *Simplified Design of Linear Power Supplies*, both published by Butterworth-Heinemann. The following paragraphs summarize both the testing and troubleshooting of power-supply/regulator circuits. This information is included so that readers not familiar with power supplies can both test the circuits described here and localize problems if the circuits fail to perform as shown. Notice that micropower/single-cell battery supplies are covered in Chapter 8.

Power-supply/regulator testing

This section describes the basic tests for all types of power-supply and regulator circuits. Both simple tests and more advanced tests are described. If the circuits pass these tests, the circuits can be used immediately. If not, the tests provide a starting point for the troubleshooting procedures that are described in the power-supply/regulator troubleshooting portion of this chapter.

Basic tests

This section is devoted to simple, practical test procedures that can be applied to a just-completed power supply during design and experimentation (or to a suspected power supply as part of troubleshooting). More advanced tests are covered in this chapter. However, the following procedures are usually sufficient for practical applications.

The basic function of an off-line power supply is to convert alternating current into direct current. In a dc/dc converter, direct current is converted to direct current, but at a different voltage (usually higher, but lower in many cases). In any event, you can check the power-supply function simply by measuring the output voltage. However, for a more thorough test of a supply, the output voltage should be measured with a load, without a load, and possibly with a partial load.

If the supply delivers the full-rated output voltage into a full-rated load, the basic supply function is met. In addition, it is often helpful to measure the regulating effect of the supply, the supply internal resistance, and the amplitude of any ripple at the supply output. The following paragraphs describe each of the basic tests.

Output tests   Figure 7-A is the basic power-supply test circuit. This arrangement permits the supply to be tested at no load, half load, and full load, depending on the position of S1. With S1 in position 1, no load is on the supply. At positions 2 and 3, there is half load and full load, respectively.

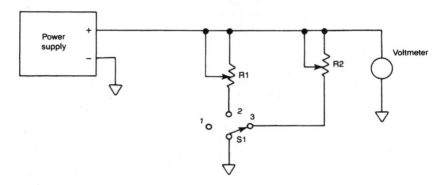

Fig. 7-A   Basic power-supply test circuit.

Using Ohm's law, $R = E/I$, R1 and R2 are chosen on the basis of output voltage and load current (maximum or half load). For example, if the supply is designed for at output of 5 V at 500 mA (full load), the value of R_2 is $5/0.5 = 10\ \Omega$. The value of R_1 is $5/0.25 = 20\ \Omega$.

Where more than one supply is to be tested, R1 and R2 should be variable, and adjusted to the correct value before testing (using an ohmmeter with the power removed). The resistors must be noninductive (not wire wound) and must dissipate the rated power without overheating. For example, using the previous values for R_1 and R_2, the power dissipation of R_1 is $5 \times 0.5 = 2.5\ W$ (use at least 5 W), and the dissipation for R_2 is $5 \times 0.25 = 1.25\ W$ (use at least 2 W).

1. Connect the equipment (Fig. 7-A).

2. Set R1 and R2 to the correct value.

3. Apply power. Set the input voltage to the correct value. Use the midrange value for the input voltage, unless otherwise specified. For example, the input voltage for a typical switching regulator/supply is between +4 V and +20 V, with certain tests (such as load and line regulation) with the input between +5.8 and +15 V. For dc/dc converters, a separate variable supply is required for the input voltage. (Figure 7-B shows such a supply.) For off-line supplies, the input

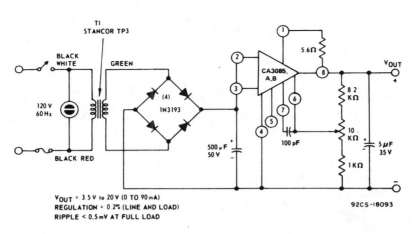

T1
STANCOR TP3

BLACK
WHITE   GREEN

120 V
60 Hz

BLACK RED

(4)
1N3193

500 μ F
50 V

CA3085,
A,B

100 pF

5.6 Ω

82
K Ω

10
K Ω

1 K Ω

5 μF
35 V

V_OUT

V_OUT = 3.5 V to 20 V (0 TO 90 mA)
REGULATION = 0.2% (LINE AND LOAD)
RIPPLE < 0.5 mV AT FULL LOAD

92CS-18093

Fig. 7-B   Adjustable off-line linear supply.

can be adjusted to the desired voltage with a variable transformer (or variac). If you must make the test with an input, use a 9-V battery for an approximate midrange value.

4. Measure the output voltage at each position of S1.
5. Calculate the current at positions 2 and 3 of S1, using Ohm's law, $I = E/R$. For example, assume that R_1 is 20 Ω and that the output voltage meter indicates 4.8 V at position 2 of S1. The actual load current is $4.8/20 = 240$ mA. If the supply output is 5 V at position 1, and it drops to 4.8 V at position 2, the supply is not producing full output with a load. This drop indicates poor regulation, possibly resulting from poor wiring design (in an experimental supply) or from component failure (when discovered as part of troubleshooting).

Load-regulation tests   Load regulation (also known as *load effect* or *output regulation*) is usually expressed as a percentage of output voltage and can be determined by:

$$\% \ regulation = \frac{(no\text{-}load \ voltage) - (full\text{-}load \ voltage)}{full\text{-}load \ voltage} \times 100$$

A low percentage of regulation is desired because it indicates that the output voltage changes very little with load changes. Use the following steps when measuring load regulation.

1. Connect the equipment (Fig. 7-A).
2. Set R2 for the correct value for a full load.
3. Apply power. Measure the output voltage at position 1 (no load) and position 3 (full load).
4. Using the equation, calculate the percentage of regulation. For example, if the no-load voltage is 5 V, and the full-load voltage is 4.999 V, the percentage of regulation is:

$$\left[\frac{(5 - 4.999)}{4.999}\right] \times 100 = 0.2\%$$

5. Notice that the power-supply output regulation is usually poor (high percentage) when the *internal resistance* is high.

Line-regulation tests Line regulation (also known as *line effect, input regulation,* or possibly *source effect* is usually expressed as a percentage of output voltage and represents the maximum allowable output voltage (with a given load) for maximum-rated input variation. For example, if the supply is designed to operate with an ac input from 110 to 120 V, and the dc output is 100 V, the output is measured (1) with an input of 120 V and (2) with an input of 110 V. If there is no change in output, input regulation is perfect (and probably impossible). If the output varies by 1 V, the output variation is 1%. The actual power-supply input regulation can be measured at full load, or half load, as desired, using the test connections shown in Fig. 7-A. However, the input voltage must be varied from maximum to minimum values (with a variac or separate dc supply) and monitored with an accurate voltmeter.

Internal-resistance tests Power-supply internal resistance is determined by the equation:

$$Internal\ resistance = \frac{(no\ load\ voltage) - (full\mbox{-}load\ voltage)}{current}$$

A low internal resistance is most desirable because this indicates that the output voltage changes very little with load.

1. Connect the equipment (Fig. 7-A).
2. Set R2 to the correct value.
3. Apply power. Measure the actual output voltage at position 1 (no load) and position 3 (full load).
4. Calculate the actual load current at position 3 (full load). For example, if R2 is adjusted to 10 Ω, and the output voltage at position 3 is 4.999 (as in the preceding example), the actual load current is:

$$\frac{4.999}{10} = 499\ mA$$

With no-load voltage, full-load voltage, and actual load current established, find the internal resistance using the equation, for example, with a no-load voltage of 5 V, a full-load voltage of 4.999, and a current of 499 mA (0.499 A), the internal resistance is:

$$\frac{(5 - 4.999)}{0.499} = 0.002\ \Omega$$

Efficiency tests   Power-supply efficiency is usually expressed as a percentage and represents output power divided by input power (times 100 to find percentage). Although the calculation is simple, you must have some means of measuring the input current as well as voltage. If you do not have an ammeter that will measure input current, use a resistor in series with the input supply. Then measure the voltage across the resistor, and calculate current ($I = E/R$). If you use a 1-Ω resistance, the result will be in amperes. A 1000-Ω series resistance will indicate milliamperes (mA). Although the steady-state input current of most switching-regulator ICs is low, the starting-current surge can be high.

Assume that the full-load output voltage is 4.999 V with a load of 75 mA and that the input is 4.5 V with an input current of 20 mA. The input power is 90 mW (4.5×0.02) and the output power is 75 mW, so the efficiency is 83.3% (75/90). This efficiency is typical for most battery-powered switching-regulator circuits.

Ripple tests   Any off-line supply, no matter how well regulated or filtered, has some ripple. A battery-operated switching supply also produces an oscillator signal (which has the same effect as ripple). No matter what the source, ripple can be measured with a meter or scope. Usually, the factor of most concern is the ratio between ripple and full-output voltage. For example, if 0.03 V of ripple is measured with a 5-V output, the ratio is 0.03/5, or 0.006, which can be converted to a percentage ($0.006 \times 100 = 0.6\%$).

1. Connect the equipment (Fig. 7-A).
2. Set R2 to the correct value. Ripple is usually measured under full-load power.
3. Apply power. Measure the dc output voltage at position 3 (full load).
4. Set the meter to measure the alternating current. Any voltage measured under these conditions is ac ripple.
5. Find the percentage of ripple, as a ratio between the two voltages (ac/dc).
6. One problem often overlooked in measuring ripple with a meter is that any ripple voltage is not a pure sine wave. Most meters provide accurate ac voltage indications only for pure sine waves. A better way to measure ripple is with a scope (as shown in Fig. 7-C) where peak values can be measured directly.
7. Adjust the scope controls to produce two or three stationary cycles of ripple on the screen. Notice that a full-wave rectifier produces two ripple "humps" per cycle, but a half-wave produces one hump per cycle.

A study of the ripple waveform can sometimes show the source of defects in a power-supply circuit. Here are some examples:

- *If the supply is unbalanced* (one rectifier passing more current than the others), the ripple humps are unequal in amplitude.
- *If there is noise or fluctuations in the supply,* especially where zener diodes are involved, the ripple humps will vary in amplitude and shape.
- *If the ripple varies in frequency,* the ac source is varying. (In switching supplies, the switching frequency is varying.)
- *If a full-wave supply produces a half-wave output,* one rectifier is not passing current.

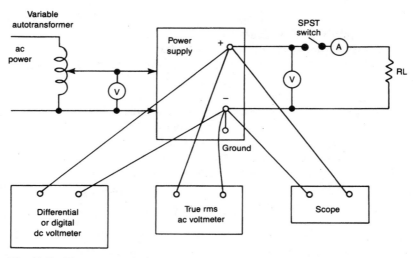

Fig. 7-C Test connections for measuring source effect, load effect, PARD (periodic and random deviation), drift, and temperature coefficient.

Advanced tests

The basic tests of the last section are generally sufficient for most practical applications (usually for the experimenters and serious hobbyists). However, you can use many other tests to measure the performance of commercial and lab power supplies. The most common (and most important) of such tests are covered in the following paragraphs.

Figure 7-C shows test connections for measurement of the five most important operating specifications of a power supply: source effect, load effect, *PARD* (periodic and random deviation), drift, and temperature coefficient. There are additional specifications, such as noise-spike measurements and transient-recovery time measurements. However, these measurements require elaborate test setups and are generally applied to lab or commercial supplies. For thorough coverage of power-supply tests and measurements, read the author's *Complete Guide to Electronic Power Supplies*, 1990, Prentice-Hall.

Test equipment All the tests described here can be performed with only four test instruments: a variable autotransformer (variac), a differential or digital ac voltmeter, a true ac voltmeter, and a scope. Of course, on those supplies that are to be battery operated, a separate variable dc supply is required (Fig. 7-B).

Using a separate supply brings up a problem. If there is any ripple or other output fluctuations in the separate (input) supply, these can pass through the supply under test, making the test supply appear defective. This problem can be checked by substituting a battery (of the same voltage) to power the supply being tested. If the ripple or other fluctuations remain, the problem is in the supply under test.

Make certain that the autotransformer (in off-line supplies) or variable supply (for battery-operated supplies) has an adequate current rating. If not, the input voltage to the supply under test might be severely distorted and the rectifying/regulating circuit might operate improperly.

The dc voltmeter should have a 1 mV (or better) accuracy. The scope should have a sensitivity of 100 μV/cm (microvolts per centimeter) and a bandwidth of at least 10 MHz. Both the scope and meter should have some means of measuring current (some form of current probe, preferably a clip-on type). In switching supplies, a scope display of the indicator current can often pinpoint circuit problems (as covered in this chapter). A nonelectronic *VOM* (volt-ohm-milliammeter), such as the classic Simpson 260 or Tripplet 630, also can be useful in switching supply tests. Some digital and other electronic meters are affected by the switching-oscillator signals.

Proper connections   For the most accurate results, the test connections should be permanent (not clip leads) and should be made to the exact point on the supply terminals. Clip-lead connections can produce measurement errors. Instead of measuring pure supply characteristics, you are measuring supply characteristics, plus the resistance between output terminals and point of connection. Even using clip leads to connect the load to the supply terminals can produce a measurement error.

Separate leads   All measurement instruments must be connected directly by separate pairs of leads to the monitoring points (Fig. 7-C). This connection avoids the subtle mutual-coupling effects that can occur between measuring instruments (unless all are returned to the low-impedance terminals of the supply). Use twisted pairs or shielded cable to avoid pickup on the measuring loads.

Load resistance   Be certain that the load resistance is adequate for the supply and test requirements. Typically, the load resistance and load voltage should permit operation of the supply at maximum-rated output voltage and current.

Pickup and ground-loop effects   Always check the test connections for possible pickup and/or ground-loop problems. As a simple test, turn off the supply and observe the scope for any unwanted signals (especially at the line frequency, typically 50/60 Hz) with the scope leads connected directly on the supply output terminals. Then connect both scope leads to either terminal (+ or −), whichever is grounded to the board—or to common ground. If there is any noise in either test condition, with the supply off, you have possible pickup and/or ground-loop effects.

Alternating-current (ac) voltmeter connections   Connect the ac voltmeter as close as possible to the input ac terminals of an off-line supply. The voltage indication is then a valid measurement of the supply input, without any error introduced by the drop present in the leads that connect the supply input to the ac line. The same is true for the dc input of battery-operated supplies. That is, measure the dc input at the supply terminals, not at the output of the variable supply or battery.

Line regulator   Do not use any form of line regulator when testing an off-line supply and when using the supply (unless it is specifically recommended for that supply). This precaution is especially important for switching supplies and regulators. A line regulator can change the shape of the output waveform in a switching supply and thus offset any improvement produced by a constant line input to the supply.

Source effect or line regulation

No matter what the test is called, the measurement is made by varying the input voltage throughout the specified range from low limit to high limit and noting the change in voltage at the supply output terminals. The test is performed with all other test conditions constant. The supply should stay within specifications for any rated output voltage, combined with any rated output current. The extreme source-effect test is with the maximum output voltage and maximum output current.

Load effect or load regulation

This test is made by closing and opening switch S1 (Fig. 7-C) and noting any change in output voltage. The test is performed with all other test conditions constant. The supply should stay within specifications for any rated output voltage, combined with any rated input voltage. The extreme load-effect test is with maximum output voltage and maximum output current.

Noise and ripple (or PARD)

In many cases, PARD (periodic and random deviation) has replaced the terms *noise* and *ripple* and represents deviation of the dc output voltage from the average value, over a specified bandwidth, with all other test conditions constant.

As an example, in Hewlett-Packard lab supplies, PARD is measured in *rms* (root means square) or peak-to-peak values over a 2-Hz to 2-MHz bandwidth. Fluctuations below 20 Hz are considered to be drift. Peak-to-peak measurements are of particular importance for applications where noise spikes can be detrimental (such as in digital logic circuits). The rms measurement is not ideal for noise because output noise spikes of short duration can be present in ripple, but not appreciably increase the rms value. Always use twisted-pair leads (for single-ended scopes) or shielded two-wire leads (for differential scopes) when making PARD or noise/ripple tests.

Drift (stability)

Drift measurements are made by monitoring the supply output on a differential or digital voltmeter over a stated measurement interval (typically eight hours, after a 30-minute warmup). In some cases, a strip chart is used to provide a permanent record. Place a thermometer near the supply to verify that the ambient temperature remains constant during the period of measurement. The supply should be at a location immune from stray air currents (away from open doors or windows and from air-conditioning vents). If practical, place the supply in an oven and hold the temperature constant. As a guideline, a well-regulated supply will drift less during the eight-hour period than during the 30-minute warmup.

Temperature coefficient (or TC)

TC (also known as *tempco*) measurements are made by placing the supply in an oven and varying the temperature over a given range, following a 30-minute warmup. The supply is allowed to stabilize at each measurement temperature. In the absence of other specifications, the temperature coefficient is the output-voltage changes that result from a 5°C change in temperature. The measuring instrument should be placed outside the oven and must have a long-term stability that is adequate to ensure that any voltmeter drift does not affect measurement accuracy.

Switching power-supply troubleshooting

The remainder of the introduction to this chapter is devoted to troubleshooting switching power-supply circuits. In general, most supply problems are the result of wiring mistakes (which you never make), defective (or inadequate) components, and possibly test errors. All these mistakes can be located by basic voltage checks, resistance checks, and point-to-point wiring checks. However, switch-mode supplies and switching regulators present problems—especially when an experimental circuit is first tested. The following notes describe some of the most common troubleshooting problems for such supplies and regulators.

Ground loops

Figure 7-D shows a typical ground-loop condition. A generator is driving a 5-V signal into 50 Ω on the experimental circuit, which results in a 100-mA current. The return path for this current divides between the ground from the generator (typically, the shield on a BNC cable) and the secondary ground loop. The secondary ground loop is created by the scope-probe ground clip (shield), and the two "third-wire" connections on the generator and scope.

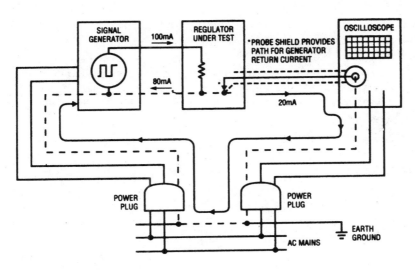

Fig. 7-D   Ground-loop errors.

Assume that 20 mA flows in the parasitic ground loop. If the scope ground lead has a resistance of 0.2 Ω, the scope will show a 4-mV bogus signal. The problem gets much worse for higher currents and for fast-signal edges, where the inductance of the scope-probe shield is important. The most practical solution is to use an isolation transformer for the scope. For a quick check, touch the scope-probe tip to the probe ground clip, with the clip connected to the experimental-circuit ground. The scope should show a flat line. Any signal displayed on the scope is a ground-loop problem or pickup problem.

Scope probe compensation

Always check that the scope probe is properly compensated when testing switching supplies. It is especially important for the ac attenuation (on a 10× probe, for example), to match the dc attenuation exactly. If not, low-frequency signals will be distorted and high-frequency signals will have the wrong amplitude. Remember that at typical switching frequencies, the waveshape might appear good because the probe acts purely capacitive, so the wrong amplitude might not be immediately obvious.

Ground-clip pickup

Do not make any test measurements on a switching regulator with a standard (alligator) ground-clip lead. Replace the alligator clip with a special soldered-in probe terminator, which you can get from many probe manufacturers. The standard ground-clip lead can act as an antenna, and it can pick up magnetic and other radiated signals. Make the test described for ground loops if you suspect pickup by the scope probe.

Measuring at the component

Make all measurements (output voltage, ripple, etc.) at the component, not at a wire that is connected to a component. This precaution is necessary because wires are not shorts. For example, switching regulators (such as those of this chapter) produce square waves or pulses. In turn, these pulses are applied to an output capacitor (in most cases). A typical regulator can produce about 2-V per inch spikes in the capacitor leads. The farther you measure from the capacitor, the greater the spike voltage.

EMI (electromagnetic interference) suppression

EMI (electromagnetic interference) is a fact of life with switching regulators. EMI takes two basic forms: conducted (which travels down input and output wiring) and radiated (which produces electric and magnetic fields). Although these fields do not usually cause regulator problems, they can create problems for surrounding circuits. The following guideline is helpful in minimizing EMI problems.

1. Avoid long high-current grounds and feedback nodes. Figure 7-E shows the right and wrong ways to make grounds and feedback connections to switching regulators. Even though low-power switching-regulator ICs are generally easy to use, you must pay some attention to PC (printed circuit) layout and routing—especially at power levels greater than 1 W or when high-speed *PWM* (pulse-width modulation) ICs are used. Trace out the high-current paths and minimize their length—especially the ground trace. Use a *star ground*, in which all grounds are brought to one point. Place any input filter capacitor physically close to the IC. Minimize any stray capacitance at the feedback (FB) pin. Return all compensation capacitors and bypass capacitors to quiet, well-filtered points (such as an analog ground pin).

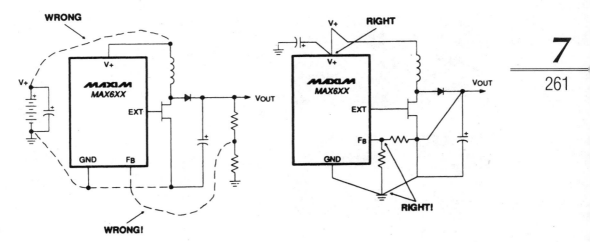

Fig. 7-E Correct and incorrect ground and feedback connections for switching regulators.

2. Use inductors or transformers with good EMI characteristics, such as toroids or pot cores. Avoid rod inductors. If you must use rod inductors, keep them in the output filter, where ripple current is low (hopefully). Use the inductor values that are shown in the circuit descriptions. Figure 7-F shows some typical current waveforms that are produced by good and bad inductors. In general, most inductor problems (other than using the incorrect size) can be traced to inadequate saturation (peak current) ratings or excessive dc resistance. If an inductor saturates, the current rises exponentially with time. If there is excessive resistance, a distinct *LR* characteristic is seen. If the waveform takes small (but strange) bends, the inductor might be producing both effects.

3. Route all traces carrying high ripple current over a ground plane to minimize radiated fields. This includes the catch-diode leads, input and output capacitor

leads, snubber leads, inductor leads, IC input and switch-pin leads, and input power leads. Keep these leads short and keep the components close to the ground plane.

4. Keep sensitive, low-level circuits as far away as possible and use field-canceling tricks, such as twisted-pair differential lines.

5. In very crucial applications, add a "spike killer" bead on the catch diode to suppress high harmonics, which can create higher transient switch voltages at switch turn-off. Check each switch waveform carefully.

6. Add an input filter if radiation from input lines could be a problem. Just a few μH (microhenrys) in the output line will allow the regular input capacitor to swallow nearly all the ripple current that is created at the regulator input.

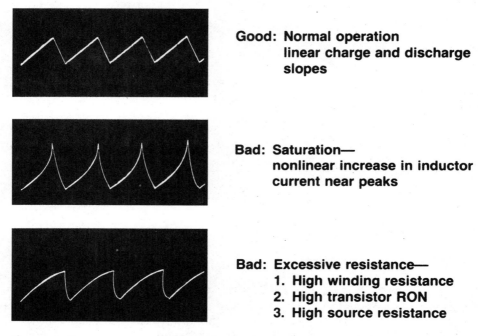

Good: Normal operation
linear charge and discharge
slopes

Bad: Saturation—
nonlinear increase in inductor
current near peaks

Bad: Excessive resistance—
1. High winding resistance
2. High transistor RON
3. High source resistance

Fig. 7-F Typical current waveforms produced by good and bad inductors.

Troubleshooting hints and tips

The following notes apply specifically when troubleshooting an inoperative or poorly performing switch-mode supply or switching regulator, particularly those that are in experimental form.

If the circuit is inoperative, look for such things as transformers wired backwards (always check the polarity dots or color codes on transformers), electrolytic capacitors wired backwards (usually you will find this out shortly after power is applied) and IC pins reversed (check the data sheet and follow the wiring that is given in the circuit schematics).

If the circuit works (no smoke, fire, or explosion), but the performance is not as expected (low efficiency, low output voltage or current, line or load regulation out of tolerance, etc.), one of the first troubleshooting steps is to display the inductor-current waveform on a scope. The most convenient way to monitor the inductor current is with a clip-on type current probe. Figure 7-F shows some typical inductor-current waveforms.

Ideally, both the charge and discharge slopes of the waveforms should be linear, as shown in Fig. 7-F. If so, but the circuit does not perform as desired, look for problems with the inductor, oscillator frequency, output capacitor value, and diode characteristics, in that order. For example, if the inductance has decreased because of saturation, the output will increase, but the inductor-current waveform will be nonlinear near the peaks, as shown. On the other hand, if the inductor resistance value is excessive, the waveform might be nonlinear during the entire charge time, as shown.

If you suspect that the inductor is faulty, try operating the inductor in a standard test circuit, such as shown in Fig. 7-G. Compare the waveforms obtained with the inductor in the test circuit against the waveforms shown in Figs. 7-H through 7-K. In each case, trace A is the voltage at the VSW pin of the LT1070, and trace B is the current (measured with a clip probe at the VSW pin). When the VSW-pin voltage is low, inductor current flows. With high inductance (Fig. 7-H), current rises slowly with no saturation. As inductance decreases, current rise is steeper (Figs. 7-I and 7-J), but there is still no saturation. However, the highest inductance (Fig. 7-K) but wound on a low-capacity core, produces extreme saturation and is unsuitable for switching-regulator use. Using this test circuit and procedure, you can narrow the selection of an inductor down to the best unit for your circuit (with regard to cost, size, etc.) Figure 7-L shows an inductor selection kit, which includes 18 inductors of various ratings and size. The kit is available from:

Pulse Engineering, Inc.
PO Box 12235
San Diego, California 92112
(619) 268-2400

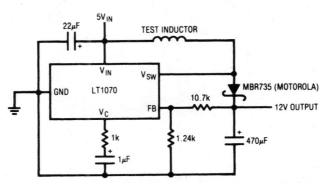

Fig. 7-G Standard test circuit for inductors used in switching circuit.

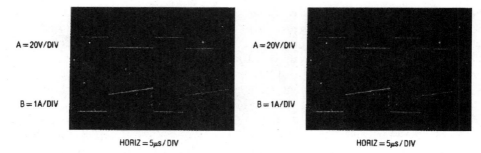

Fig. 7-H Waveforms for 450-nH high-capacity core.

Fig. 7-I Waveforms for 170-nH high-capacity core.

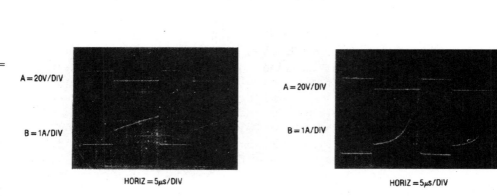

Fig. 7-J Waveforms for 55-μH *high-capacity core.*

Fig. 7-K Waveforms for 500-μH *low-capacity core.*

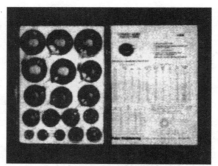

Fig. 7-L Model 845 inductor-selection kit.

If the input voltage appears to dip, it is possible that the input leads (from the battery to the switching IC, for example) are too long when connected in experimental form. (This problem should not occur when the circuit is in PC form). Switching regulators draw current from the input supply in pulses. Long input wires can cause dips in experimental form, try adding input capacitance (perhaps a 1000-μF or higher capacitance close to the IC regulator). If this doesn't cure the problem, it might be necessary to add some input capacitance when the circuit is in final form.

If a nonbattery input supply simply does not come up and switching does not start, with the right components all properly connected, it is possible that the input supply cannot deliver the necessary start-up current. Switching regulators have negative input resistance at start-up and draw high current. The high current can latch some input supplies to a low or off condition. In a battery-input supply, it is possible that the batteries are not "stiff" (not capable of delivering a large momentary current at start-up).

If efficiency is low (much more power going into the supply than power coming out), suspect the inductors (or transformers). Core or copper losses might be the problem. Of course, the problem could be an accumulation of all losses (inductor, capacitor, diode, etc.), which results in an inefficient supply/regulator circuit.

If the switch timing varies, check for excessive ripple at the output and at any feedback or compensation pins of the IC regulator. If the switch timing varies from cycle to cycle, try connecting a capacitor (about 1000 to 3000 pF) across the output capacitor (and/or at the feedback/compensation pin) to ground. If any of these capacitors eliminate the variation in switch timing, you have located the problem area.

If you definitely have high output ripple or noise spikes, suspect the output capacitor. Capacitors have a capacitance value (in μF or pF) and an *ESR* (equivalent series resistance) expressed in ohms. An increase in capacitance value decreases ripple, but an increase in ESR increases ripple. It is possible that your capacitor has high ESR, even though the capacitance value is correct, when there is ripple of unknown origin.

If the IC blows up in an experimental circuit (with the correct components, all properly connected, no shorted outputs, and no reversed electrolytics or battery polarities), it is possible that start-up surges are causing momentary large switching voltages. (These voltages also can occur when good experimental circuits are converted to PC form). Generally, this indicates that one or more components were on the borderline (such as excessive leakage in the output capacitor or catch diode).

If the IC runs hot, suspect a problem in the heatsink (not properly connected to the IC regulator). For example, a TO-220 package has a thermal resistance of about 50 to 55°C/W with no heatsink. A 5-V, 3-A output (15-W) regulator with a typical 10% switch loss will dissipate more than 1.5 W in the IC. This energy causes a 75°C temperature rise or 100°C case temperature at 25°C room temperature (which is hot!). In an experimental circuit, simply soldering the TO-220 tab to an enlarged copper pad on the PC board will reduce thermal resistance to about 25°C/W.

If you have poor load or line regulation, check in the following order:

- An output capacitor with high ESR (particularly if the capacitor is outside the feedback loop)
- A ground-loop error in the scope (Fig. 7-D)
- Improper connections of the output-divider resistors to current-carrying lines. (Figure 7-E shows both correct and incorrect connections for switching regulators)
- Excessive output ripple; switching frequency too high. (It is assumed that you have checked the inductor for saturation, as shown in Fig. 7-F).

If efficiency is poor, do not overlook the diodes. If the diode is not fast enough, or is not rated for the current level, efficiency will drop—even though the circuit might work. If the diode is hopelessly underrated, it will be destroyed. Most switching-regulator diodes are Schottky or ultrafast diodes. The major advantage of Schottky versus ultrafast is efficiency. As a guideline, when the output voltage is below 12 V, a Schottky will provide about 5% improvement in efficiency compared to an ultrafast. If the voltage is greater than 12 V, the advantage is less significant. Of course, the breakdown rating of any diode must exceed the input voltage, with allowances made for input surges.

Linear supply troubleshooting

The first step in troubleshooting the supply of Fig. 7-B is to test the circuit using both the basic and advanced tests. With a 90-mA load, you should be able to set the output voltage between 3.5 and 20 V, using the 10-kΩ pot (potentiometer) at pin 6 of the CA3085 positive voltage regulator.

If you find no voltage, check the neon lamp when the power switch is closed (right after you make sure the power cord is plugged in!). If the lamp is off, suspect the fuse.

If the lamp is on, check for ac at the transformer secondary (about 24 V across the diodes). If ac is absent, suspect the transformer. If it is present, check for dc between pin 3 of the CA3085 and ground. If it is absent, suspect problems with the diodes. It also is possible that the 500-μF capacitor is shorted or leaking badly.

If there is dc at pins 2/3 (about 25 V), but there is no output (pin 8) or you cannot adjust the output across the range, suspect the 5-μF output capacitor, the 100-pF compensation capacitor, and the CA3085 (in that order).

If you get the correct voltage output (across the range), but there is excessive ripple, or if line/load regulation is out of tolerance (alleged to be 0.2%), suspect the CA3085 (although leaking filter capacitors are always a possibility).

Dual pre-regulated supply troubleshooting

Again, the first step in troubleshooting the supply of Fig. 7-M is to test the circuit. Notice that you should get both a +12-V and a −12-V output, neither of which is

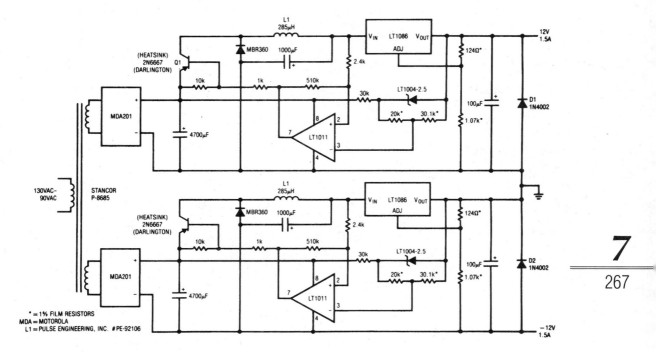

Fig. 7-M Dual pre-regulated off-line linear supply.

adjustable. (The adjustment pin on the three-terminal regulators is connected to a fixed voltage divider network in both cases.) If you get an output voltage, but not at the correct level, check the values of the 124-Ω and 1.07-kΩ resistors.

If you get +12 V, but not −12 V (or vice versa), you have eliminated half of the circuits as suspects. The same is true if you get excessive ripple or if the two supplies do not show substantially the same line/load regulation. (If both supplies are equally bad, suspect the transformer.)

Assume that the +12-V supply is bad, but the −12-V supply is good (−12-V output with a 1.5-A load, when the line is varied between 90 and 130 Vac). If there is no +12 V, check for ac at the input of the MDA201 and dc at the output. If the ac is absent or is abnormal, suspect the transformer wiring. If there is ac, but not dc, suspect the MDA201 (or the 4700-μF capacitor).

If there is dc from the MDA201, compare this to the dc from the MDA201 in the −12-V supply. While you are at it, compare the voltage at V_{IN} of both LT1086 regulators. If V_{IN} for the bad +12-V LT1086 is not substantially the same as for the −12-V regulator, suspect Q1 and the associated parts (such as L1, the LT1011, the MRB360, and the 1000-μF capacitor).

If the dc is the same for both V_{IN} terminals, but the +12-V output is absent or abnormal, suspect the 100-μF output capacitor, diode D1, the LT1086, and the LT1004 zener (in that order). If the zener is leaking badly (or is completely dead), you will get a different voltage at the V_{IN} terminals of the LT1086.

Linear supply troubleshooting

Power-supply circuit titles and descriptions

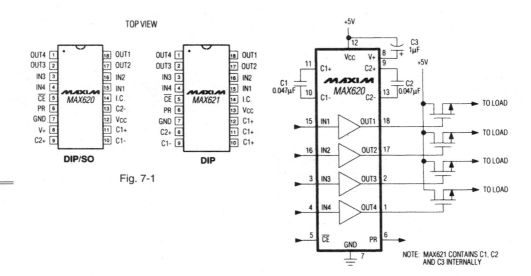

Fig. 7-1

High-side MOSFET driver

Figure 7-1 shows a MAX620 connected to power high-side switching and control circuits. The charge pump delivers a regulated output voltage that is 11 V higher than V_{CC} to the drivers. In turn, the drivers translate a TTL/CMOS input signal to a noninverted output that swings from ground to the high-side voltage. The continuous driver-output current is 25 mA, with a typical quiescent current of 70 μA. The MAX620 requires three external charge-pump capacitors. The MAX621 has internal capacitors. MAXIM NEW RELEASES DATA BOOK, 1992, P. 4-19.

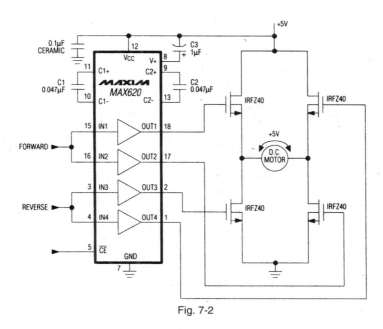

Fig. 7-2

H-bridge dc motor controller

Figure 7-2 shows a MAX620 driving an H-bridge switch that controls the direction of a +5-Vdc motor. By toggling between the forward and reverse inputs, each driver-output pair turns on the associated pair, which passes current through the motor, causing rotation in the desired direction. To prevent all four MOSFETs from switching on at once, update the forward/reverse inputs before clocking $\overline{CE}$ low, and do not assert forward and reverse simultaneously. Do not use a supply that will cause the gate drive to exceed the absolute maximum gate-to-source voltage of the low-side switch. MAXIM NEW RELEASES DATA BOOK, 1992, P. 4-27.

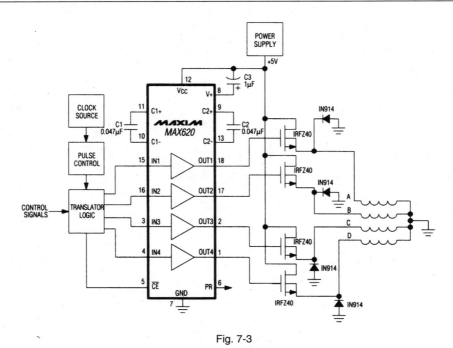

Fig. 7-3

Four-phase stepper-motor drive system
Figure 7-3 shows a MAX620 connected to form a complete stepper-motor drive.
TTL/CMOS signals from the logic network are translated to high-side levels that
drive four N-channel power MOSFETs, supplying current to each of four stepper-
motor phases. The diodes provide a discharge current path for the stepper-motor
windings. MAXIM NEW RELEASES DATA BOOK, 1992, P. 4-28.

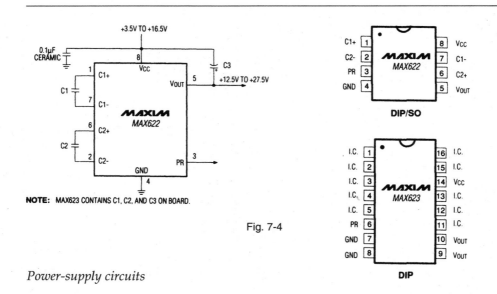

NOTE: MAX623 CONTAINS C1, C2, AND C3 ON BOARD.

Fig. 7-4

Power-supply circuits

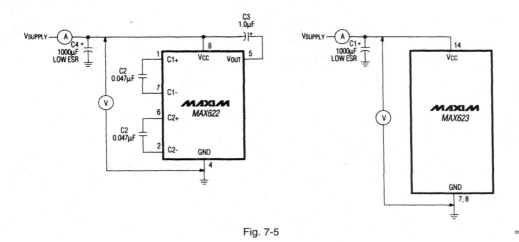

Fig. 7-5

High-side power supplies

Figure 7-4 shows a MAX622 connected to generate a regulated output voltage that is 11 V greater than the input supply voltage. The primary use is to power high-side switching and control circuits. The output current is 25 mA, with a typical quiescent current of 70 µA. Figure 7-5 shows quiescent supply-current test circuits. The MAX623 has internal charge-pump capacitors. The logic-level PR (power ready) output indicates when the high-side voltage reaches the proper level. MAXIM NEW RELEASES DATA BOOK, 1992, P. 4-31, 4-37.

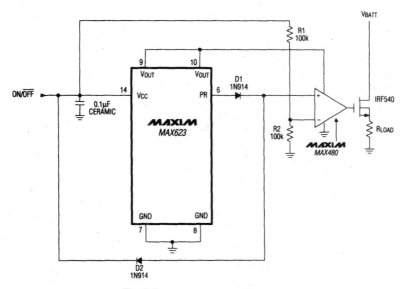

Fig. 7-6

Power-supply circuit titles and descriptions

Single-load switch
Figure 7-6 shows a MAX623 connected as a comparator-switch. The switch is turned on by applying VBATT to the ON/OVERLINE{OFF} input and turned off by pulling the input to ground. MAXIM NEW RELEASES DATA BOOK, 1992, P. 4-38.

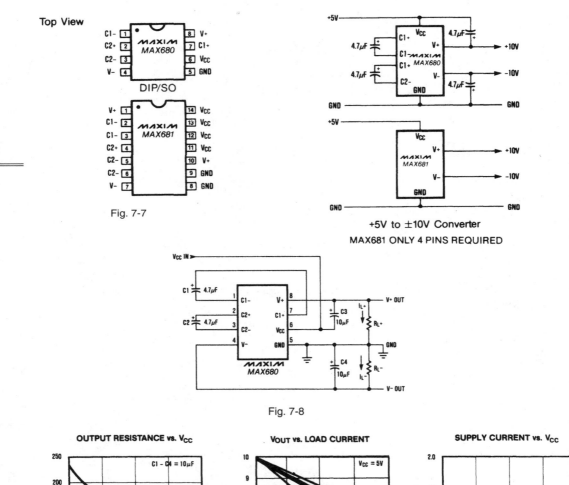

Fig. 7-7

+5V to ±10V Converter
MAX681 ONLY 4 PINS REQUIRED

Fig. 7-8

Fig. 7-9

Power-supply circuits

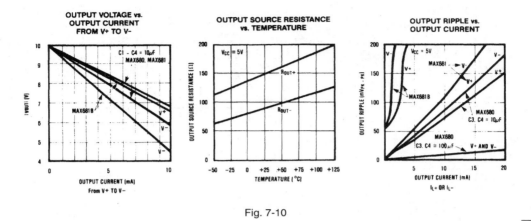

Fig. 7-10

Voltage converters
Figure 7-7 shows MAX680 and MAX681 converters used to provide ±10-V outputs from a +5-V input. The input voltage range is +2.0 V to +6.0 V. Voltage-conversion efficiency is 95%, with 85% power-conversion efficiency. The supply current is 500 μA. The output source impedances are typically 150 Ω, providing useful output currents up to 10 mA. No external capacitors are required for the MAX681. Figure 7-8 shows a test circuit. Figures 7-9 and 7-10 show typical test results. MAXIM NEW RELEASES DATA BOOK, 1992, P. 4-141.

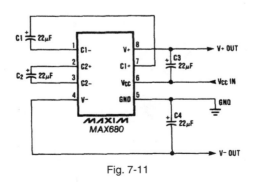

Fig. 7-11

Positive and negative voltage converter

Figure 7-11 shows a MAX680 used as a positive and negative voltage converter. Capacitors C1 and C3 must be rated at 6 V or greater. Capacitors C2 and C4 must be rated at 12 V or greater. If the MAX680 is used for low-current applications, C1 and C2 can be reduced to 1 μF, and capacitors C3 and C4 can be reduced to 4.7 μF. MAXIM NEW RELEASES DATA BOOK, 1992, P. 4-144.

7

274

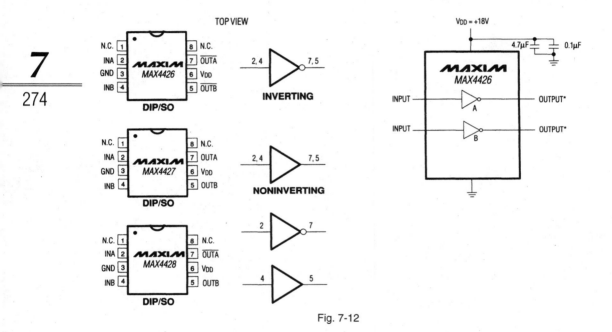

Fig. 7-12

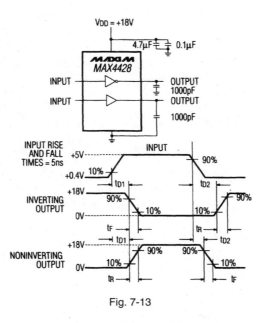

Fig. 7-13

High-speed 1.5-A MOSFET drivers

Figure 7-12 shows the basic connections for the MAX4426/27/28 drivers. Such ICs rapidly charge and discharge the gate capacitance of the largest MOSFETs to within millivolts of the supply. The typical on resistance is 4 Ω, the delay times are 10 ns for t_{D1} and 25 ns for t_{D2}, the peak output current is 1.5 A, rise and fall times are typically 20 ns with 1000-pF loads, the operating range is 4.5 V to 18 V, and the power consumption is 1.8 mA with a logic-1 output and 200 µA with a logic-0 input. The ICs are TTL/CMOS compatible and will withstand greater than 500-mA reverse current. Figure 7-13 is a test circuit. MAXIM HIGH-RELIABILITY DATA BOOK, 1993, P. 4-5, 4-6.

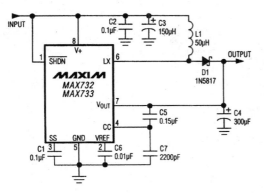

PART	INPUT SUPPLY RANGE (V)	OUTPUT VOLTAGE (V)	GUARANTEED OUTPUT CURRENT (mA)
MAX732	4.5 to 9.3	+12	150
	6.0 to 9.3	+12	200
MAX733	4.5 to 11.0	+15	100
	6.0 to 11.0	+15	200

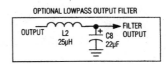

NOTE: PIN NUMBERS REFER TO 8-PIN PACKAGES.

Fig. 7-14

Power-supply circuit titles and descriptions

Continued

TOP VIEW

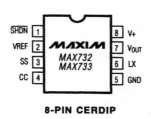

8-PIN CERDIP

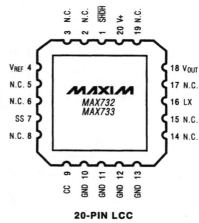

20-PIN LCC

N.C. = NO CONNEC

Fig. 7-15

7

276

Step-up PWM regulators

Figure 7-14 shows the MAX732 and MAX733 connected as step-up PWM (pulse-width modulation) regulators. Typical efficiency for the MAX731 is 82% to 87%, with 85% to 95% for the MAX733. Figure 7-15 shows the pin configurations. MAXIM HIGH-RELIABILITY DATA BOOK, 1993, P. 4-65.

TOP VIEW

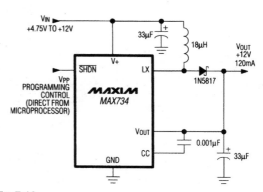

DIP/SO

Fig. 7-16

Power-supply circuits

Figure 7-16 shows a MAX734 connected as a flash-memory programming supply. Typical efficiency is 85%. The IC also has a logic-controlled shutdown pin that allows direct microprocessor control. MAXIM HIGH-RELIABILITY DATA BOOK, 1993, P. 4-65.

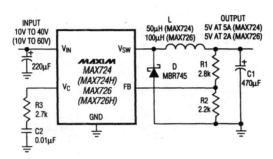

Fig. 7-17

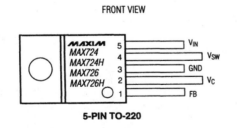

5-PIN TO-220

CASE IS CONNECTED TO GROUND.
CONTACT FACTORY FOR STRAIGHT PINS.
Pin Configurations continued on last page.

Fig. 7-18

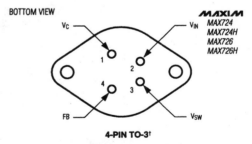

4-PIN TO-3†

CASE IS CONNECTED TO GROUND.
†CONTACT FACTORY FOR AVAILABILITY.

Fig. 7-19

Surface-Mount Components (for designs typically below 2A)

Inductors:	Sumida Electric - CDR125 Series
	USA: Phone (708) 956-0666
	Japan: Phone (03) 3607-5111
	FAX (03) 3607-5428
	Coiltronics - CTX series
	USA: Phone (305) 781-8900
	FAX (305) 782-4163
Capacitors:	Matsuo - 267 series
	USA: Phone (714) 969-2491
	FAX (714) 960-6492
	Japan: Phone (06) 332-0871
	Sprague - 595D series
	USA: Phone (603) 224-1961
	FAX (603) 224-1430
Diodes:	Motorola - MBRS series
	USA: (602) 244-6900
	Nihon - NSQ series
	USA: Phone (805) 867-2555
	FAX (805) 867-2698

Through-Hole Components

Inductors:	Sumida - RCH-110 series
	(see above for phone number)
	Cadell-Burns - 7070, 7300, 6860, and 7200 series
	USA: Phone (516) 746-2310
	FAX (516) 742-2416
	Renco - various series
	USA: Phone (516) 586-5566
	FAX (516) 586-5562
	Coiltronics - various series
	(see above for phone number)
Capacitors:	Nichicon - PL series low-ESR electrolytics
	USA: Phone (708) 843-7500
	FAX (708) 843-2798
	United Chemi-Con - LXF series
	USA: Phone (708) 696-2000
	FAX (708) 640-6311
	Sanyo - OS-CON low-ESR organic semiconductor
	USA: Phone (619) 661-6322
	Japan: Phone (0720) 70-1005
	FAX (0720) 70-1174
Diodes:	General Purpose - 1N5820-1N5825
	Motorola - MBR and MBRD series
	(see above for phone number)

Fig. 7-20

Step-down PWM switch-mode regulators
Figure 7-17 shows the MAX724/26 connected as basic step-down converters. Figures 7-18 and 7-19 show the pin configurations. Figure 7-20 shows component suppliers. MAXIM EVALUATION KIT DATA BOOK, 1994, P. 3-125, 3-131, 3-139.

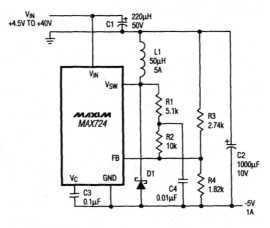

D1 - MOTOROLA MBR745
C1 - NICHICON UPL1C221MRH6
C2 - NICHICON UPL1A102MRH6
L1 - COILTRONICS CTX25-5-52
ALL RESISTORS HAVE 1% TOLERANCE

Fig. 7-21

Positive-to-negative dc-dc inverter
Figure 7-21 shows the MAX724 connected to provide −5 V at a 1-A output with a
+4.5-V to +40-V input. See Figs. 7-18 and 7-19 for pin configurations, and Fig.
7-20 for component suppliers. MAXIM EVALUATION KIT DATA BOOK, 1994, P. 3-138.

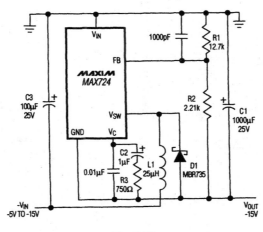

Fig. 7-22

Negative step-up dc-dc converter

Figure 7-22 shows the MAX724 connected to provide a −15-V output with a −5-V to −15-V input. See Figs. 7-18 and 7-19 for pin configurations, and Fig. 7-20 for component suppliers. MAXIM EVALUATION KIT DATA BOOK, 1994, P. 3-138.

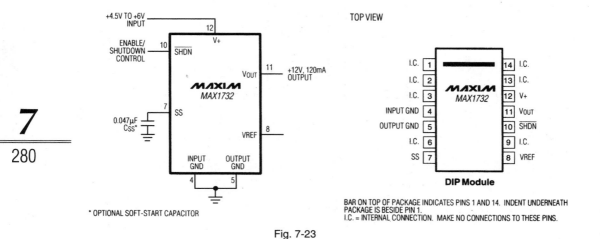

TOP VIEW

BAR ON TOP OF PACKAGE INDICATES PINS 1 AND 14. INDENT UNDERNEATH PACKAGE IS BESIDE PIN 1.
I.C. = INTERNAL CONNECTION. MAKE NO CONNECTIONS TO THESE PINS.

* OPTIONAL SOFT-START CAPACITOR

Fig. 7-23

Flash-memory programming module (120 mA)

Figure 7-23 shows a MAX1732 connected as a flash-memory programming module. This function is similar to that performed by the circuit of Fig. 7-16, but with only one external capacitor required. Quiescent current is 1.7 mA, with 70-μA shutdown current. MAXIM NEW RELEASES DATA BOOK, 1994, P. 4-15.

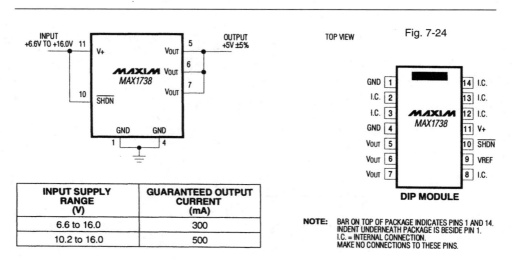

TOP VIEW   Fig. 7-24

NOTE: BAR ON TOP OF PACKAGE INDICATES PINS 1 AND 14.
INDENT UNDERNEATH PACKAGE IS BESIDE PIN 1.
I.C. = INTERNAL CONNECTION.
MAKE NO CONNECTIONS TO THESE PINS.

INPUT SUPPLY RANGE (V)	GUARANTEED OUTPUT CURRENT (mA)
6.6 to 16.0	300
10.2 to 16.0	500

Power-supply circuits

Continued

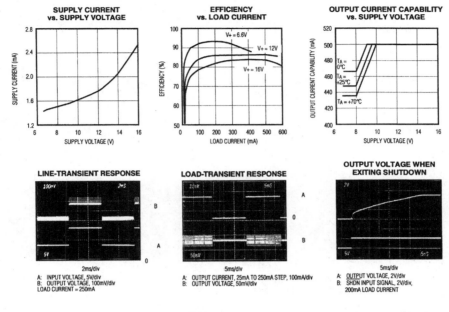

(Using *Typical Operating Circuit*, T$_A$ = +25°C, V+ = 12V, unless otherwise noted.)

SUPPLY CURRENT vs. SUPPLY VOLTAGE

EFFICIENCY vs. LOAD CURRENT

OUTPUT CURRENT CAPABILITY vs. SUPPLY VOLTAGE

LINE-TRANSIENT RESPONSE

2ms/div

A: INPUT VOLTAGE, 5V/div
B: OUTPUT VOLTAGE, 100mV/div
LOAD CURRENT = 250mA

LOAD-TRANSIENT RESPONSE

5ms/div

A: OUTPUT CURRENT, 25mA TO 250mA STEP, 100mA/div
B: OUTPUT VOLTAGE, 50mV/div

OUTPUT VOLTAGE WHEN EXITING SHUTDOWN

5ms/div

A: OUTPUT VOLTAGE, 2V/div
B: SHDN INPUT SIGNAL, 2V/div,
    200mA LOAD CURRENT

Pin Description

PIN	NAME	FUNCTION
1,4	GND	Ground. Connect both pins to ground.
2,3,8, 12,13,14	I.C.	Internal Connection. Make no connections to these pins.
5,6,7	V$_{OUT}$	Output Voltage, +5V. The output is **not** isolated from the input supply voltage. Connect all V$_{OUT}$ pins together.
9	VREF	Reference Voltage Output, +1.23V. Drives up to 100µA external load. If externally loaded, bypass this pin to GND with 0.01µF.
10	SHDN	Shutdown. Connect to V+ for normal operation. Connect to GND to shut down the device. When shut down, the output falls to 0V.
11	V+	Supply Voltage Input

Fig. 7-25

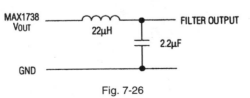

MAX1738 V$_{OUT}$ — 22µH — FILTER OUTPUT

2.2µF

GND

Fig. 7-26

+5-V, 500-mA step-down converter module
Figure 7-24 shows a MAX1738 2.5-W converter module that requires no external components. Figure 7-25 shows the pin descriptions and operating characteristics. The no-load current is 1.7 mA, with 60-µA shutdown current. Ripple is a typical ±30 mV, which can be improved using the optional output filter of Fig. 7-26. MAXIM NEW RELEASES DATA BOOK, 1994, P. 4-23, 4-25, 4-26.

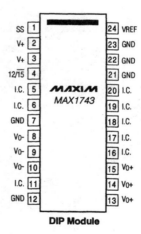

INPUT 4.5V
TO 5.5V

1	SS	VREF	24
2	V+	GND	23
3	V+	GND	22
4	12/$\overline{15}$	GND	21
5	I.C.	I.C.	20
6	I.C.	I.C.	19
7	GND	I.C.	18
8	Vo-	I.C.	17
9	Vo-	I.C.	16
10	Vo-	Vo+	15
11	I.C.	Vo+	14
12	GND	Vo+	13

-12V OR -15V
OUTPUT

OUTPUT
+12V OR +15V

*CONNECT PIN 4 TO V+ FOR ±12V OPERATION OR TO GND FOR ±15V OPERATION.

TOP VIEW

SS	1	24	VREF
V+	2	23	GND
V+	3	22	GND
12/$\overline{15}$	4	21	GND
I.C.	5	20	I.C.
I.C.	6	19	I.C.
GND	7	18	I.C.
Vo-	8	17	I.C.
Vo-	9	16	I.C.
Vo-	10	15	Vo+
I.C.	11	14	Vo+
GND	12	13	Vo+

DIP Module

Fig. 7-27

(V+ = 5.0V, T_A = +25°C, unless otherwise noted.)

75mA LOAD TRANSIENT

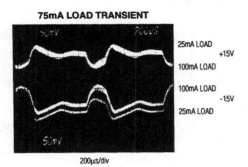

200µs/div

OUTPUT VOLTAGE vs. LOAD CURRENT (±15V MODE)

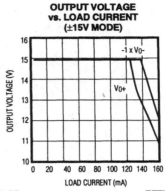

EFFICIENCY vs. LOAD CURRENT (±15V MODE)

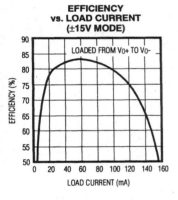

OUTPUT VOLTAGE vs. LOAD CURRENT (±12V MODE)

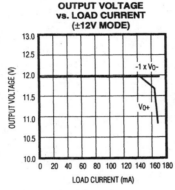

EFFICIENCY vs. LOAD CURRENT (±12V MODE)

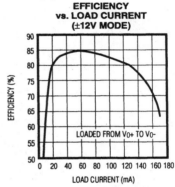

Pin Description

PIN	NAME	FUNCTION
1	SS	Soft-Start Pin. During power-up, the current into the V+ pin will surge to 1.5A for approximately 2ms while the outputs reach regulation. Adding an optional 0.1µF capacitor from SS to GND prevents current surge, but lengthens the time until the outputs reach regulation to approximately 60ms.
2, 3	V+	+5V Input
4	12/$\overline{15}$	Pin-Strap Input for selecting ±12V or ±15V. Tie 12/$\overline{15}$ to V+ to get a ±12V output; tie 12/$\overline{15}$ to GND to get a ±15V output.
5, 6, 11, 16-20	I.C.	Internal Connection. Make no connection to these pins.
7, 12, 21, 22, 23	GND	Ground
8, 9, 10	V_{O-}	Negative Output Voltage. -12V when 12/$\overline{15}$ = V+ and -15V when 12/$\overline{15}$ = 0V. This output is short-circuit protected.
13, 14, 15	V_{O+}	Positive Output Voltage. +12V when12/$\overline{15}$ = V+ and +15V when 12/$\overline{15}$ = 0V. This output is **not** short-circuit protected. Do not short V_{O+} to any potential less than V+.
24	VREF	+2.0V Reference Voltage Output. Pulling VREF up to V+ puts the MAX1743 in a low-current standby mode with V_{O+} a Schottky diode drop below V+ and V_{O-} at GND. See Figure 1.

Fig. 7-28

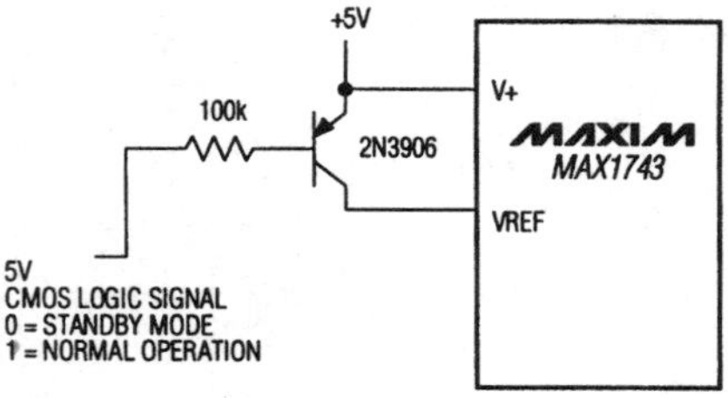

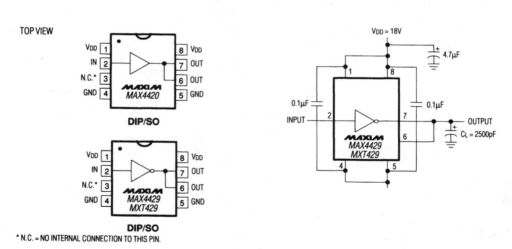

Fig. 7-29

+5-V to ±12-V/±15-V converter module
Figure 7-27 shows a MAX1743 3-W converter module that requires no external components. The module supplies 125 mA at ±12 V or 100 mA at ±15 V. Pin strapping selects ±12-V or ±15-V operation. Figure 7-28 shows the pin descriptions and operating characteristics. Figure 7-29 shows an optional stand-by mode circuit. MAXIM NEW RELEASES DATA BOOK, 1994, P. 4-27, 4-29, 4-30.

* N.C. = NO INTERNAL CONNECTION TO THIS PIN.

Fig. 7-30

Continued

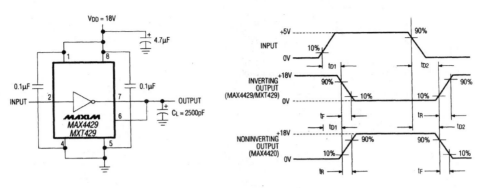

Fig. 7-31

High-speed, 6-A single-MOSFET drivers
Figure 7-30 shows the MAX4429/MXT429 connected to translate TTL/CMOS inputs to high-voltage/high-current outputs. The output impedance is 1.5 Ω with a 6-A current. Delay time is 40 ns, with 25-ns rise/fall times (into a 2500-pF load). Figure 7-31 shows the timing test circuit. The supply range is 4.5 V to 18 V, with an output swing to within 25 mV of V_{DD} and ground. MAXIM NEW RELEASES DATA BOOK, 1994, P. 4-31, 4-35.

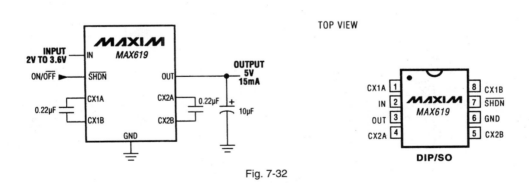

Fig. 7-32

Charge-pump voltage converter
Figure 7-32 shows a MAX619 connected to provide a 5-V, 15-mA output, with 2-V to 3.6-V input, using two inexpensive capacitors. The quiescent current is 150 μA (max) with 10-μA (max) shutdown current. MAXIM NEW RELEASES DATA BOOK, 1994, P. 4-37.

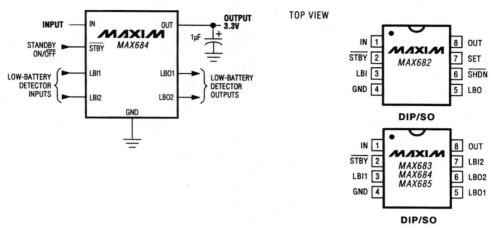

Fig. 7-33

Low-dropout P-channel linear regulator

Figure 7-33 show a MAX684 linear regulator. The output voltages are: MAX682 = 3.3 V, MAX683 = 5.0 V, MAX684 = 3.3 V, and MAX685 = 3.0 V. The supply range is 2.7 V to 11.5 V, the maximum dropout is 300 mV at 200 mA output, the maximum quiescent current is 15 µA, the shutdown current is 1 µA (maximum). Maxim New Releases Data Book, 1994, p. 4-49.

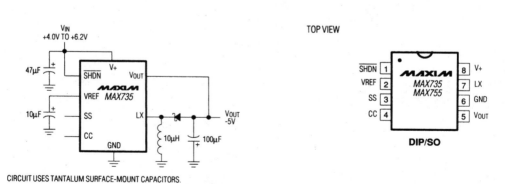

CIRCUIT USES TANTALUM SURFACE-MOUNT CAPACITORS.

Fig. 7-34

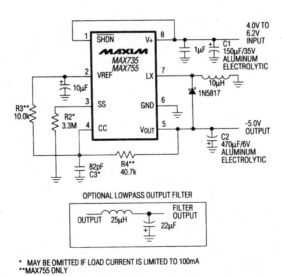

Fig. 7-35

Fig. 7-36

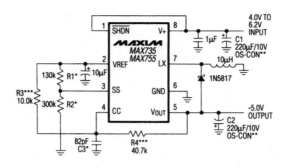

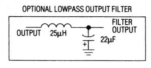

OPTIONAL LOWPASS OUTPUT FILTER

* MAY BE OMITTED IF LOAD CURRENT IS LIMITED TO 100mA
** OS-CON CAPACITORS LIMITED TO -55°C TO +105°C OPERATION.
   FOR OPERATION UP TO +125°C, OTHER CAPACITORS MUST BE SPECIFIED.
*** MAX755 ONLY

Fig. 7-37

PRODUCTION METHOD	INDUCTORS	CAPACITORS
Surface Mount	Sumida CD54-100 (10µH)	Matsuo 267 series
Miniature Through Hole	Sumida RCH855-100M (10µH)	Sanyo Os-Con series low-ESR organic semiconductor
Low-Cost Through Hole	Renco RL 1284 (10µH)	Nichicon PL series low-ESR electrolytics United Chemicon LXF series

Matsuo USA (714) 969-2491 FAX (714) 960-6492
Matsuo Japan (06) 332-0871
Nichicon (708) 843-7500 FAX (708) 843-2798
Renco (516) 586-5566 FAX (516) 586-5562
Sanyo Os-Con USA (619) 661-6322
Sanyo Os-Con Japan (0720) 70-1005 FAX (0720) 70-1174
Sumida USA (708) 956-0666
Sumida Japan (03) 3607-5111 FAX (03) 3607-5428
United Chemi-Con (708) 696-2000 FAX (708) 640-6311

Fig. 7-38

Continued

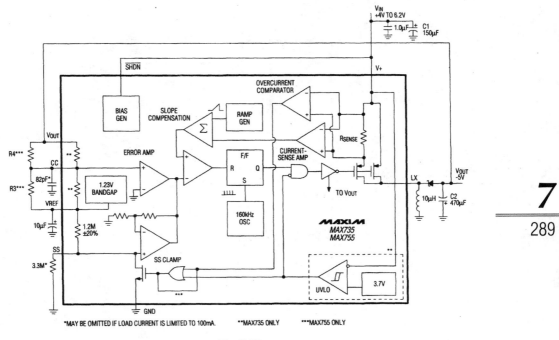

Fig. 7-39

−5-V/adjustable inverting PWM regulators

Figure 7-34 shows a MAX735 connected in a basic circuit to provide a 200-mA, −5-V output with a +4.0-V to +6.2-V input. Figure 7-35 shows the corresponding circuit using surface-mount components. Figure 7-36 shows the circuit using through-hole components (at commercial temperature ranges). Figure 7-37 shows the circuit using through-hole components (for all temperature ranges). Figure 7-38 shows component suppliers and Fig. 7-39 shows the block diagram with typical external components. The MAX755 operates from a +2.7-V to +9-V input and generates an adjustable negative output, depending on the values of R3 and R4. The relationship is:

$$R_4 = \frac{V_{OUT}}{1.23\ V}\ R_3$$

Resistor R3 can be any value from 10 kΩ to 20 kΩ, V_{OUT} is limited to 11.7 V $-V_{IN}$. The quiescent current is 1.6 mA, with a 10-μA shutdown current. MAXIM NEW RE-LEASES DATA BOOK, 1994, P. 4-111, 4-115, 4-116, 4-117.

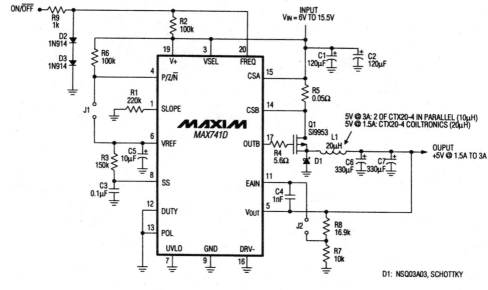

Fig. 7-40

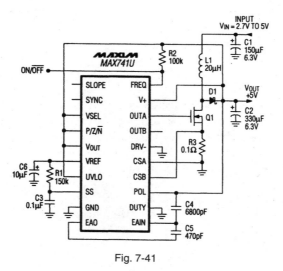

Fig. 7-41

Current-mode SMPS controllers

Figure 7-40 shows the MAX741D connected to provide +5 V at 1.5 A or 3.0 A with a 6-V to 15.5-V input. Figure 7-41 shows the MAX741D connected to provide +5 V at 1 A from a +3-V input. Both configurations are SMPS (switch-mode power supplies). The data sheet shows how the MAX741 can be programmed to provide different outputs with supply voltages from 2.7 V. MAXIM NEW RELEASES DATA BOOK, 1994, P. 4-131.

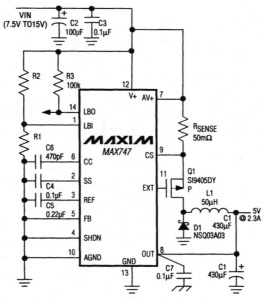

Fig. 7-42

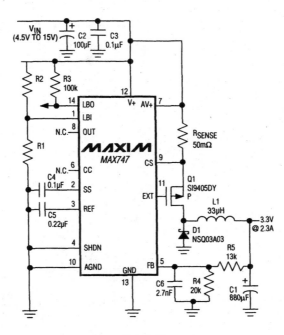

Fig. 7-43

Power-supply circuit titles and descriptions

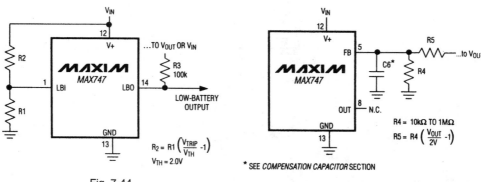

Fig. 7-44

Fig. 7-45

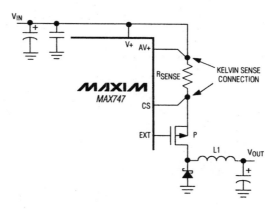

Fig. 7-46

Step-down P-channel controllers

Figure 7-42 shows the MAX747 connected to provide +5 V at 2.3 A with a 7.5-V to 15-V input. Figure 7-43 shows the MAX747 connected to provide +3.3 V at 2.3 A with a 4.5-V to 15-V input. Figure 7-44 shows the calculations required for the low-battery detector circuit. LBO goes low when $V+$ is equal to or less than VTRIP. LBO is high impedance in the shutdown mode. Figure 7-45 shows the calculations required to adjust the output from 2 V to 14 V. The value of C6 is calculated using:

$$C_6 = \frac{(C_1)\,(ESRC1)}{R_4 \; || \; R_5}$$

where C_1 is the value of C1 in µF, and ESRC1 (effective series resistance of C1 at 100 kHz) is in ohms. Figure 7-46 shows the recommended Kelvin connections for the current-sense resistor. MAXIM NEW RELEASES DATA BOOK, 1994, P. 4-140, 4-143, 4-145.

Power-supply circuits

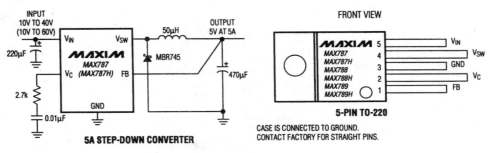

Fig. 7-47

5-A step-down PWM regulator
Figure 7-47 shows the MAX787 connected to provide an output of +5 V at 5 A, with an input of 10 V to 40 V. The MAX787/88/89 provide 5-A outputs at 5 V, 3.3 V, and 3 V, respectively. Quiescent current is 8.5 mA. The ICs are also available in TO-3 packages. MAXIM NEW RELEASES DATA BOOK, 1994, P. 4-245.

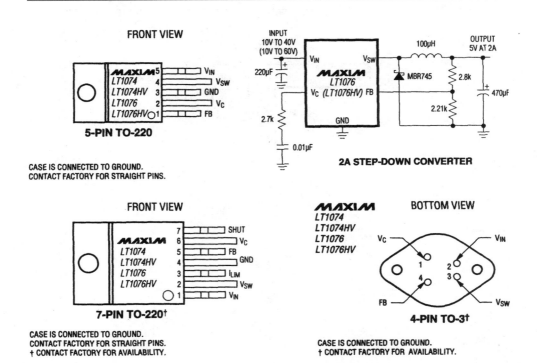

Fig. 7-48

Continued

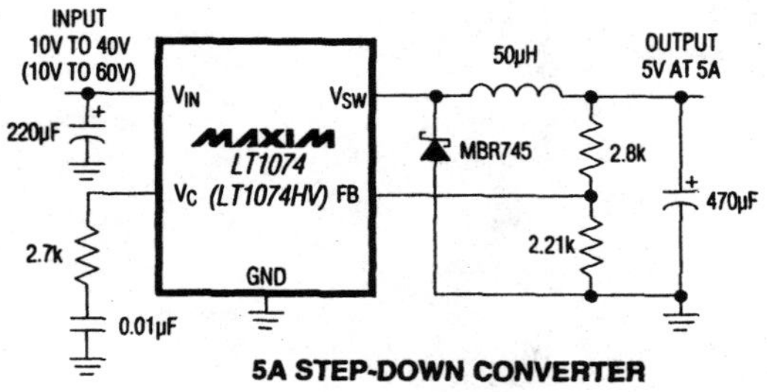

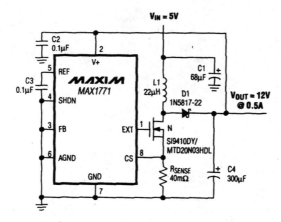

Fig. 7-49

2-A/5-A step-down PWM regulators

Figure 7-48 shows an LT1076 connected to provide an output of +5 V at 2 A, with an input of 10 V to 40 V. Figure 7-49 shows an LT1074 connected to provide an output of +5 V at 5 A, with an input of 10 V to 40 V. Both ICs have a quiescent current of 8.5 mA. MAXIM NEW RELEASES DATA BOOK, 1994, P. 4-251, 4-254.

Fig. 7-50

Continued

PRODUCTION	INDUCTORS	CAPACITORS	TRANSISTORS	DIODES

Surface Mount	Sumida
	  CD54 series
	  CDR125 series
	Coiltronics
	  CTX20 series
	Coilcraft
	  DO3316 series
	  DO3340 series
		Matsuo
		  267 series
		Sprague
		  595D series
		AVX
		  TPS series
			Siliconix
			  Si9410DY
			  Si9420DY (high voltage)
			Motorola
			  MTP3055EL
			  MTD20N03HDL
			  MMFT3055ELTI
			  MTD6N1O
			  MMBT8099LTI
			  MMBT8599LTI
				Central Semiconductor
				  CMPSH-3
				  CMPZ5240
				Nihon
				  EC11 FS1 series (high-
				  speed silicon)
				Motorola
				  MBRS1100T3
				  MMBZ5240BL

Through Hole	Sumida
	  RCH855 series
	  RCH110 series
		Sanyo
		  OS-CON series
		Nichicon
		  PL series
				Motorola
				  1N5817–1N5822
				  MUR115 (high voltage)
				  MUR105 (high-speed
				  silicon)

7
_____
295

SUPPLIER	PHONE	FAX
AVX	USA: (803) 448-9411	(803) 448-1943
Central Semiconductor	USA: (516) 435-1110	(516) 435-1824
Coilcraft	USA: (708) 639-6400	(708) 639-1469
Coiltronics	USA: (407) 241-7876	(407) 241-9339
Matsuo	USA: (714) 969-2491 Japan: 81-6-337-6450	(714) 960-6492 81-6-337-6456
Motorola	USA: (800) 521-6274	(602) 952-4190
Nichicon	USA: (708) 843-7500	(708) 843-2798
Nihon	USA: (805) 867-2555	(805) 867-2556
Sanyo	USA: (619) 661-6835 Japan: 81-7-2070-1005	(619) 661-1055 81-7-2070-1174
Siliconix	USA: (800) 554-5565	(408) 970-3950
Sprague	USA: (603) 224-1961	(603) 224-1430
Sumida	USA: (708) 956-0666 Japan: 81-3-3607-5111	(708) 956-0702 81-3-3607-5144

Fig. 7-51

12-V high-efficiency controller (boot-strapped)
Figure 7-50 shows a MAX1771 connected to provide 12-V output at 0.5 A, with a 5-V input. Figure 7-51 shows component suppliers. Maxim New Releases Data Book, 1995, p. 4-20.

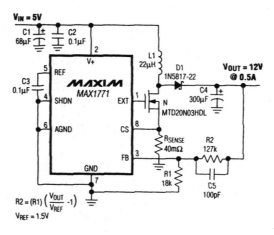

Fig. 7-52

12-V high-efficiency controller (non-boot-strapped)
Figure 7-52 shows the MAX1771 connected to provide 12-V output at 0.5 A, with a 5-V input. See Fig. 7-51 for component suppliers. This non-boot-strapped version of the Fig. 7-50 controller requires less supply current, but with the full 5-V input. Use the Fig. 7-50 controller for input less than 5 V. MAXIM NEW RELEASES DATA BOOK, 1995, P. 4-20.

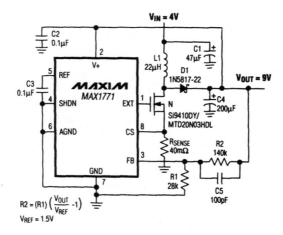

Fig. 7-53

9-V high-efficiency controller
Figure 7-53 shows the MAX1771 connected to provide 9-V output with a 4-V input. See Fig. 7-51 for component suppliers. MAXIM NEW RELEASES DATA BOOK, 1995, P. 4-20.

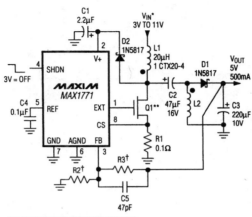

SEE TEXT FOR FURTHER COMPONENT INFO
* V$_{IN}$ MAY BE LOWER THAN INDICATED IF THE SUPPLY IS NOT
  REQUIRED TO START UNDER FULL LOAD
** MOTOROLA MMFT3055ELTI
† FOR 5V: R2 = 200kΩ, R3 = 470kΩ
     3.3V: R2 = 100kΩ, R3 = 20kΩ

Fig. 7-54

Step-up/down for 5-V/3.3-V output
Figure 7-54 shows the MAX1771 connected to provide either 5-V or 3.3-V output at 0.5 A, with a 3-V to 11-V input. See Fig. 7-51 for component suppliers. MAXIM NEW RELEASES DATA BOOK, 1995, P. 4-25.

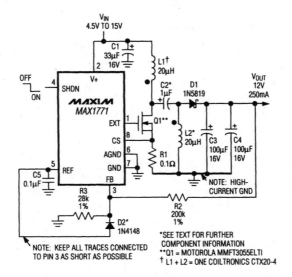

Fig. 7-55

12-V buck-boost from a 4.5-V to 15-V input
Figure 7-55 shows the MAX1771 connected to provide a 12-V output at 250 mA, with a 4.5-V to 15-V input. See Fig. 7-51 for component suppliers. MAXIM NEW RELEASES DATA BOOK, 1995, P. 4-26.

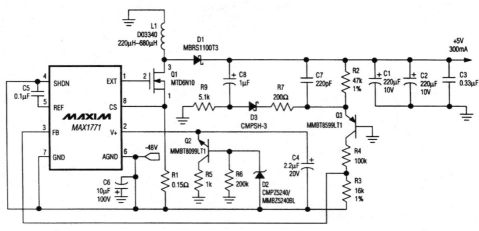

Fig. 7-56

−48-V input to 5-V output (without transformer)
Figure 7-56 shows the MAX1771 connected to provide a 5-V output at 300 mA, with a −48-V input. No transformer is required. The conversion efficiency is typically 82%. See Fig. 7-51 for component suppliers. MAXIM NEW RELEASES DATA BOOK, 1995, P. 4-27.

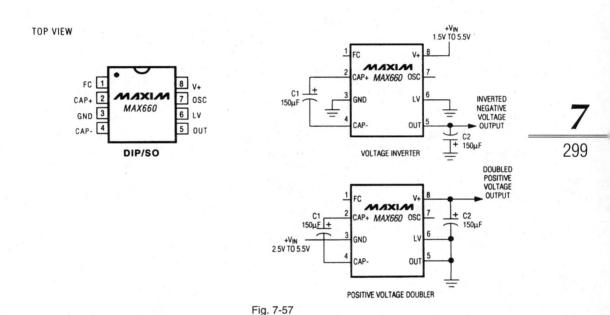

Fig. 7-57

CMOS monolithic voltage converter
Figure 7-57 shows a MAX660 connected to provide either voltage inversion or positive voltage doubling. The conversion efficiency is a typical 88% at 100 mA, with a typical loss of 0.65 V. The typical output impedance is 6.5 Ω. The operating current is 120 μA. MAXIM NEW RELEASES DATA BOOK, 1995, P. 4-73.

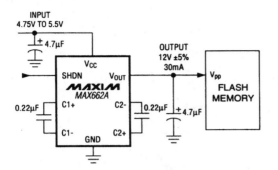

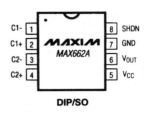

TOP VIEW

Fig. 7-58

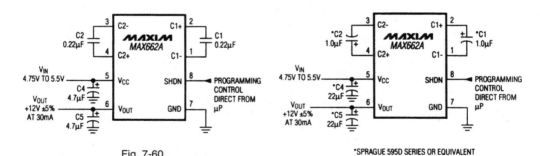

Supplier	Phone Number	Fax Number	Capacitor	Capacitor Type*
Murata Erie	(814) 237-1431	(814) 238-0490	GRM42-6Z5U224M50	0.22µF Ceramic (SM)
			RPE123Z5U105M50V	1.0µF Ceramic (TH)
Sprague Electric	(603) 224-1961 (207) 324-4140	(603) 224-1430 (207) 324-7223	595D475X9016A7	4.7µF Tantalum (SM)
			595D105X9016A7	1.0µF Tantalum (SM)

*Note: (SM) denotes surface-mount component, (TH) denotes through-hole component.

Fig. 7-59

Fig. 7-60

Fig. 7-61

*SPRAGUE 595D SERIES OR EQUIVALENT

Flash-memory programming supply (30 mA)
Figure 7-58 shows a MAX662A connected to provide a +12-V, 30-mA output for programming byte-wide flash memories. Figure 7-59 shows component suppliers. Figures 7-60 and 7-61 show connections for programming commercial-temperature range, and extended/military-range applications, respectively. Quiescent current is 185 µA, with 0.5-µA shutdown current. The IC has a logic-controlled shutdown pin that allows direct microprocessor control. MAXIM NEW RELEASES DATA BOOK, 1995, P. 4-75, 4-79.

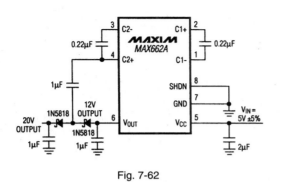

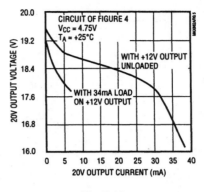

Fig. 7-62

Fig. 7-63

+12-V and +20-V dual supply
Figure 7-62 shows the MAX662A flash-memory supply (Figs. 7-58 through 7-61)
used as a dual power supply. Figure 7-63 shows the output-current characteristics.
MAXIM NEW RELEASES DATA BOOK, 1995, P. 4-80.

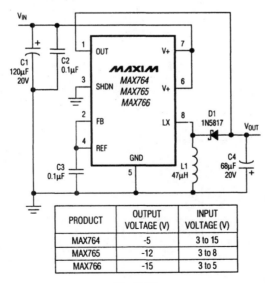

PRODUCT	OUTPUT VOLTAGE (V)	INPUT VOLTAGE (V)
MAX764	-5	3 to 15
MAX765	-12	3 to 8
MAX766	-15	3 to 5

Fig. 7-64

Power-supply circuit titles and descriptions

PRODUCTION METHOD	INDUCTORS	CAPACITORS	DIODES
Surface Mount	Sumida CD75/105 series Coiltronics CTX series Coilcraft DT/D03316 series	Matsuo 267 series Sprague 595D/293D series AVX TPS series	Nihon EC10QS02L (Schottky) EC11FS1 (high-speed silicon)
Miniature Through-Hole	Sumida RCH895 series	Sanyo OS-CON series (very low ESR)	Motorola 1N5817, 1N5818, (Schottky) MUR105 (high-speed silicon)
Low-Cost Through-Hole	Renco RL1284 series	Nichicon PL series	

SUPPLIER	PHONE		FAX
AVX	USA:	(803) 448-9411	(803) 448-1943
Coilcraft	USA:	(708) 639-6400	(708) 639-1469
Coiltronics	USA:	(407) 241-7876	(407) 241-9339
Matsuo	USA: Japan:	(714) 969-2491 81-6-337-6450	(714) 960-6492 81-6-337-6456
Motorola	USA:	(800) 521-6274	(602) 952-4190
Nichicon	USA: Japan:	(708) 843-7500 81-7-5231-8461	(708) 843-2798 81-7-5256-4158
Nihon	USA: Japan:	(805) 867-2555 81-3-3494-7411	(805) 867-2556 81-3-3494-7414
Renco	USA:	(516) 586-5566	(516) 586-5562
Sanyo OS-CON	USA: Japan:	(619) 661-6835 81-7-2070-1005	(619) 661-1055 81-7-2070-1174
Sprague Electric Co.	USA:	(603) 224-1961	(603) 224-1430
Sumida	USA: Japan:	(708) 956-0666 81-3-3607-5111	(708) 956-0702 81-3-3607-5144

Fig. 7-65

−5-V/−12-V/−15-V high-efficiency inverter

Figure 7-64 shows the MAX764/65/66 connected to provide inverter operation with fixed output voltages. Figure 7-65 shows component suppliers. MAXIM NEW RELEASES DATA BOOK, 1995, P. 4-121, 4-123.

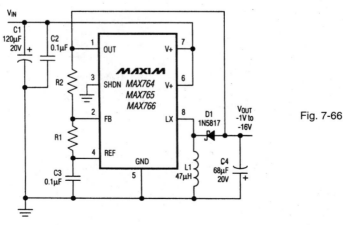

Fig. 7-66

Power-supply circuits

High-efficiency inverter with adjustable output
Figure 7-66 shows the MAX764/65/66 connected to provide inverter operation with an adjustable output (-1.0 V to -16 V). For adjustable operation, let $R_1 = 150$ kΩ, then find the value of R2 using:

$$R_2 = R_1 \left(\frac{V_{OUT}}{V_{REF}} \right)$$

Where $V_{REF} = 1.5$ V. See Fig. 7-65 for component suppliers. MAXIM NEW RELEASES DATA BOOK, 1995, P. 4-121.

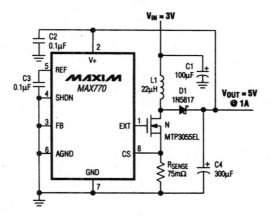

Fig. 7-67

PRODUCTION	INDUCTORS	CAPACITORS	TRANSISTORS	DIODES
Surface Mount	Sumida CD54 series CDR125 series Coiltronics CTX20 series	Matsuo 267 series Sprague 595D series	N-FET Siliconix Si9410DY Si9420DY (high voltage) Motorola MTP3055EL MTD20N03HDL	Nihon EC10 series
Through Hole	Sumida RCH855 series RCH110 series Renco RL1284-18	Sanyo OS-CON series Nichicon PL series United Chemi-Con LXF series	NPN Zetex ZTX694B	Motorola 1N5817–1N5822 MUR115 (high voltage)

SUPPLIER	PHONE	FAX
Coiltronics	USA: (407) 241-7876	(407) 241-9339
Matsuo	USA: (714) 969-2491 Japan: 81-6-337-6450	(714) 960-6492 81-6-337-6456
Nichicon	USA: (708) 843-7500	(708) 843-2798
Nihon	USA: (805) 867-2555	(805) 867-2556
Renco	USA: (516) 586-5566	(516) 586-5562
Sanyo	USA: (619) 661-6835 Japan: 81-7-2070-6306	(619) 661-1055 81-7-2070-1174
Sumida	USA: (708) 956-0666 Japan: 81-3-3607-5111	81-3-3607-5144
United Chemi-Con	USA: (714) 255-9500	(714) 255-9400
Zetex	USA: (516) 543-7100 UK: 44-61-627-4963	(516) 864-7630 44-61-627-5467

Fig. 7-68

Power-supply circuit titles and descriptions

5-V preset-output high-efficiency controller
Figure 7-67 shows a MAX770 connected to provide 5-V output at 1 A, with a 3-V input. Figure 7-68 shows component suppliers. MAXIM NEW RELEASES DATA BOOK, 1995, P. 4-150, 4-149.

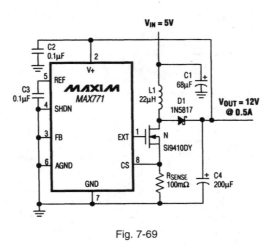

Fig. 7-69

12-V preset-output high-efficiency controller (boot-strapped)
Figure 7-69 shows a MAX771 connected to provide 12-V output at 0.5 A, with a 5-V input. See Fig. 7-68 for component suppliers. MAXIM NEW RELEASES DATA BOOK, 1995, P. 4-150.

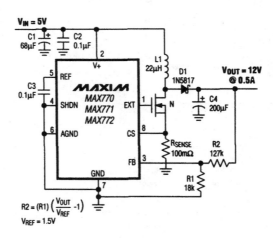

Fig. 7-70

12-V preset-output high-efficiency controller (non-boot-strapped)
Figure 7-70 shows a MAX770/71/72 connected to provide 12-V output at 0.5 A, with a 5-V input. See Fig. 7-68 for component suppliers. This non-boot-strapped version of the Fig. 7-69 circuit requires less supply current, but with the full 5-V input. Use the Fig. 7-69 controller for input voltages less than 5 V. MAXIM NEW RELEASES DATA BOOK, 1995, P. 4-15.

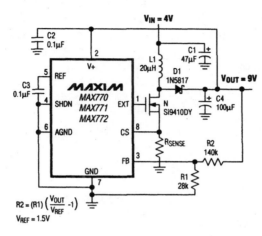

Fig. 7-71

9-V high-efficiency controller (lower power)
Figure 7-71 shows the MAX770/71/72 connected to provide 9-V output with a 4-V input. See Fig. 7-68 for component suppliers. These IC controllers are similar to the MAX1771 (Fig. 7-53), but with lower power capability. MAXIM NEW RELEASES DATA BOOK, 1995, P. 4-150.

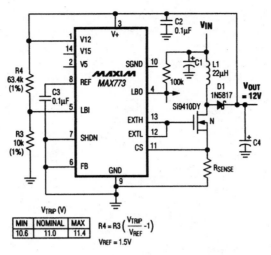

Fig. 7-72

12-V preset-output controller with power monitor

Figure 7-72 shows a MAX773 connected to provide 12-V output, with a power-monitor (or low battery) function. See Fig. 7-68 for component suppliers. MAXIM NEW RELEASES DATA BOOK, 1995, P. 4-151.

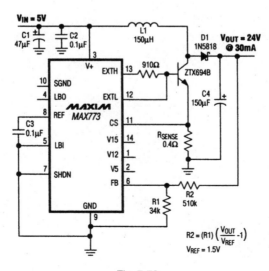

Fig. 7-73

24-V preset-output controller

Figure 7-73 shows a MAX773 connected to provide 24-V output at 30 mA, with a 5-V input. See Fig. 7-69 for component suppliers. MAXIM NEW RELEASES DATA BOOK, 1995, P. 4-151.

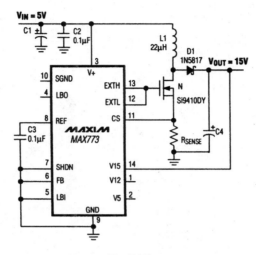

Fig. 7-74

15-V preset-output controller

Figure 7-74 shows a MAX773 connected to provide 15-V output, with a 5-V input. This circuit is non-boot-strapped for minimum current consumption. See Fig. 7-68 for component suppliers. MAXIM NEW RELEASES DATA BOOK, 1995, P. 4-151.

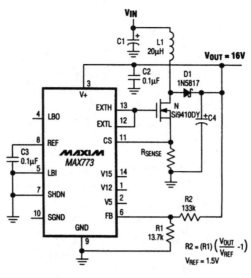

Fig. 7-75

16-V preset-output controller
Figure 7-75 shows a MAX773 connected to provide 16-V output, with a 5-V input.
This circuit is boot-strapped for operation with inputs less than 5 V. See Fig. 7-68
for component suppliers. MAXIM NEW RELEASES DATA BOOK, 1995, P. 4-151.

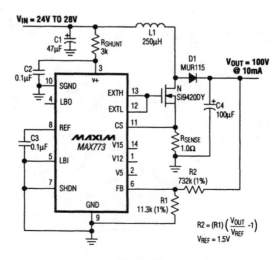

Fig. 7-76

Power-supply circuits

Continued

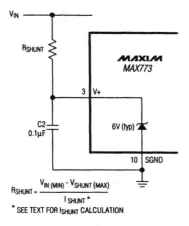

$$R_{SHUNT} = \frac{V_{IN\,(MIN)} - V_{SHUNT\,(MAX)}}{I_{SHUNT}\,^*}$$
* SEE TEXT FOR I_{SHUNT} CALCULATION

Fig. 7-77

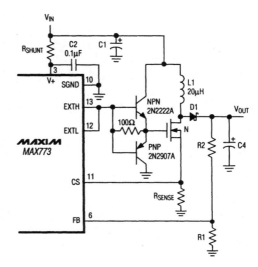

Fig. 7-78

100-V output with shunt regulation
Figure 7-76 shows the MAX773 connected to provide 100-V output at 10 mA, with 24-V to 28-V input. Figure 7-77 shows the calculations for selecting the R_{SHUNT} value. R_{SHUNT} should be selected so that I_{SHUNT} is greater than 1 mA, but less than 20 mA. If the calculated shunt regulator current exceeds 20 mA, or if the shunt current exceeds 5 mA, and less shunt-regulator current is desired, use the circuit of Fig. 7-78. This provides increased drive and reduced shunt current when driving N-FETs with large gate capacitances. Use an I_{SHUNT} of 3 mA. This provides adequate biasing current for the circuit, although higher shunt currents can be used. Notice that the shunt regulator is not disabled in the shutdown mode, and continues to draw the calculated shunt current. To prevent the shunt regulator from drawing current in the shutdown mode, place a switch in series with the shunt resistor. See Fig. 7-68 for component suppliers. MAXIM NEW RELEASES DATA BOOK, 1995, P. 4-152, 4-154.

Power-supply circuit titles and descriptions

=8=

Battery-power and micropower circuits

This chapter is devoted to circuits that can be operated from a battery (often a single 1.5-V cell) and draw a minimum of current (micropower). Testing and troubleshooting for these circuits is the same as for corresponding circuits in other chapters and are not duplicated here. (The majority of the circuits in this chapter are power-supply circuits, with testing and troubleshooting, as described in Chapter 7.) Where it is not obvious, reference is made (in the circuit descriptions) to the appropriate chapter for testing and troubleshooting. For additional information on micropower and battery circuits, read the author's *Simplified Design of Micropower and Battery Circuits*, 1996, published by Butterworth-Heinemann.

Battery-power/micropower circuit titles and descriptions

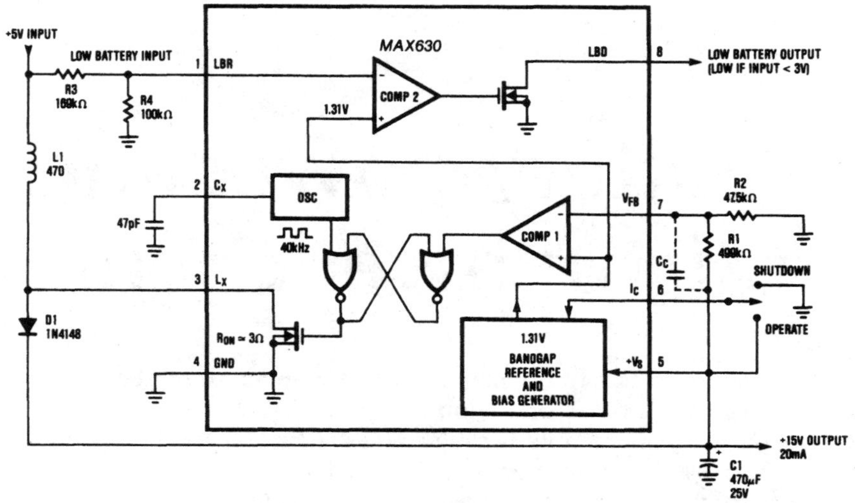

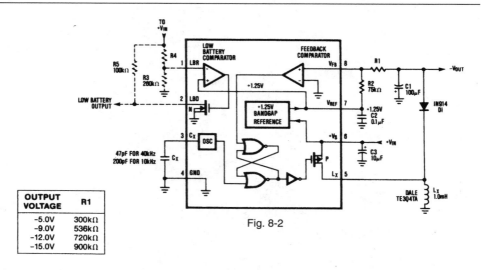

Fig. 8-1

Micropower step-up switching regulator

Figure 8-1 shows a MAX630 connected to provide −15-V output at 20 mA. The input can be from 2.0 V to 16.5 V, with 70-μA typical operating current, 1-μA maximum quiescent current, and a 85% (typical) efficiency. MAXIM NEW RELEASES DATA BOOK, 1992, P. 4-49.

OUTPUT VOLTAGE	R1
−5.0V	300kΩ
−9.0V	536kΩ
−12.0V	720kΩ
−15.0V	900kΩ

Fig. 8-2

Battery-power and micropower circuits

MANUFACTURER	TYPICAL PART #	DESCRIPTION
MOLDED INDUCTORS		
Dale	IHA-104	500μH, 0.5 ohms
Caddell-Burns	6860-19	330μH, 0.33 ohms
TRW	LL-500	500μH, 0.75 ohms
POTTED TOROIDAL INDUCTORS		
Dale	TE-3Q4TA	1mH, 0.82 ohms
TRW	MH-1	600μH, 1.9 ohms
Torotel Prod.	PT 53-18	500μH, 5 ohms
FERRITE CORES AND TOROIDS		
Allen Bradley	T0451S100A	Tor. Core, 500nH/T²
Siemens	B64290-K38-X38	Tor. Core, 4μH/T²
Magnetics	555130	Tor. Core, 53nH/T²
Stackpole	57-3215	Pot Core, 14mm x 8mm
Magnetics	G-41408-25	Pot Core, 14 x 8, 250nH/T²

Note 1: This list does not constitute an endorsement by Maxim Integrated Products and is not intended to be a comprehensive list of all manufacturers of these components.

Fig. 8-3

Micropower inverting switching regulator
Figure 8-2 shows a MAX634 connected to provide −5-V, −9-V, −12-V, or −15-V outputs, depending on the value of R1. Figure 8-3 shows coil and core manufacturers. The input can be from +3 V to +16.5 V, with 100 μA typical operating current and 85% efficiency. MAXIM NEW RELEASES DATA BOOK, 1992, P. 4-69, 4-70.

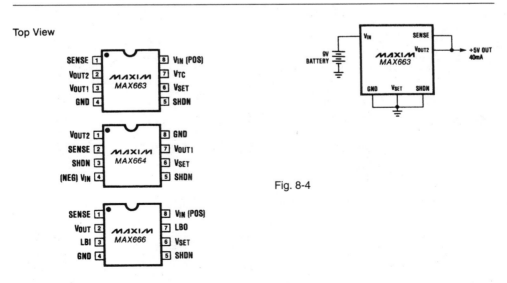

Fig. 8-4

Micropower voltage regulator
Figure 8-4 shows a MAX663 connected to provide a +5-V output at 40 mA from a 9-V battery. The maximum quiescent current is 12 μA. MAXIM NEW RELEASES DATA BOOK, 1992, P. 4-125.

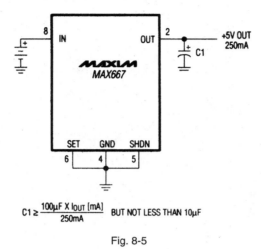

$$C1 \geq \frac{100\mu F \times I_{OUT} \text{ [mA]}}{250mA} \quad \text{BUT NOT LESS THAN } 10\mu F$$

Fig. 8-5

+5-V low-dropout voltage regulator
Figure 8-5 shows a MAX667 connected to provide a fixed +5-V output with minimum components and an input from +3.5 V to +16.5 V. Normal-mode quiescent current is a typical 20 μA, with 0.2-μA shutdown quiescent. MAXIM NEW RELEASES DATA BOOK, 1992, P. 4-137.

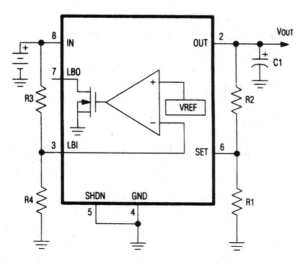

Fig. 8-6

Battery-power and micropower circuits

Programmable low-dropout voltage regulator
Figure 8-6 shows the MAX667 connected to provide a programmable output from +1.3 V to +16 V, with a low-battery function. Let R1 be 1 MΩ, and select R2 for a desired output using: $R_2 = R_1 \times (V_{OUT}/V_{SET} - 1)$, where $V_{SET} = 1.225$ V. Let R4 be 2.4 MΩ, and select R3 for a desired threshold of the low-battery detector using: $R_3 = R_4 \times (V_{BATT}/V_{LBI} - 1)$, where V_{BATT} is the desired threshold of the low-battery detector, and R3/R4 are the LBI input divider resistors. If V_{OUT} is 5 V, a 5.5-V low-battery threshold can be set using 8.2 MΩ for R3 and 2.4 MΩ for R4. MAXIM NEW RELEASES DATA BOOK, 1992, P. 4-137.

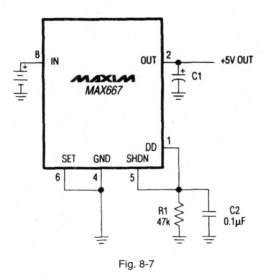

Fig. 8-7

Quiescent-current reduction below dropout
Figure 8-7 shows how the no-load quiescent current of the MAX667 (Figs. 8-5, 8-6) can be reduced to about 160 μA when the dropout voltage is reached. Notice that this circuit increases the dropout voltage by about 0.1 V. MAXIM NEW RELEASES DATA BOOK, 1992, P. 4-138.

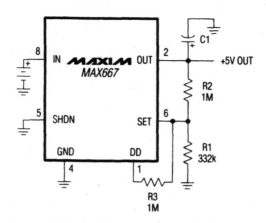

Fig. 8-8

Correction for minimum quiescent current near dropout
Figure 8-8 shows how the quiescent current of the MAX667 (Figs. 8-5, 8-6) can be
further reduced near the dropout voltage (compared to the circuit of Fig. 8-7).
MAXIM NEW RELEASES DATA BOOK, 1992, P. 4-138.

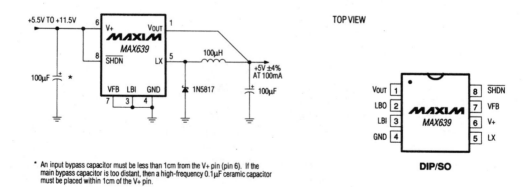

* An input bypass capacitor must be less than 1cm from the V+ pin (pin 6). If the
main bypass capacitor is too distant, then a high-frequency 0.1 μF ceramic capacitor
must be placed within 1cm of the V+ pin.

Fig. 8-9

High-efficiency +5-V step-down regulator
Figure 8-9 shows a MAX639 connected to provide +5-V output at 100 mA with
inputs from +5.5 V to +11.5 V. Maximum quiescent current is 20 μA, with a typical
dropout of 0.5 V at (100-mA load). Efficiency is greater than 90%. MAXIM HIGH
RELIABILITY DATA BOOK, 1993, P. 4-31.

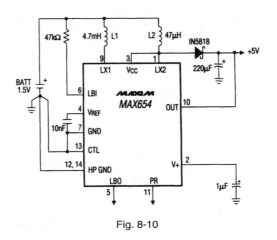

Fig. 8-10

Low-voltage (single-cell) step-up converter
Figure 8-10 shows a MAX654 connected to provide a +5-V output at 40 mA with a 1.5-V input. The guaranteed start-up voltage is 1.15 V. The quiescent current shutdown is 80 μA. MAXIM HIGH RELIABILITY DATA BOOK, 1993, P. 4-37.

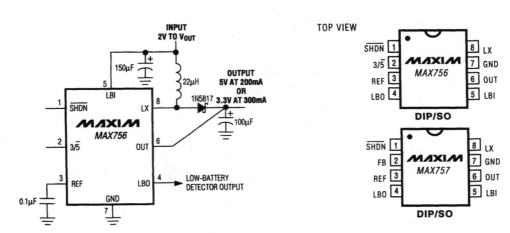

Fig. 8-11

3.3-V/5-V step-up converter
Figure 8-11 shows a MAX756 connected to provide a 5-V output at 200 mA or 3.3 V at 300 mA (depending on the status of 3/5 pin 2) with an input from 1.1 V to 5.5 V. Quiescent current is 60 μA, with a 20-μA shutdown current, and an efficiency of 87% (at 200 mA). MAXIM NEW RELEASES DATA BOOK, 1994, P. 4-173.

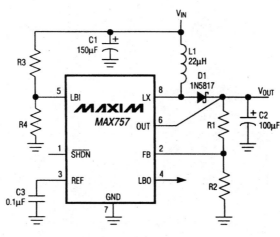

Fig. 8-12

PRODUCTION METHOD	INDUCTORS	CAPACITORS
Surface-Mount	Sumida CD54-220 (22µH) Coiltronics CTX20-1	Matuso 267 series
Miniature Through-Hole	Sumida RCH654-220	Sanyo Os-Con Os-Con series Low-ESR organic semiconductor
Low-Cost Through-Hole	Renco RL 1284-22 Coilcraft PCH-27-223	Maxim MAXC001 150µF, low-ESR electrolytic Nichicon PL series Low-ESR electrolyic United Chemi-Con LXF series

Coilcraft	USA:	(708) 639-6400
Coiltronics	USA:	(305) 781-8900
Matsuo	USA:	(714) 969-6291, FAX (714) 960-6492
	Japan:	(06) 332-0871
Nichicon	USA:	(708) 843-7500, FAX (708) 843-2798
Renco	USA:	(516) 586-5566, FAX (516) 586-5562
Sanyo Os-Con	USA:	(619) 661-6835
	Japan:	(0720) 70-1005, FAX (0720) 70-1174
Sumida	USA:	(708) 956-0666
	Japan:	(03) 3607-5111, FAX (03) 3607-5428
United Chemi-Con	USA:	(708) 696-2000, FAX (708) 640-6311

Fig. 8-13

Adjustable step-up converter
Figure 8-12 shows a MAX757 connected to provide an adjustable output from 2.7 V to 5.5 V, with a low-battery function. Figure 8-13 shows component suppliers. Let R2 be 100 kΩ, and select R1 for a desired output using:

$$R_1 = R_2 \times \left[\frac{(V_{OUT}}{V_{REF})} + 1 \right]$$

Where V_{REF} is 1.25 V. Let R4 100 kΩ, and select R3 for a desired threshold of the low-battery detector using:

$$R_3 = \left[\frac{V_{IN}}{V_{REF}} - 1 \right] (R_4)$$

where V_{REF} is 1.25 V. MAXIM NEW RELEASES DATA BOOK, 1994, P. 4-178.

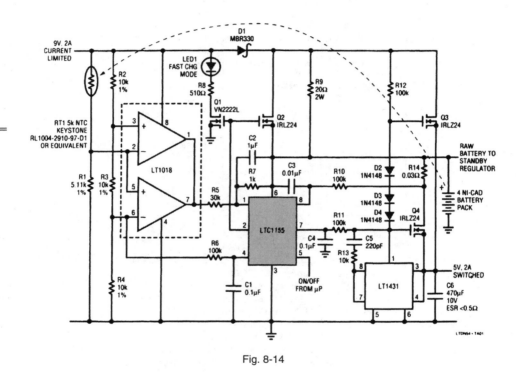

Fig. 8-14

Four-cell NiCad regulator/charger
Figure 8-14 shows an LTC1155 dual-power MOSFET driver connected to provide control of a four-cell charger/regulator. The LTC1155 has the ability to deliver 12-V of gate drive to two N-channel power MOSFETs when powered from a 5-V supply with no external component required. The circuit is suited for a notebook-computer supply. (A four-cell NiCad battery pack can be used to power a 5-V notebook computer.) Q3 and Q4 must be provided with heatsinks. The regulator is switched off by the microprocessor when the battery voltage drops below 4.6 V. The standby current for the 5-V 2-A regulator is less than 10 µA. The regulator is switched on again when the battery voltage rises during charging. LINEAR TECHNOLOGY, DESIGN NOTE 51, P. 1.

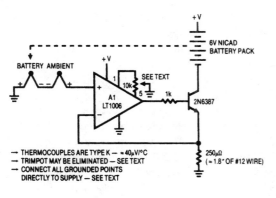

— THERMOCOUPLES ARE TYPE K — ≈ 40µV/°C
— TRIMPOT MAY BE ELIMINATED — SEE TEXT
— CONNECT ALL GROUNDED POINTS
   DIRECTLY TO SUPPLY — SEE TEXT

Fig. 8-15

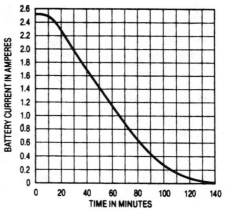

Fig. 8-16

WIRE GAUGE	µΩ/INCH
10	83
11	100
12	130
13	160
14	210
15	265
16	335
17	421
18	530
19	670
20	890
21	1000
22	1300
23	1700
24	2100
25	2700

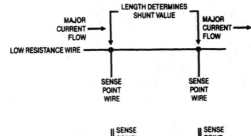

Fig. 8-17

Battery-power/micropower circuit titles and descriptions

Thermally based NiCad charger

Figure 8-15 shows a simple thermally based charging circuit for NiCad batteries. Figure 8-16 shows the charge characteristics. Thermocouples sense both cell and ambient temperature simultaneously. A1 provides the amplification necessary for microvolt-level thermocouple signals. The 10-kΩ trimpot is set to introduce enough input-offset so that the A1 output swings positive, turning on the transistor. Figure 8-17 shows a simple, inexpensive way to construct low-resistance shunts using a small length of wire or a PC trace. The type and length of wire determines the shunt resistance (which can be altered to produce the desired charging characteristics). Figure 8-17 also shows resistance-versus-length characteristics for various wire sizes, as well as the details for both wire and PC shunts. In both cases (PC or wire), the shunt should have separate connections for sensing (Kelvin style) so that high current does not affect the readings. LINEAR TECHNOLOGY, APPLICATION NOTE 37, P. 4.

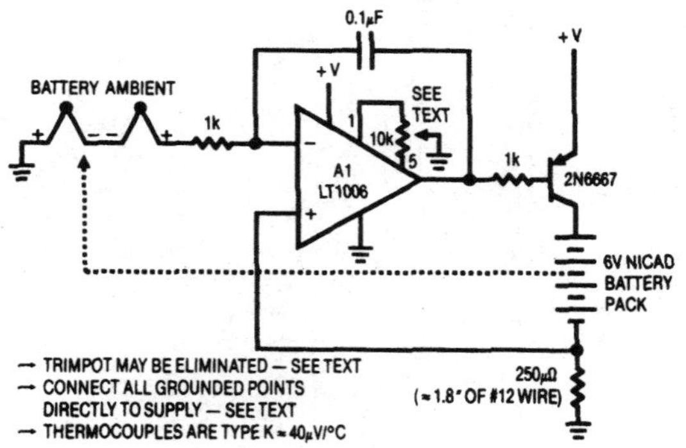

Fig. 8-18

NiCad charger for grounded batteries

Figure 8-18 shows a thermally based NiCad charger for use with batteries that are common to ground. The transistor is connected as a common emitter, so the inputs to A1 are reversed. However, operation is the same as for the Fig. 8-15 circuit. Notice that in both the Fig. 8-15 and 8-18 circuits, the trimpot can be eliminated by specifying an LT1006 set (at manufacture) to the desired offset value. High-quality grounds must be used, and all ground returns must be brought directly back to the supply common terminal. LINEAR TECHNOLOGY, APPLICATION NOTE 37, PAGE 2.

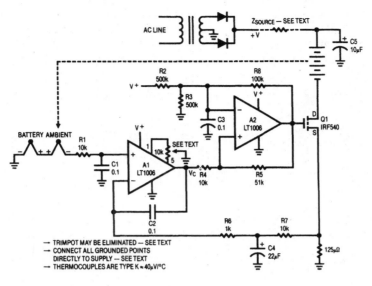

Fig. 8-19

Switch-mode thermally based NiCad charger

Figure 8-19 shows a thermally based NiCad charger operated in the switching mode. The circuits of Figs. 8-15 and 8-18 both force the transistor to dissipate some power, and the resulting heat might be a problem in a small enclosure (typical of micropower circuits). The circuit of Fig. 8-19 relies on the source impedance of the wall transformer to limit current through Q1 and the battery pack. The source impedance can be set when specifying the transformer. LINEAR TECHNOLOGY, APPLICATION NOTE 37, P. 3.

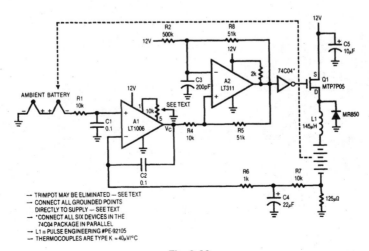

Fig. 8-20

Battery-power/micropower circuit titles and descriptions

Thermally based NiCad charger with low-impedance charging source
Figure 8-20 shows a circuit where the charging source has low impedance. In this circuit, the output is essentially a step-down switching regulator. The 74VO4s provide phase inversion and drive for Q1 (a P-channel MOSFET). LINEAR TECHNOLOGY, APPLICATION NOTE 37, P. 4.

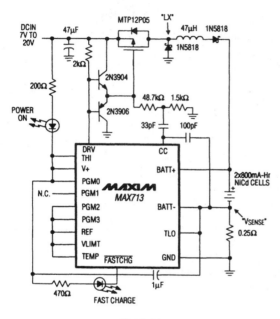

Fig. 8-21

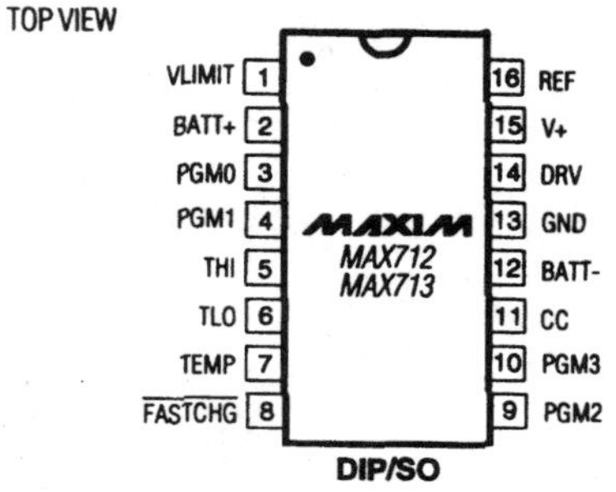

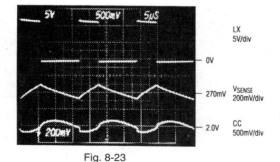

Fig. 8-23

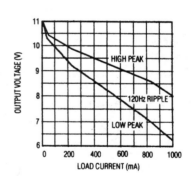

TOP VIEW

Fig. 8-22

Fig. 8-24

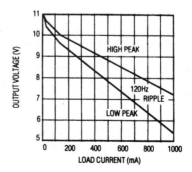

Fig. 8-25

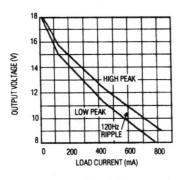

Fig. 8-26

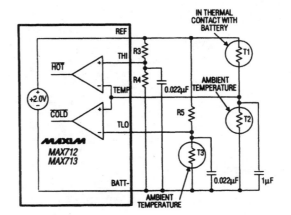

NOTE: FOR ABSOLUTE TEMPERATURE CHARGE CUTOFF, T2 AND T3 MAY BE REPLACED WITH STANDARD RESISTORS

Fig. 8-27

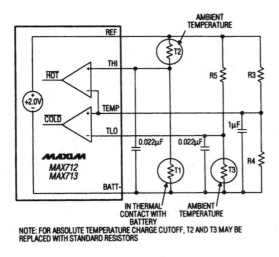

NOTE: FOR ABSOLUTE TEMPERATURE CHARGE CUTOFF, T2 AND T3 MAY BE REPLACED WITH STANDARD RESISTORS

Fig. 8-28

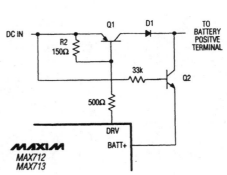

Fig. 8-29

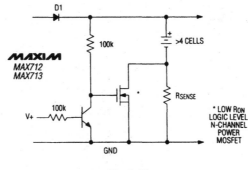

Fig. 8-30

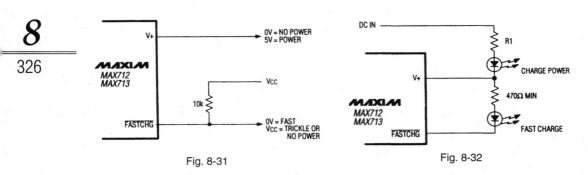

Fig. 8-31   Fig. 8-32

Fast-charge controllers for NiCad/NiMH

Figure 8-21 shows a MAX713 connected to provide a fast charge for two cells at 1 A. Higher charge currents and greater numbers of cells can be accommodated by changing R_{SENSE} and the PGM0-PGM3 connections (as described in the data sheet). Figure 8-22 shows the pin connections and Fig. 8-23 shows the circuit waveforms. Notice that the circuit of Fig. 8-21 cannot service a load while charging. The Fig. 8-21 circuit is usually used with an ac-to-dc wall cube (transformer, bridge rectifier, and capacitor in a plug-in package). Figures 8-24, 8-25, and 8-26 show the characteristics of three consumer-product wall cubes. When selecting a wall cube, be sure that the lowest dip in the in the wall-cube voltage during fast-charge is at least 1 V higher than the maximum battery voltage. Figure 8-27 shows the connections for controlling charge cutoff with negative-temperature-coefficient (NTC) thermistors. Use the same model of thermistor for T1 and T2 so that both have the same nominal resistance. The voltage at TEMP is 1 V (referred to Batt−) when the battery is at ambient temperature. Some battery packs might come with a temperature-detecting thermistor connected to the negative terminal of the battery pack. In this case, use the connections shown in Fig. 8-28. T2 and T3 can be replaced with standard resistors if absolute temperature-charge cutoff is acceptable. The absolute maximum voltage rating for the BATT+ input voltage must be limited by external circuits when DC IN is not applied, as shown in Fig. 8-29. R_{SENSE} causes a small efficiency loss during battery use. The efficiency loss is significant only if R_{SENSE} is much greater than the internal resistance of the battery pack. The circuit of Fig. 8-30 can be used to shunt R_{SENSE} whenever power is removed from the charger. Figure 8-31 shows a circuit used to indicate charger status, with logic-level outputs. Figure 8-32 shows an LED drive circuit that indicates charger status. MAXIM NEW RELEASES DATA BOOK, 1994, P. 4-60, 4-61, 4-63, 4-64.

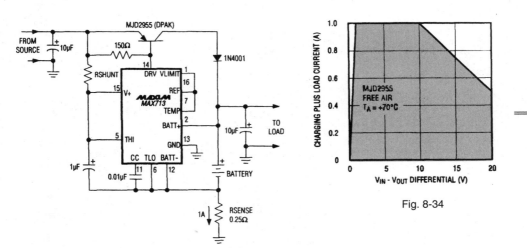

Note: See the MAX712/MAX713 data sheet for additional pin-strap connections to program the number of cells and the timer (PG0–PG3).

Fig. 8-33

Fig. 8-34

Fast-charger with linear-regulator current source
Figure 8-33 shows the MAX713 connected to provide a fast charge (with linear regulation) to both NiCad and NiMH batteries. This circuit solves two closely related problems found in portable power supplies. The circuit charges the battery and switches over from battery power to ac power when an external ac/dc adapter is plugged in. The MAX713 supplies the system load current (while the battery is being charged) by sensing and dynamically regulating the battery current. Figure 8-34 shows the operating area of the circuit. The MAX713 must be programmed for the desired number of cells; charging time, using pin straps, is described in the data sheet. MAXIM BATTERY MANAGEMENT CIRCUIT COLLECTION, 1994, P. 2.

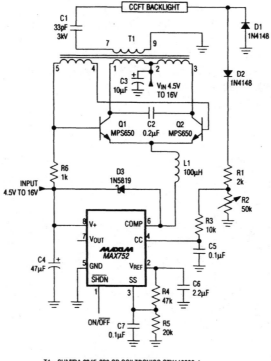

Fig. 8-35

CCFT/CCFL backlight supplies

Figure 8-35 shows a MAX752 connected as a power supply for cold-cathode fluorescent tubes (CCFT or CCFL, whichever you prefer). CCFT/CCFLs require high-voltage ac power, taken from a source that is not part of the main power supply. This is because the supply should be located physically close to the display (to prevent losses resulting from cable capacitance). Typically, the CCFT/CCFLs need about 2 W of 400-Vac power (that must reach near 1200 V upon startup) to arc and turn on the lamp. In this circuit, the MAX752 boost-regulator IC acts as a switching-regulator current source to feed the tail of a traditional Royer-type self-oscillating dc/dc converter. The Royer circuit drives a 33:1 transformer that steps up the battery voltage to ac high voltages (near 1200 V). Capacitance C_2 and the primary inductance form a resonant circuit, which provides a low-EMI sine-wave drive signal to the lamp. A half-wave rectified signal proportional to tube current is returned to the MAX752 feedback input. This maintains the CCFT/CCFL current at a constant level. Display brightness is adjusted by R2. MAXIM ENGINEERING JOURNAL, VOL. 3, 1994, P. 51.

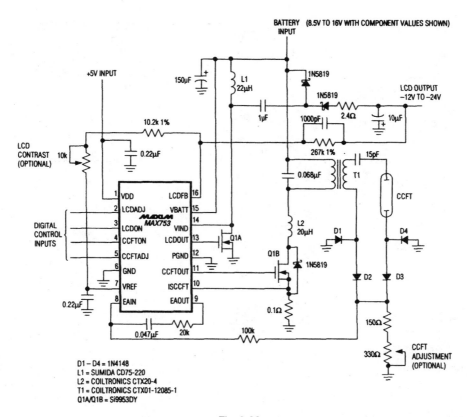

BATTERY (8.5V TO 16V WITH COMPONENT VALUES SHOWN)
INPUT

LCD OUTPUT
−12V TO −24V

D1 – D4 = 1N4148
L1 = SUMIDA CD75-220
L2 = COILTRONICS CTX20-4
T1 = COILTRONICS CTX01-12085-1
Q1A/Q1B = Si9953DY

Fig. 8-36

Combined contrast and backlight supply for LCD displays

Figure 8-36 shows a MAX753 connected to provide both the contrast and backlight for LCD displays. The dc contrast voltage (−12 V to −24 V) is generated with a hybrid of boost regulator, plus charge pump, under supervision of an on-board five-bit D/A converter, thus permitting microprocessor control of brightness. The circuit operates over a 6-V to 20-V range, draws 3-mA quiescent current, and is capable of 3-W output. MAXIM ENGINEERING JOURNAL, VOL. 3, 1994, P. 53.

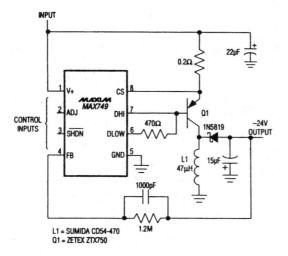

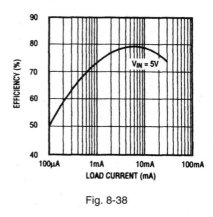

Fig. 8-38

Fig. 8-37

L1 = SUMIDA CD54-470
Q1 = ZETEX ZTX750

Stand-alone LCD contrast (bias) supply (digitally adjusted)
Figure 8-37 shows a MAX749 connected as an LCD-contrast supply with a digitally adjusted negative output. Figure 8-38 shows the efficiency curve. The MAX749 has an on-board five-bit D/A converter that adjusts the LCD contrast voltage (from ⅛ full-scale to full-scale) through a serial interface. No adjustment pots are needed. The circuit operates over a 2-V to 6-V range, draws 310-μA quiescent current, and is capable of 25-mA output. MAXIM BATTERY MANAGEMENT CIRCUIT COLLECTION, 1994, P. 46.

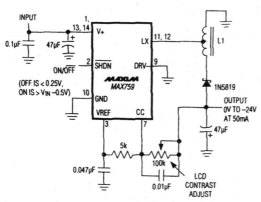

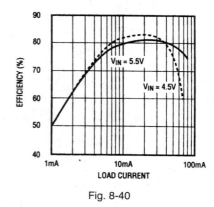

Fig. 8-40

L1 = COILTRONICS CTX100-1 OR MAGNETICS, INC. "KOOL-MU" 77030-A7
30 TURNS AND 30 TURNS 26 AWG

Fig. 8-39

Battery-power and micropower circuits

Negative LCD-contrast supply

Figure 8-39 shows a MAX759 connected as an LCD-contrast supply with a -24-V output using an autotransformer. Figure 8-40 shows the efficiency curve. The MAX759 has a large low-saturation, internal P-channel MOSFET. The use of an autotransformer trades off higher peak switch currents for reduced stress voltage on the switch transistor. The circuit operates over a 4-V to 6-V range (with a maximum input/output differential of 30 V), draws 3.7-mA quiescent current, and is capable of 50-mA output. Mount L1 close to the IC to minimize PC-trace inductance in the L_X lead. MAXIM BATTERY MANAGEMENT CIRCUIT COLLECTION, 1994, P. 47, 48.

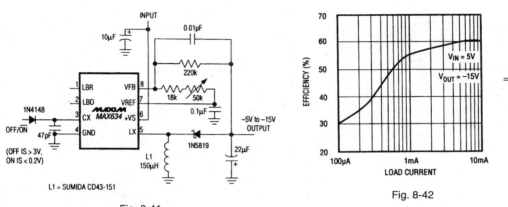

L1 = SUMIDA CD43-151

Fig. 8-41

Fig. 8-42

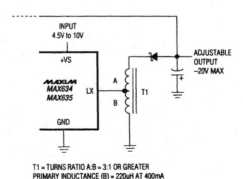

T1 = TURNS RATIO A:B = 3:1 OR GREATER
PRIMARY INDUCTANCE (B) = 220μH AT 400mA

Fig. 8-43

Micropower LCD-contrast supply
Figure 8-41 shows a MAX634 connected as a micropower LCD-contrast supply
with an adjustable −5-V to −15-V output. Figure 8-42 shows the efficiency curve.
This "flea-power" circuit is suitable for small multiplexed LCD displays (such as
those on cellular phones). The circuit operates over a 4-V to 6-V range (with a
maximum input/output differential of 24 V), draws 500-µA quiescent current,
and is capable of 10-mA output. Figure 8-43 shows a way to extend the maximum
input/output of 24-V by substituting an autotransformer for the inductor. MAXIM
BATTERY MANAGEMENT CIRCUIT COLLECTION, 1994, P. 48.

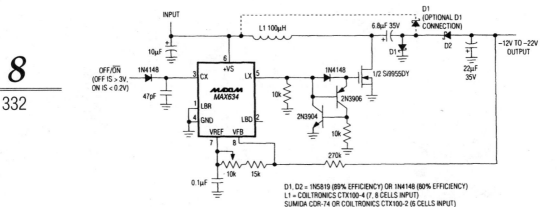

Fig. 8-44

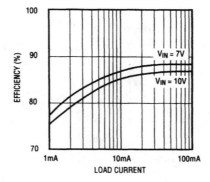

Fig. 8-45

Charge-pump LCD-contrast supply

Figure 8-44 shows a MAX634 connected as an LCD-contrast supply using the charge-pump principle for higher efficiency. Figure 8-45 shows the efficiency curves. This circuit is particularly useful where the LCD supply must be operated directly from the battery. (The circuit is a hybrid in that the switching regulator boosts the input to a high positive output voltage, with the actual negative output voltage being generated by a charge-pump tap on the switching node.) The circuit operates over a 4-V to 16-V (four- to eight-cell) range, draws 330 μA of quiescent current, and is capable of a 30-mA output. The optional D1 connection (dotted line) is used when the battery voltage exceeds the absolute value of the output voltage. If this does not occur, connect D1 to ground. Maxim Battery Management Circuit Collection, 1994, p. 49, 50.

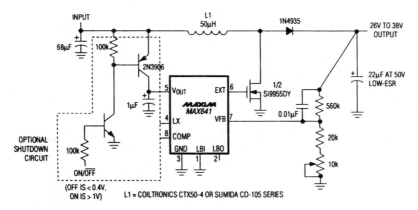

Fig. 8-46

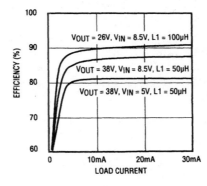

Fig. 8-47

LCD-contrast supply for color displays
Figure 8-46 shows a MAX641 connected as an LCD-contrast supply with positive 26-V to 38-V output (primarily for color displays). Figure 8-47 shows the efficiency curves. The circuit operates over a 4.5-V to 15-V (five- to eight-cell) range, draws 550 µA (with 5-V in and 26-V output) or 1-mA (with 9.6-V input and 38-V output) quiescent current, and is capable of 30-mA output. Load regulation is 0.06%/mA, line regulation is 0.16%/V, and output noise is 200 mVp-p. The input range makes the circuit suitable for +5-V regulated input, or for a direct-battery connection. For input voltages above 7 V, substitute a 100-µH inductor (Coiltronix CTS100-4) to maintain low output noise and ripple. MAXIM BATTERY MANAGEMENT CIRCUIT COLLECTION, 1994, P. 50.

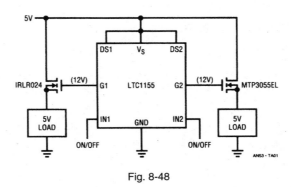

Fig. 8-48

Dual high-side switch driver
Figure 8-48 shows an LTC1155 connected as a high-side MOSFET switch driver (Chapter 2). The circuit generate 12-V from a 5-V supply to fully enhance (switch) logic-level N-channel MOSFET switches, with no external components required. The supply current is typically 85 µA with the switch fully enhanced, and 8 µA with the LTC1155 in standby (both inputs off). This combination of a low-drop N-channel MOSFET switch and micropower driver is an efficient means of controlling power in complex loads. Switch efficiencies in the 99% range are practical. LINEAR TECHNOLOGY, APPLICATION NOTE 53, P. 1.

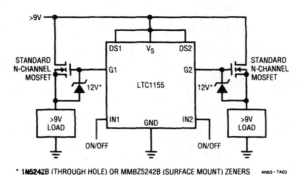

Fig. 8-49

Dual high-side switch driver with extended voltage range
Figure 8-49 shows an LTC1155 connected as a switch driver for voltages greater than 9 V. The circuit is the same as that shown in Fig. 8-48, with the addition of the zeners. LINEAR TECHNOLOGY, APPLICATION NOTE 53, P. 2.

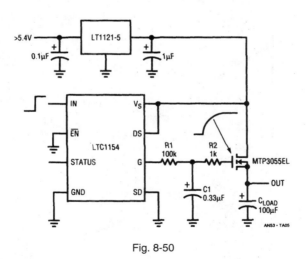

Fig. 8-50

Slew-rate reduction for high-capacity loads
Figure 8-50 shows a method for reducing the slew rate in the circuits of Figs. 8-48 and 8-49. Supply circuits in battery-operated equipment are often bypassed with large capacitors to reduce transients. If not properly switched, such filter capacitors can produce glitches at start-up. For example, the 100-μF load

capacitance shown could easily produce a start-up current of 10 A (well beyond the capability of the regulator), and possibly prevent start-up. The circuit of Fig. 8-50 reduces start-up current to about 15 mA. LINEAR TECHNOLOGY, APPLICATION NOTE 53, P. 3.

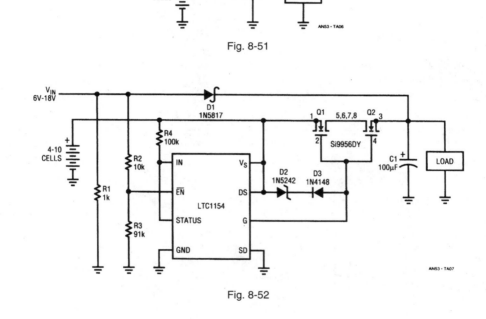

Fig. 8-51

Fig. 8-52

Bidirectional MOSFET-switch controllers
Figures 8-51 and 8-52 show manual and automatic bidirectional switches, respectively. Such circuits are used when the supply voltage is higher or lower than the load voltage, when powered by a secondary source. In the manual circuit of Fig. 8-51, S1 disconnects the battery from both the load and source when V_{IN} (from a wall-unit power supply) is connected. This permits the load voltage to fluctuate above or below the battery voltage, without forcing current into (or out of) the battery. In the far more practical automatic circuit of Fig. 8-52, Q1 and Q2 fully disconnect the battery from the load immediately after the wall-unit supply is connected to V_{IN}. The two diodes in Q1 and Q2 are also connected back to back, and no current can flow through the switch when the gate drive is removed. The

LTC1154 $\overline{\text{EN}}$ (enable) input senses when the wall-unit voltage exceeds 3 V, and inverts the switch action so that Q1/Q2 are turned off when the wall-unit supply is disconnected. LINEAR TECHNOLOGY, APPLICATION NOTE 53, P. 3.

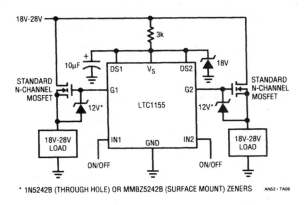

* 1N5242B (THROUGH HOLE) OR MMBZ5242B (SURFACE MOUNT) ZENERS   AN53 · TA08

Fig. 8-53

Switch driver for 18-V to 28-V operation
Figure 8-53 shows an LTC1155 connected for 18-V to 28-V switch-driver operation—even though the LTC1154/55/56 family is designed for 4.5-V to 18-V operation. The supply pin (V_S) is clamped to 18 V. These drivers typically produce 36-V of drive from a 18-V supply. This fully enhances N-channel MOSFET switches operating from 18 V to 28 V. The 12-V zener clamps are added to ensure that the maximum MOSFET V_{GS} is not exceeded. LINEAR TECHNOLOGY, APPLICATION NOTE 53, P. 4.

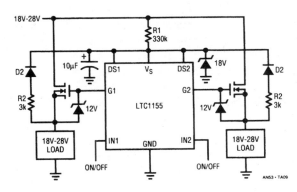

Fig. 8-54

Battery-power/micropower circuit titles and descriptions

Switch driver with micropower standby mode (18-V to 28-V)

Figure 8-54 shows an LTC1155 connected for switch-driver operation with micropower standby. This circuit is similar to that shown in Fig. 8-53, but with R1 increased from 3 kΩ to 330 kΩ, and a boot-strap R2/D2 network added. These modifications reduce the Fig. 8-54 standby current to less than 30 μA. The extra supply current is provided only when the switch is turned on. The supply current drops back to 30 μA when the switch is turned off, and the LTC1155 is returned to standby. LINEAR TECHNOLOGY, APPLICATION NOTE 53, P. 4.

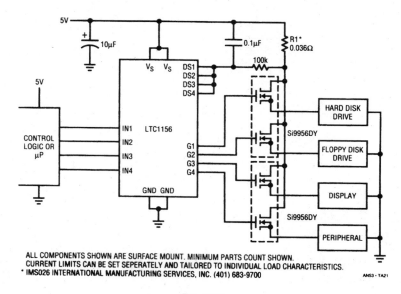

ALL COMPONENTS SHOWN ARE SURFACE MOUNT. MINIMUM PARTS COUNT SHOWN.
CURRENT LIMITS CAN BE SET SEPERATELY AND TAILORED TO INDIVIDUAL LOAD CHARACTERISTICS.
* IMS026 INTERNATIONAL MANUFACTURING SERVICES, INC. (401) 683-9700   AN53 - TA21

Fig. 8-55

Notebook-computer power management

Figure 8-55 shows an LTC1156 connected to control power (and protect) for components of a notebook computer (or similar device). Each load in the circuit is activated by the microprocessor only when that component is required to process or display information. When not needed (as determined by microprocessor control), the individual systems are placed in the standby mode where quiescent current is reduced to microamp levels. The standby current of the LTC1156 is typically 16 μA with all four inputs turned off. LINEAR TECHNOLOGY, APPLICATION NOTE 53, P. 10.

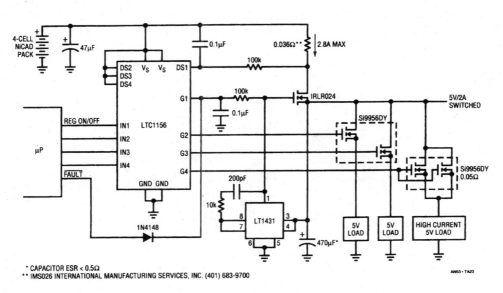

* CAPACITOR ESR < 0.5Ω
** IMS026 INTERNATIONAL MANUFACTURING SERVICES, INC. (401) 683-9700

AN53 • TA23

Fig. 8-56

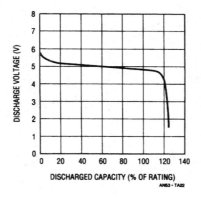

AN53 • TA22

Fig. 8-57

Four-cell NiCad power management

Figure 8-56 shows one channel of an LTC1156 used to regulate the output of a four-cell NiCad battery pack to power a notebook or palmtop computer. Figure 8-57 shows typical four-cell NiCad discharge characteristics. As long as the input voltage to the regulator is sufficient to produce 5 V at the output, the regulator limits at 5 V. When the battery-pack voltage drops below 5 V, the MOSFET is fully enhanced, and acts as a direct connection between the battery and the computer circuits. A battery-voltage monitor in the microprocessor decides when the battery voltage drops

Battery-power/micropower circuit titles and descriptions

below 4.6 V, and housekeeping is performed (data storage, etc.) before the batteries are completely discharged. The other three channels of the LTC 1156 act as switches (under microprocessor control) to power the remaining sections of the computer. The number of switches can be increased by adding more LTC1155 or LTC1156 circuits as needed. LINEAR TECHNOLOGY, APPLICATION NOTE 53, P. 11.

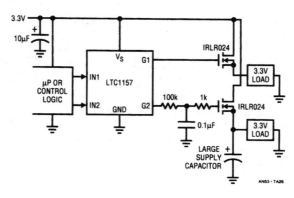

Fig. 8-58

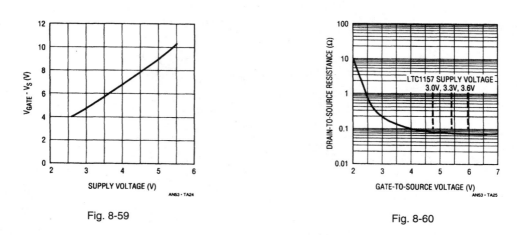

Fig. 8-59

Fig. 8-60

High-side switches for 3.3-V operation

Figure 8-58 shows an LTC1157 connected to provide high-side switching (Chapter 2) for two 3.3-V loads. Figure 8-59 shows the gate-voltage (above supply) characteristics for the LTC1157, which is a dual, low-voltage MOSFET driver specifically designed for operation between 2.7 V and 5.5 V. The LTC1157 internal charge pump boosts the gate-drive voltage 5.4 V above the 3.3-V positive supply (or 8.7 V above ground). This voltage fully enhances a logic-level N-channel

MOSFET for 3.3-V high-side switching. Figure 8-60 is a graph of $R_{DS(ON)}$ versus V_{GS} for a typical logic-level N-Channel MOSFET switch. LINEAR TECHNOLOGY, APPLICATION NOTE 53, P. 11, 12.

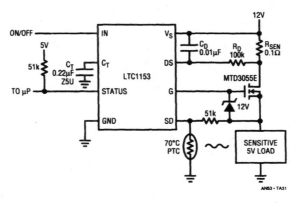

Fig. 8-61

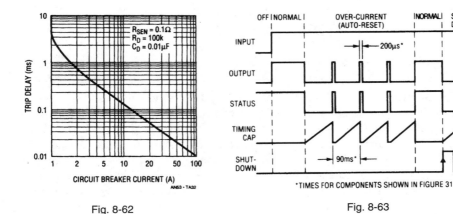

Fig. 8-62

Fig. 8-63

Electronic circuit breaker

Figure 8-61 shows an LTC1153 connected for thermal shutdown. Figures 8-62 and 8-63 show the characteristics and timing diagram, respectively. The LTC1153 interrupts power to a sensitive load in the event of an overcurrent condition, and remains tripped for a period of time programmed by external capacitor CT. The switch is then automatically reset, and the load is momentarily reconnected. If the load current is still too high, the switch is shut down again. This cycle continues until the overcurrent condition is removed, thus protecting both the sensitive load and the power MOSFET. Dc trip current is set by R_{SEN} at 1 A. The trip-delay time is set by R_D and C_D, and is shorter for increasing current (similar to a mechanical circuit breaker), as shown in Fig. 8-62. The circuit breaker is automatically reset (retried) every 200 ms until the overload is removed. An open-drain STATUS output warns the host microprocessor whenever the circuit is tripped. A shutdown (SD) input interfaces with a PTC thermistor to sense overtemperature conditions (over 70°C) and trip the circuit breaker. LINEAR TECHNOLOGY, APPLICATION NOTE 53, P. 15, 16.

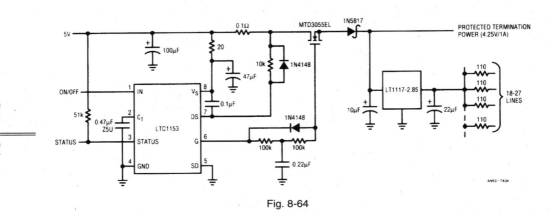

Fig. 8-64

SCSI terminal power protection
Figure 8-64 shows how an LTC1153 can be used to protect the termination power for an SCSI (small computer serial interface) circuit. With the values shown, the dc current is limited to 1 A with a trip-delay time of 1 ms. The breaker trips if the cable or connector is accidentally shorted and will retry every second until the short circuit is removed. The termination power then returns to normal, and the interface is reconnected. The microprocessor can continuously monitor the status of the termination power (via the STATUS pin) and take further action if the fault condition persists. A gate-voltage ramp is slowed to smoothly start large capacitive loads (soft start). The circuit also includes a power-supply filter to ensure that the supply (V_s) is maintained above 3.5 V until the gate is fully discharged during a short circuit. LINEAR TECHNOLOGY, APPLICATION NOTE 53, P. 16.

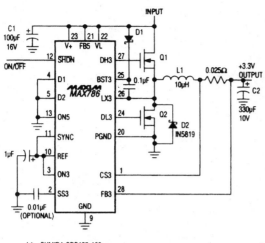

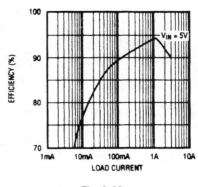

Fig. 8-66

L1 = SUMIDA CDR125-100
Q1,Q2 = SILICONIX Si9410DY
D1 = CENTRAL SEMICONDUCTOR CMPSH-3 OR 1N5817
C1,C2 = AVX TPS OR SPRAGUE 595D
OTHER PINS ARE NO CONNECTS

Fig. 8-65

High-power 5-V to 3.3-V supply
Figure 8-65 shows a MAX786 connected to provide 3.3 V from a 5-V supply. Figure 8-66 shows the efficiency curve. The circuit is suitable as a converter (possibly on a daughter card) to upgrade an existing desktop system with a new 3.3-V microprocessor. A synchronous rectifier is required. The input voltage range is 4.5 V to 6 V, with a quiescent current of 780 µA and a maximum load current capability (V_{IN} = 4.5 V) of 3 A. MAXIM BATTERY MANAGEMENT CIRCUIT COLLECTION, 1994, P. 43.

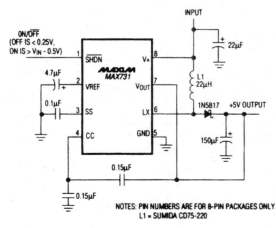

NOTES: PIN NUMBERS ARE FOR 8-PIN PACKAGES ONLY.
L1 = SUMIDA CD75-220

Fig. 8-67

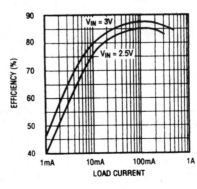

Fig. 8-68

Battery-power/micropower circuit titles and descriptions

Low-power 3.3-V to 5-V supply
Figure 8-67 shows a MAX731 connected to provide 5 V from a 3.3-V supply. Figure 8-68 shows the efficiency curves. This PWM boost regulator is designed for low-noise, battery-powered applications, such as cellular phones or sub-notebook computers without rotating disk drives (where the maximum +5-V load is 2 W or less). The input voltage range is 1.4 V to 5 V, with a quiescent current of 2 mA, and a maximum load current of 350 mA. The no-load start-up voltage is 1.8 V. The switching frequency is 170 kHz, and the shutdown current is 35 μA. MAXIM BATTERY MANAGEMENT CIRCUIT COLLECTION, 1994, P. 41.

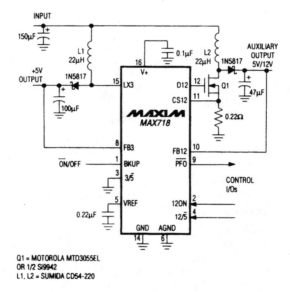

Fig. 8-69

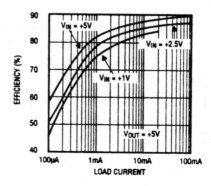

Fig. 8-70

Battery-power and micropower circuits

Dual 3.5-V to 5-V/12-V supply

Figure 8-69 shows a MAX718 controller connected to provide 5 V and 12 V from a 3.3-V supply. Figure 8-70 shows the efficiency curves. This circuit is designed for those applications that require a 3.3-V to 5-V converter to supply power for peripheral equipment, and a +12-V supply for flash-memory programming. The circuit delivers 5 V at 400 mA, and 12 V at 120 mA when powered from a 3.3-V (±10%) supply. The 12-V output can be programmed to 5 V, under logic control without external switches or pullup. The input voltage is 0.9 V to V_{OUT}, with 1.4-V no-load start-up. Quiescent current is 140 μA for 5 V, and 500 μA for both 5 V and 12 V. The switching frequency is 0.5 MHz. MAXIM BATTERY MANAGEMENT CIRCUIT COLLECTION, 1994, P. 42.

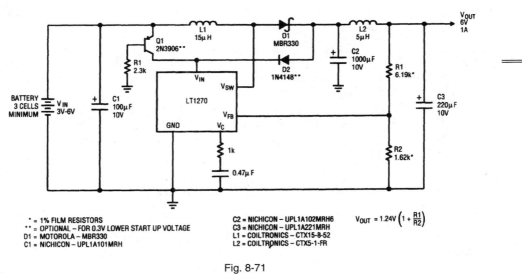

* = 1% FILM RESISTORS
** = OPTIONAL – FOR 0.3V LOWER START UP VOLTAGE
D1 = MOTOROLA – MBR330
C1 = NICHICON – UPL1A101MRH

C2 = NICHICON – UPL1A102MRH6
C3 = NICHICON – UPL1A221MRH
L1 = COILTRONICS – CTX15-8-52
L2 = COILTRONICS – CTX5-1-FR

$V_{OUT} = 1.24V \left(1 + \frac{R1}{R2}\right)$

Fig. 8-71

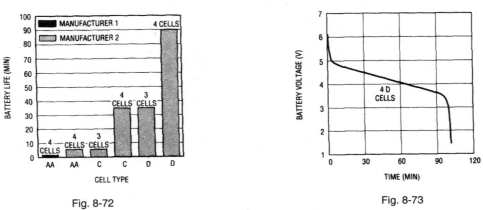

Fig. 8-72

Fig. 8-73

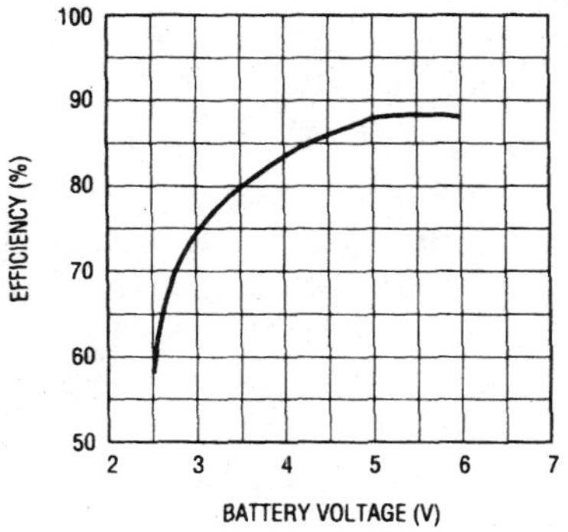

EFFICIENCY (%)

Fig. 8-74

Alkaline-battery switching regulator
Figure 8-71 shows an LT1270 connected to provide a 6-V, 1-A output using three or four alkaline-battery cells (instead of NiCads). The output voltage can be changed by the selection of resistors R1 and R2, using the V_{OUT} equation (as is the case with most switching regulators). Figure 8-72 shows alkaline-battery-life characteristics for a 6-W load. Figure 8-73 shows alkaline-battery-discharge characteristics for the same 6-W load. Figure 8-74 shows the efficiency for various alkaline-battery voltages. LINEAR TECHNOLOGY, DESIGN NOTE 41, P. 2.

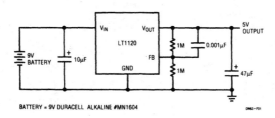

BATTERY = 9V DURACELL ALKALINE #MN1604

Fig. 8-75

Battery-power and micropower circuits

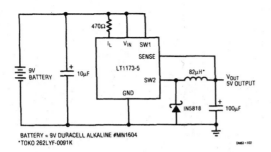

BATTERY = 9V DURACELL ALKALINE #MN1604
*TOKO 262LYF-0091K

Fig. 8-76

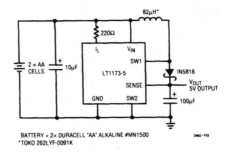

BATTERY = 2× DURACELL "AA" ALKALINE #MN1500
*TOKO 262LYF-0091K

Fig. 8-77

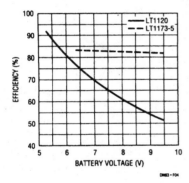

Fig. 8-78

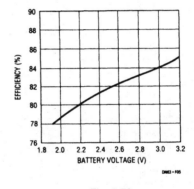

Fig. 8-79

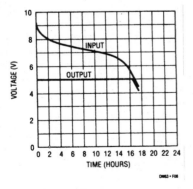

Fig. 8-80

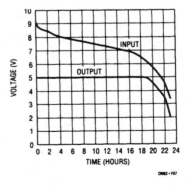

Fig. 8-81

Battery-power/micropower circuit titles and descriptions

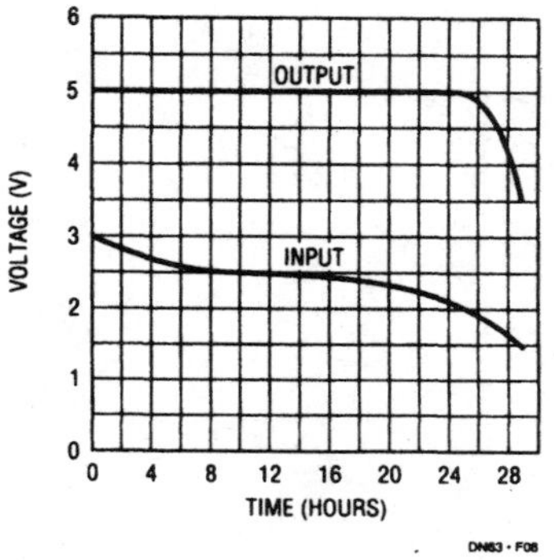

Continued

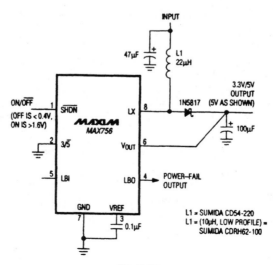

Fig. 8-82

Replacing a 9-V battery with two AA cells
Figure 8-75 shows an LT1120 linear regulator used with a 9-V battery to provide a 5-V, 30-mA output. Figure 8-76 shows an LT1173-5 switching regulator used with a 9-V battery to provide the same output. Figure 8-77 shows the same LT1173-5 used with two AA cells to provide the same output (step-up). Figure 8-78 shows efficiency for the step-down circuits. Figure 8-79 shows efficiency for the step-up (two-AA cell) circuit. The battery life for the three circuits is shown in Figs. 8-80, 8-81, and 8-82, respectively. A study of these curves shows that efficiency remains higher, and battery life is longer, with the two AA cells, for the same voltage and current output. LINEAR TECHNOLOGY, DESIGN NOTE 63, P. 1, 2.

Fig. 8-83

Battery-power and micropower circuits

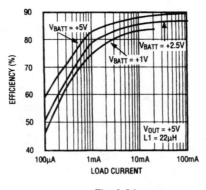

Fig. 8-84

3.3/5 V from two to three cells

Figure 8-83 shows a MAX756 connected to provide a 3.3-V (or 5-V) output at 400 mA, with a 0.9-V to V_{OUT} input range, and a start-up of 1.4 V. Quiescent current (V_{IN} = 3 V) is 60 μA in the 3.3-V mode and 140 μA in the 5-V mode. Figure 8-84 show the efficiency curves. The circuit draws only 20 μA in the shutdown mode. This is particularly important in those micropower applications where the batteries must always remain connected (when there is no on/off switch). Unfortunately, the shutdown current is not always provided on all datasheets! MAXIM BATTERY MANAGEMENT CIRCUIT COLLECTION, 1994, P. 10.

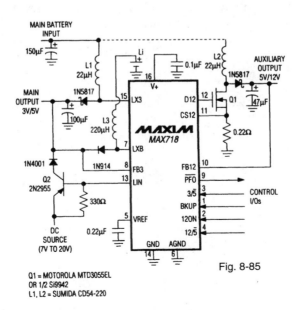

Fig. 8-85

Q1 = MOTOROLA MTD3055EL
OR 1/2 Si9942
L1, L2 = SUMIDA CD54-220

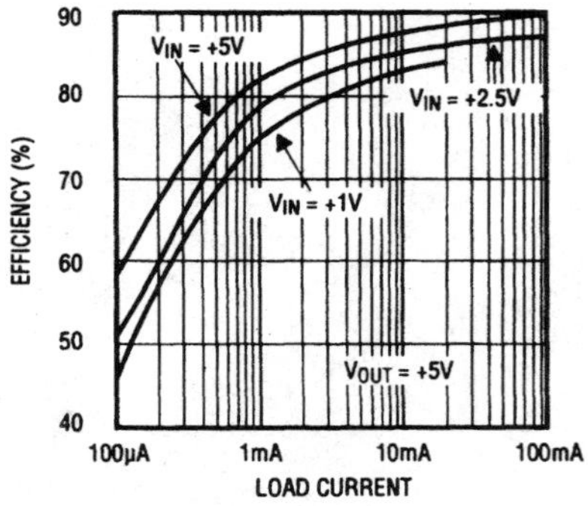

Fig. 8-86

3.3/5 V and 5/12 V from two to three cells

Figure 8-85 shows a MAX718 dual regulator connected to provide a main 3.3-V/5-V output and an auxiliary 5-V/12-V output, with a 0.9-V to V_{OUT} input range, and a start-up of 1.4 V. Quiescent current ($V_{IN} = 3$ V) is 140 μA for the main 5-V output, and 500 μA with both outputs. The maximum load-current capability of the main supply is 400 mA, and 120 mA for the auxiliary supply. Figure 8-86 shows the efficiency curves. The circuit can generate 3.3 V and 5 V, 3.3 V, and 12 V, or 5 V and 12 V, depending on the state of the logic inputs at pins 3 and 4. Two regulated output voltages are generated from one of three input-voltage sources: an ac/dc wall-cube adapter (7 V or 20 V), a main two-cell or three-cell battery, or a backup lithium battery. The dotted-line connection at the top of L2, which powers the auxiliary supply, is normally hard-wired to the main battery, but it can also be connected to the main output when power comes from the ac wall adapter. The main switching regulator automatically shuts off to save the batteries when the ac/dc wall cube is plugged in. MAXIM BATTERY MANAGEMENT CIRCUIT COLLECTION, 1994, P. 11.

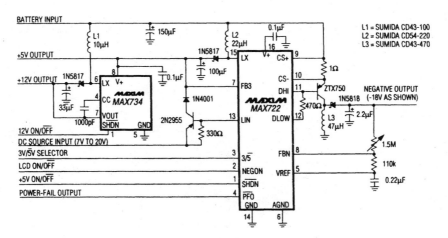

Fig. 8-87

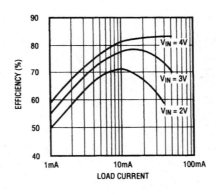

Fig. 8-88

3.3/5 V, 12 V, and −18 V from two to three cells

Figure 8-87 shows a MAX722 palmtop SMPS (surface-mount power supply), and a MAX734 regulator connected to provide a main 3.3-V/5-V output, a +12-V output for flash programming, and a −18-V output for LCD bias—all with a 1.8-V to 5.5-V input range. Quiescent current (when $V_{IN} = 3$ V, and the 12-V line is off) is 350 μA. With a V_{IN} of 2 V, the maximum load capability for the 5-V output is 200 mA, and the +12-V output, 40 mA. When V_{IN} is increased to 2.5 V, the 5-V output provides 275 mA, and the +12-V output, 60 mA. Figure 8-88 shows the efficiency curves for the +12-V output. If the main output is set at 3.3 V (pin 3 of the MAX722 high), connect the MAX734 $V+$ pin to +12 V (instead of +5 V, as shown) to get the extra gate drive for the MAX734 MOSFET. MAXIM BATTERY MANAGEMENT CIRCUIT COLLECTION, 1994, P. 12.

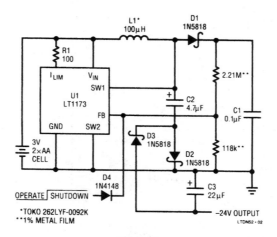

Fig. 8-89

−24 V from two to three cells

Figure 8-89 shows an LT1173 regulator connected to provide −24 V for LCD biasing using two AA cells. The 3-V input is converted to +24 V by the MOSFET switch in U1, inductor L, diode D1, and capacitor C1. The U1 switch pin (SW1) then drives a charge pump composed of C2, C3, D2, and D3 to generate the −24 V. Line regulation is less than 0.2% from 3.3-V to 2.0-V inputs. Load regulation, although it suffers somewhat because the −24-V output is not directly regulated, measures 2% from a 1-mA to 7-mA load. The circuit will deliver 7 mA from a 2.0-V input at 73% efficiency, and will also operate from a 5-V supply. LINEAR TECHNOLOGY, DESIGN NOTE 51, P. 1.

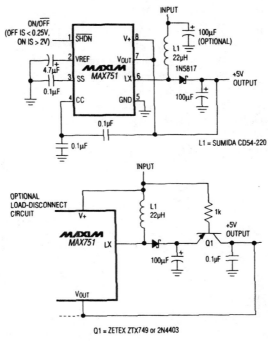

Fig. 8-90

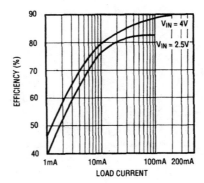

Fig. 8-91

Battery-power/micropower circuit titles and descriptions

5 V from two to three cells (with low noise)
Figure 8-90 shows a MAX751 connected to provide a 5-V output, with a 2-V to 5-V input range, but with low noise. This circuit provides a fixed-frequency PWM alternative to the pulse-skipping control scheme usually found in low-voltage switching-regulator ICs. (The noise generated by switching regulators is a major problem in many portable products, such as cellular phones and medical instruments.) The trade-offs for low-noise operation are increased quiescent supply current and lower efficiency for light loads. The quiescent current with a V_{IN} of 3 V is 1.2 mA, with a maximum load current of 100 mA (with V_{IN} = 2.7 V). The circuit has a shutdown current of 30 μA and a fixed oscillator frequency of 170 kHz. The no-load start-up is 1.2 V. Figure 8-91 shows the efficiency curves. MAXIM BATTERY MANAGEMENT CIRCUIT COLLECTION, 1994, P. 13.

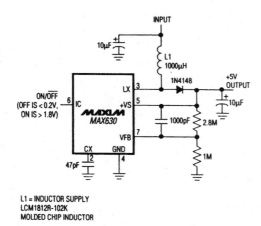

L1 = INDUCTOR SUPPLY
LCM1812R-102K
MOLDED CHIP INDUCTOR

Fig. 8-92

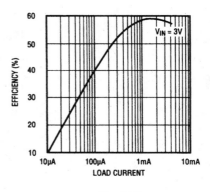

Fig. 8-93

5 V from two to three cells (low-cost micropower)
Figure 8-92 shows a MAX630 connected to provide a low-cost, micropower 5-V output, with a 1.6-V to 5-V input range. Figure 8-93 shows the efficiency curve. The quiescent current is 160 μA, with a start-up of 2 V, a maximum load current of 5 mA, and a shutdown current of 1 μA. This current is most useful in micropower applications where cost, not efficiency, is the main concern. (Efficiency can be improved by substituting a Schottky rectifier for the 1N4148, and a low-resistance inductor for L1, at the expense of higher cost.) The circuit is boot-strapped ($+V_S$ connected to the +5-V output). In those applications where minimum start-up voltage is essential, connect the $+V_S$ pin directly to the input. Unfortunately, removing the boot-strap connection is done at the expense of low-voltage load-current capability. MAXIM BATTERY MANAGEMENT CIRCUIT COLLECTION, 1994, P. 14.

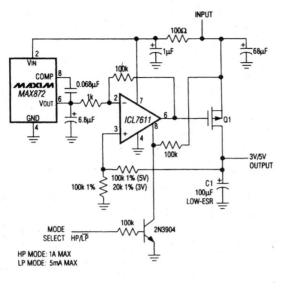

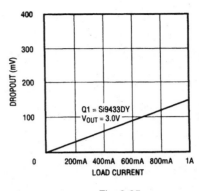

Fig. 8-95

Fig. 8-94

Q1: SILICONIX Si9433DY OR SMD10P05L

3 V/3.3 V from three cells with linear regulation

Figure 8-94 shows a MAX872 voltage reference and an ICL7611 micropower op amp connected to form a linear regulator for a 3-V/3.3-V supply. This circuit is particularly effective with NiCad and NiMH batteries. The end of life for such cells is about 1 V, so a linear regulator (with very low dropout) can be used in place of a switching regulator or charge pump. The dropout characteristics for the circuit are shown in Fig. 8-95, and depend primarily on the characteristics of Q1. When the circuit is used with low voltage, such as a three-cell battery, Q1 must have a gate-threshold voltage below that of the lowest battery voltage. For example, the $R_{DS(ON)}$ for the Si9433 is guaranteed at a V_{GS} of 2.7 V. The circuit will operate at input voltages from 3 V to 15 V. The quiescent current (V_{IN} = 6.5 V) is 40 μA when the circuit is operated in the low-power mode, but it increases to 70 μA in the high-power mode. The maximum load current is 1 A in high-power and 5 mA in low-power. The high- and low-power modes are selected by logic at the MODE SELECT input. High-power mode is selected when the input is high. MAXIM BATTERY MANAGEMENT CIRCUIT COLLECTION, 1994, P. 15.

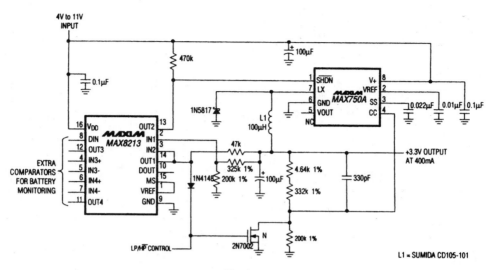

Fig. 8-96

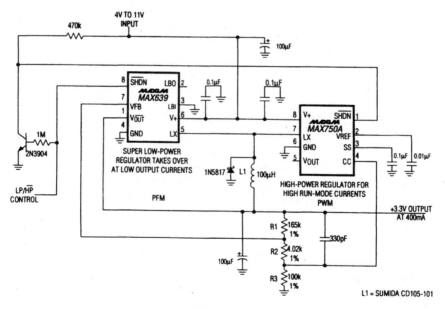

Fig. 8-97

Battery-power and micropower circuits

3.3 V from four to six cells
Figures 8-96 and 8-97 show two circuits that provide 3.3 V with four- to six-cell inputs. Both circuits can be switched (by an LP/HP control signal) between high- and low-power modes to accommodate equipment that operates in two modes. The circuit of Fig. 8-96 has a quiescent current of 60 μA (V_{IN} = 4.8 V) in the LP mode and 1.6 mA in the HP mode. Maximum load current (V_{IN} = 4 V) in LP is 10 mA, and 400 mA in HP. Efficiency (V_{IN} = 4.8 V) in LP is 72%, with a 1-mA load, and 92% in HP, with a 100-mA load. The circuit of Fig. 8-97 has a quiescent current (V_{IN} = 4.8 V) in the LP mode of 25 μA and 1.6-mA in HP. The maximum load current (V_{IN} = 4 V) in LP is 50 mA, and 400 mA in HP. Efficiency (V_{IN} = 4.8 V) in LP is 86%, with a 1-mA load, and 92% in HP, with a 100-mA load. MAXIM BATTERY MANAGEMENT CIRCUIT COLLECTION, 1994, P. 17.

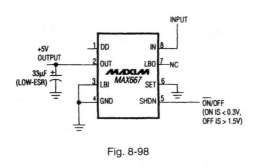

Fig. 8-98

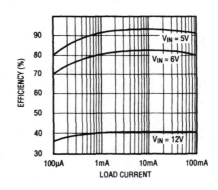

Fig. 8-99

5 V from four cells (linear regulation)
Figure 8-98 shows a MAX667 linear regulator connected to provide a 5-V output from four cells. The input voltage range is 4 V to 16.5 V. The quiescent current (V_{IN} = 6 V) is 10 μA, and the maximum load current (V_{IN} = 6 V) is 250 mA. The dropout voltage is 100 mV with a 100-mA load. Figure 8-99 shows the efficiency curves. MAXIM BATTERY MANAGEMENT CIRCUIT COLLECTION, 1994, P. 21.

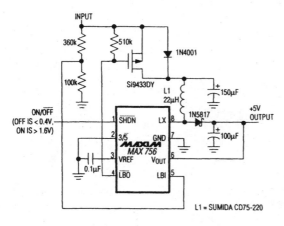

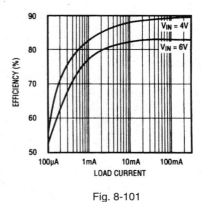

Fig. 8-101

Fig. 8-100

5 V from four cells (preregulation)
Figure 8-100 shows a MAX756 connected to provide a 5-V output from four cells. This circuit uses a diode and a PFET (in parallel) as a pre-regulator for the MAX756 boost regulator. The input voltage range is 2 V to 6.2 V, with a no-load start-up of 1.5 V. The quiescent current (V_{IN} = 5 V) is 70 μA, with a maximum load current of 400 mA. The shutdown current is 55 μA, and battery life is 15.5 hours (using four alkaline AA cells with a 100-mA load). Figure 8-101 shows the efficiency curves. MAXIM BATTERY MANAGEMENT CIRCUIT COLLECTION, 1994, P. 22.

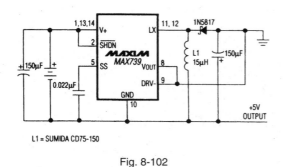

Fig. 8-102

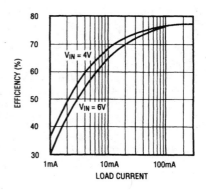

Fig. 8-103

5 V from four cells (inverter or flyback)
Figure 8-102 shows a MAX739 connected to provide a 5-V output from four cells. This circuit uses the MAX739 as an inverter with a battery-referenced output. The input voltage range is 3.8 V to 11 V, with a no-load startup of 4 V maximum. Quiescent current (V_{IN} = 5 V) is 1.8 mA, with a maximum load current of 200 mA. Shutdown current is 1 μA, and battery life is 13.5 hours (using four alkaline AA cells with a 100-mA load). Figure 8-103 shows the efficiency curves. MAXIM BATTERY MANAGEMENT CIRCUIT COLLECTION, 1994, P. 23.

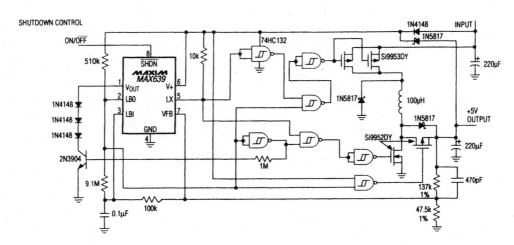

Fig. 8-104

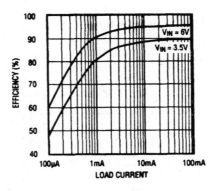

Fig. 8-105

5 V from four cells (switchable step-up/step-down)

Figure 8-104 shows a MAX639 connected to provide a 5-V output from four cells. This circuit switches from step-down to step-up mode as the battery output falls below 5 V. The circuit operates over an input-voltage range of 3 V to 6.5 V, with a quiescent current (V_{IN} = 5.5 V) of 50 μA, and 110 μA for a V_{IN} of 4.5 V. The maximum load current is 200 mA with a V_{IN} of 3.75 V. Battery life is 17.2 hours (using four alkaline AA cells with 100-mA load). Figure 8-105 shows the efficiency curves. MAXIM BATTERY MANAGEMENT CIRCUIT COLLECTION, 1994, P. 24.

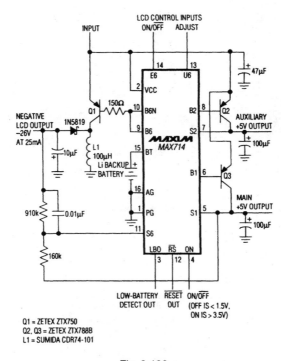

Q1 = ZETEX ZTX750
Q2, Q3 = ZETEX ZTX788B
L1 = SUMIDA CDR74-101

Fig. 8-106

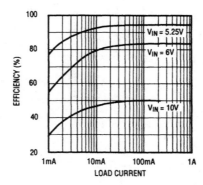

Fig. 8-107

Battery-power and micropower circuits

Multiple-output linear regulator (five cells)
Figure 8-106 shows a MAX714 connected to provide multiple outputs from five cells. Figure 8-107 shows the efficiency curves. This circuit operates over an input-voltage range of 5.05 V to 11 V, with a quiescent current of 300 μA (V_{IN} = 6 V), a maximum load current (V_{IN} = 6 V) of 1 A, and a standby current of 35 μA (V_{IN} = 6 V). The MAX714 is part of a family (MAX714/15) that contains supervisory functions, such as low-battery detection. The MAX714 shown here generates two 5-V outputs (main output at 1 A, with a 2-A peak and auxiliary output at 100 mA). The circuit also generates a negative LCD bias voltage (-10 V to -26 V), controlled by an internal D/A converter. MAXIM BATTERY MANAGEMENT CIRCUIT COLLECTION, 1994, P. 27.

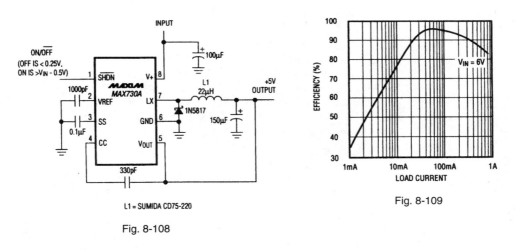

Fig. 8-109

Fig. 8-108

5 V from five cells (imitation linear)
Figure 8-108 shows a MAX730A step-down PWM regulator connected to provide a 5-V output from 5 cells. Figure 8-109 shows the efficiency curve. This circuit has high efficiency because it imitates a low-dropout linear regulator and "rides down" a falling battery voltage. The circuit operates over an input-voltage range of 5.2 V to 11 V, with a quiescent current of 1.4 mA (V_{IN} = 6 V), a maximum load current of 500 mA (V_{IN} = 6 V), and an efficiency of 95% (V_{IN} = 6 V, with a 100-mA load). MAXIM BATTERY MANAGEMENT CIRCUIT COLLECTION, 1994, P. 28, 29.

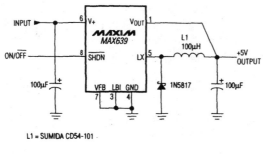

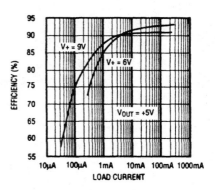

Fig. 8-110

Fig. 8-111

3.3 V/5 V from 9-V transistor-radio battery
Figure 8-110 shows a MAX639 step-down PFM regulator connected to provide 5 V from a 9-V transistor-radio battery (or any other six-cell battery). Figure 8-111 shows the efficiency curves. The circuit operates over an input-voltage range of 5.5 V to 11.5 V, with a quiescent current of 10 μA (V_{IN} = 9 V) and a maximum load current of 100 mA for V_{IN} = 5.5 V, and 175 mA for V_{IN} = 8 V. If a 3.3-V output is required instead of 5 V, use the companion MAX640 regulator. MAXIM BATTERY MANAGEMENT CIRCUIT COLLECTION, 1994, P. 29.

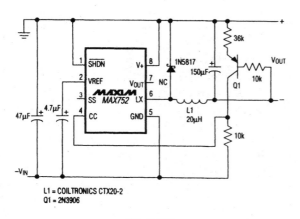

L1 = COILTRONICS CTX20-2
Q1 = 2N3906

Fig. 8-112

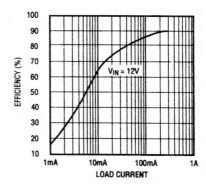

Fig. 8-113

+5 V from a negative-input voltage

Figure 8-112 shows a MAX752 step-up regulator connected to provide a +5-V output from a −6-V to −15-V input (six- to eight-cell battery). Quiescent current with a V_{IN} of −6 V is 1.5 mA, and maximum load current with a V_{IN} of −6 V is 500 mA. Figure 8-113 shows the efficiency curves. The circuit works only in situations where the negative battery terminal need not be tied to ground (in systems that do not require multiple-output voltages taken from the battery). MAXIM BATTERY MANAGEMENT CIRCUIT COLLECTION, 1994, P. 33.

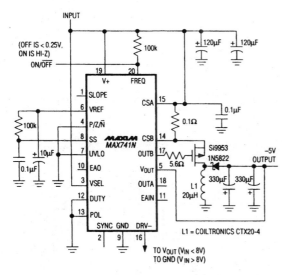

Fig. 8-114

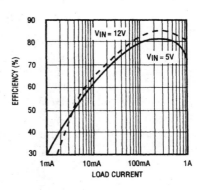

Fig. 8-115

−5 V from +5 V, or four to eight cells

Figure 8-114 shows a MAX714 PWM regulator connected to provide −5 V from a four- to eight-cell battery (or a fixed +5-V source). Figure 8-115 shows the efficiency curves. The input voltage range is 4 V to 15 V, with a quiescent current of 3 mA (V_{IN} = 5 V), a maximum load current of 1 A (V_{IN} = 4.75 V), and 1.25 A (V_{IN} = 12 V), and a shutdown current of 30 μA. MAXIM BATTERY MANAGEMENT CIRCUIT COLLECTION, 1994, P. 56.

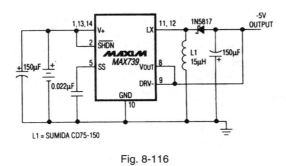

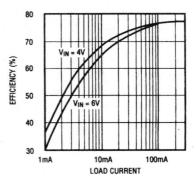

Fig. 8-116

Fig. 8-117

−5 V from +5 V, or five to eight cells

Figure 8-116 shows a MAX739 PWM regulator connected to provide −5 V from a five- to eight-cell battery (or a fixed +5-V source). Figure 8-117 shows the efficiency curves. The input voltage range is 3.8 V to 11 V, with a no-load start-up of 4 V maximum, a quiescent current of 1.8 mA (V_{IN} = 5 V), a maximum load current of 200 mA (V_{IN} = 5 V), and a 1-μA shutdown current. MAXIM BATTERY MANAGEMENT CIRCUIT COLLECTION, 1994, P. 57.

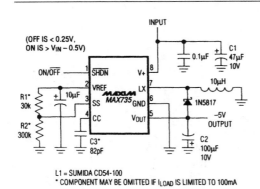

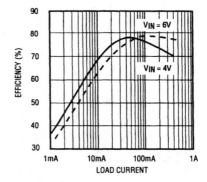

Fig. 8-118

Fig. 8-119

−5 V from + 5 V at medium power (low noise)

Figure 8-118 shows a MAX735 PWM regulator connected to provide −5 V at medium power. Figure 8-119 shows the efficiency curves. The input voltage range is 4 V to 6.2 V, with a quiescent current of 1.6 mA (V_{IN} = V), a maximum load current of 200 mA (V_{IN} = 5 V), and a shutdown current of 10 μA. MAXIM BATTERY MANAGEMENT CIRCUIT COLLECTION, 1994, P. 58.

Battery-power and micropower circuits

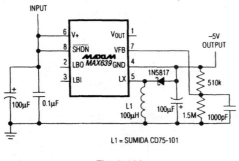

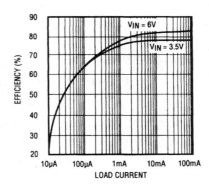

Fig. 8-120

Fig. 8-121

−5 V from +5 V at low power

Figure 8-120 shows a MAX639 PFM regulator connected to provide −5 V at low power. Figure 8-121 shows the efficiency curves. The input voltage range is 1.2 V to 6 V, with a no-load start-up of 1 V, a quiescent current of 35 μA (V_{IN} = 5 V), a maximum load current of 80 mA (V_{IN} = 4.75 V), and a shutdown current of 10 μA. MAXIM BATTERY MANAGEMENT CIRCUIT COLLECTION, 1994, P. 59.

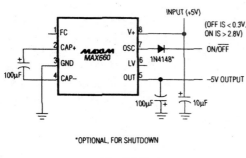

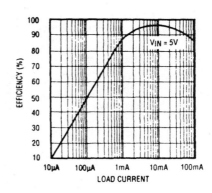

Fig. 8-122

Fig. 8-123

−5 V from a charge pump (low power)

Figure 8-122 shows a MAX660 charge pump connected to provide −5 V at low power. Figure 8-123 shows the efficiency curve. The input voltage range is 1.5 V to 5.5 V, output impedance is 6.5 Ω, quiescent current is 100 μA (V_{IN} = 5 V), maximum load current is 100 mA (V_{IN} = 4.75 V), and shutdown current is 10 μA. The circuit is low noise, with a fixed-frequency oscillation (between 10 and 45 kHz), and with an unregulated output (output voltage follows input variations). MAXIM BATTERY MANAGEMENT CIRCUIT COLLECTION, 1994, P. 60.

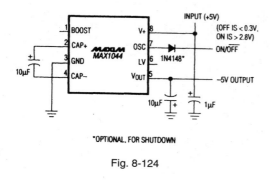

Fig. 8-124

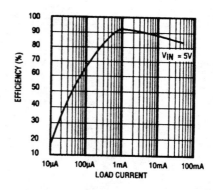

Fig. 8-125

−5 V from a charge pump (micropower)
Figure 8-124 shows a MAX1044 charge pump connected to provide −5 V at micropower. Figure 8-125 shows the efficiency curve. The input voltage range is 1.5 V to 10 V, output impedance is 65 Ω, quiescent current is 50 μA (V_{IN} = 5 V), maximum load current is 10 mA (V_{IN} = 4.75 V), and shutdown current is 1.5 μA. This circuit is a scaled-down version of the Fig. 8-122. The MAX1044 is basically identical to the MAX660, except for a somewhat higher input-voltage range, a 10-times-reduced output-current, and lower cost. The fixed-frequency oscillator range is 8 to 65 kHz. MAXIM BATTERY MANAGEMENT CIRCUIT COLLECTION, 1994, P. 61.

Battery-power and micropower circuits

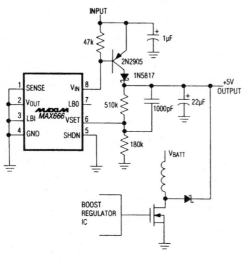

Fig. 8-126

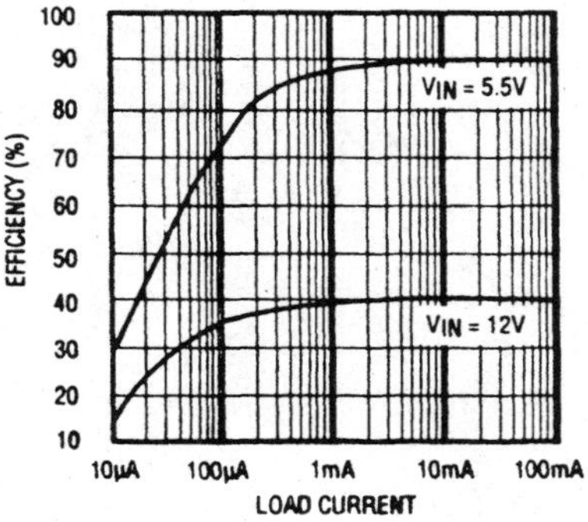

Fig. 8-127

Low-dropout linear regulator

Figure 8-126 shows a MAX666 linear regulator connected to provide a 5-V output. This circuit is useful for uninterruptible battery-backup applications, as well as for low-power circuits that use batteries for main power and optionally take power from wall-cube ac adapters. The input-voltage range is 5.4 V to 16.5 V, quiescent current (V_{IN} = 10 V) is 20 μA, dropout voltage (with 100-mA load) is 400 mV, maximum load current (V_{IN} = 6 V) is 500 mA, and shutdown current is 5 μA. Figure 8-127 shows the efficiency curves. Notice that the pass transistor is external and can be sized to handle the required power dissipation. If the dissipation is low enough, the MAX666 internal 50-mA power transistor can replace the 2N2905. The MAX667 should also be considered in such cases. (The MAX667 has a larger, 250-mA pass transistor and lower dropout voltage.) MAXIM BATTERY MANAGEMENT CIRCUIT COLLECTION, 1994, P. 74.

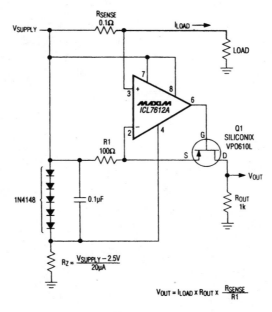

Fig. 8-128

High-side current-sense amplifier

Figure 8-128 shows an ICL7612A micropower op amp connected to provide sensing of high-side current. The operating voltage range is 4.5 V to 48 V, quiescent current (V_{IN} = 5 V) is 20 µA, and the gain factor is 1 V/A. The circuit senses current in the positive battery lead, allowing the battery's negative terminal to be directly connected to circuit ground. The circuit output is a ground-referenced output voltage that is directly proportional to the current flowing in the low-value sense resistor. The output is typically added to the input of an A/D converter or integrating V/F converter. Because the output current is a true-current source, the output can be referenced to any level within the supply limits. The value of R1 should be kept in the range of 100 × R_{SENSE} to 1000 × R_{SENSE}. MAXIM BATTERY MANAGEMENT CIRCUIT COLLECTION, 1994, P. 75.

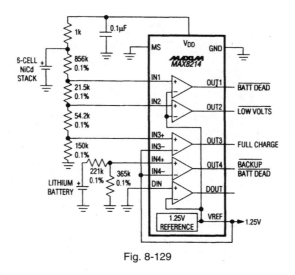

Fig. 8-129

System-voltage monitor

Figure 8-129 shows a MAX8214 voltage monitor connected to monitor the condition of a six-cell NiCad stack and a lithium backup battery. Larger portable systems often require several voltage-monitoring comparators to detect the status of main and backup batteries, as well as for power-fail monitoring and ac adapter detection. This circuit performs such monitoring and draws only 3 μA per comparator. The input-voltage range for the monitored voltage is 1.25 V to 100 V, and 2.7 V to 11 V for the IC. Quiescent current (V_{IN} = 5 V) is 16 μA, and threshold accuracy is 11% maximum. When the NiCad stack is fully charged, output 3 goes high. When the NiCad stack voltage drops to a low (but still usable) level, output 2 goes low. When the NiCad stack is dead, output 1 goes low. Output 4 goes low when the lithium battery is dead. MAXIM BATTERY MANAGEMENT CIRCUIT COLLECTION, 1994, P. 77.

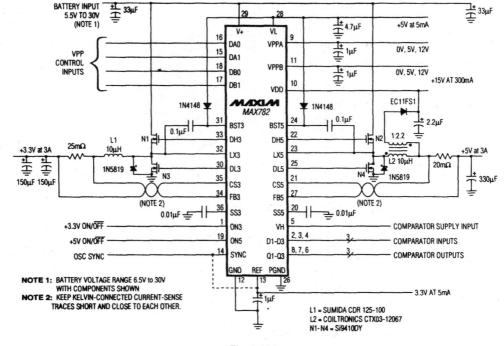

Fig. 8-130

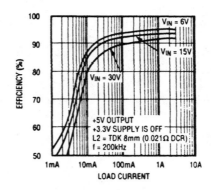

Fig. 8-131

3.3 V, 5 V, and 12 V from six to twelve cells

Figure 8-130 shows a MAX782 notebook SMPS connected to provide 3.3 V, 5 V, and 12 V from six to twelve battery cells. Figure 8-131 shows the efficiency curves. The input voltage range is 5.5 V to 30 V, quiescent current (V_{IN} = 15 V) is 420 μA, maximum load current (5-V output, V_{IN} = 6) and 3.3-V output, V_{IN} = 6 V) is 3 A, and the shutdown current is 70 μA. An evaluation kit is available from Maxim. MAXIM BATTERY MANAGEMENT CIRCUIT COLLECTION, 1994, P. 37.

Battery-power and micropower circuits

=9=

Analog/digital and digital/analog circuits

This chapter is devoted to circuits that convert analog voltages or signals to digital form (A/D or ADC, whichever you prefer) and convert digital signals to analog form (D/A or DAC). Because these circuits are essentially digital, all the testing and troubleshooting information of Chapter 1 applies to A/D and D/A circuits, whether they contain ICs or a combination of ICs and discrete components. For example, power, ground, reset, chip-select, clock, and input-output signals must be checked as with any IC. Those who are not familiar with A/Ds and D/As should read the author's *Simplified Design of Data Converters*, 1997, Butterworth-Heinemann.

A/D converter testing and troubleshooting

Analog-to-digital converters can be tested by applying precision voltages at the input and monitoring the output for corresponding digital values. For example, in the A/D converter of Fig. 9-A, a fixed voltage between 0 and +10 V can be applied to pin 2 of A1, and the corresponding digital value can be read out at pins 5 through 14 of the DAC-10, or at the lines between the DAC-10 and the 2504 SAR (successive approximation register). The lines should go to +5 V for a logic 1 and to ground or 0 V for a logic 0.

Notice that there is a serial digital output at pin 2 of the 2504 SAR. This output is best monitored on a scope. The rate at which conversions are performed is controlled by the 1- to 2-MHz clock input at pin 15 of the 2504. Pin 14 of the 2504 must receive a start-conversion input signal (typically from the system microprocessor) to initiate each conversion cycle. At the end of the conversion cycle, the conversion-complete pin 3 produces an output to the microprocessor (indicating status, conversion-complete, or conversion-not-complete).

If the output readings of the circuit are slightly off, try correcting the problem by adjusting R7 at the low end (all logic 0 when the analog input is 0 V) or try trimming the +5-V reference at the high end (all logic-1 when the analog input is +10 V). If the

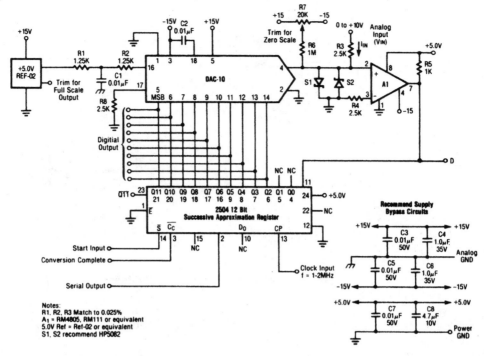

Fig. 9-A   Typical A/D converter.

outputs are absent or are way off, suspect the DAC-10, comparator A1, or the 2504. Also notice that the accuracy of this circuit depends on the precision of R1, R2, and R3.

D/A converter testing and troubleshooting

Digital-to-analog converters can be tested by applying digital input and monitoring the output for corresponding voltages. For example, in the D/A converter of Fig. 9-B, the inputs at pins 4 through 11 of the RM/RC4888 can be connected to ground (for a 0) or to 5 V (for a 1), and the output can be monitored with a precision voltmeter at pin 13.

If the output voltage is slightly off, try correcting the problem using the calibration procedures. For example, with all of the digital inputs at 0 (pins 4 through 11 grounded), adjust the offset pot until V_{OUT} is 0.0000. Then, with all digital inputs at 5 V (logic 1), adjust the gain pot until V_{OUT} is 9.9609. It might be necessary to work between these two adjustments until all output voltages are within tolerance.

Calibration Procedure:
1. Set inputs to all zeros
2. Adjust offset until V_{OUT} equals 0V
3. Set inputs to all ones
4. Adjust gain until V_{OUT} equals correct full scale value
   *Optional — reduces reference noise
   **Optional — improves settling time (see table for values)

Format	Output Scale	MSB DB7	DB6	DB5	DB4	DB3	DB2	DB1	LSB DB0	I_0 (mA)	$\bar{I}_0$ (mA)	V_{OUT}
Straight Binary: Unipolar With True Input Code. True Zero Output	Positive Full Scale	1	1	1	1	1	1	1	1	3.999	0.000	9.9609
	Positive Full Scale – LSB	1	1	1	1	1	1	1	0	3.984	0.001	9.9219
	LSB	0	0	0	0	0	0	0	1	0.0001	3.984	0.0391
	Zero Scale	0	0	0	0	0	0	0	0	0.000	3.999	0.0000
Complementary Binary: Unipolar With Complementary Input Code. True Zero Output	Positive Full Scale	0	0	0	0	0	0	0	0	0.000	3.999	9.9609
	Positive Full Scale – LSB	0	0	0	0	0	0	0	1	0.001	3.984	9.9219
	LSB	1	1	1	1	1	1	1	0	3.984	0.001	0.0391
	Zero Scale	1	1	1	1	1	1	1	1	3.999	0.000	0.0000

Fig. 9-B   Eight-bit straight-binary D/A converter.

Analog/digital and digital/analog circuit titles and descriptions

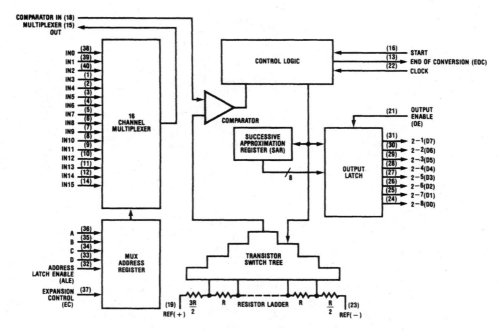

Fig. 9-1

Address D C B A				Expansion Control	Selected Channel
0	0	0	0	1	IN0
0	0	0	1	1	IN1
0	0	1	0	1	IN2
0	0	1	1	1	IN3
0	1	0	0	1	IN4
0	1	0	1	1	IN5
0	1	1	0	1	IN6
0	1	1	1	1	IN7
1	0	0	0	1	IN8
1	0	0	1	1	IN9
1	0	1	0	1	IN10
1	0	1	1	1	IN11
1	1	0	0	1	IN12
1	1	0	1	1	IN13
1	1	1	0	1	IN14
1	1	1	1	1	IN15
X	X	X	X	0	NONE

Fig. 9-2

Analog/digital and digital/analog circuits

Continued

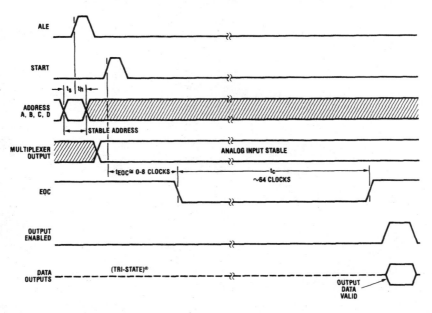

Fig. 9-3

8-bit ADC with on-chip 16-channel multiplexer

Figure 9-1 shows an ADC0816/17 specifically designed for data-acquisition applications. Figures 9-2 and 9-3 show the analog input-selection code and timing diagram, respectively. In addition to a standard 8-bit SAR-type ADC, these ICs also contain a 16-channel analog multiplexer (Chapter 2) with 4-bit latched address inputs. As a result, the ICs include much of the circuitry required to build an 8-bit-accurate, medium-throughput data-acquisition system.

Although similar to other ADCs used in data-acquisition systems, these ICs have externally available multiplexer output and A/D-comparator inputs. This feature is useful when connecting signal-processing circuits to the ADC. Also, these ICs have an expansion-control pin to allow addition of more multiplexers, producing more input channels. The ADC8016 is identical to the ADC0817, except for the accuracy. The ADC0816 is the more accurate device, having a total unadjusted error of ±½ LSB. The ADC0817 has a total unadjusted error of ±1 LSB (and is, as you might expect, less expensive!). NATIONAL SEMICONDUCTOR, APPLICATION NOTE 258, P. 590, 591.

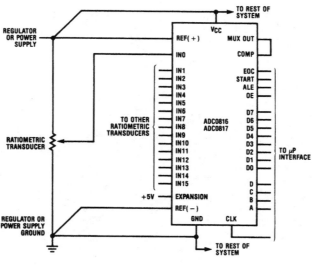

Fig. 9-4

Ratiometric conversion

Figure 9-4 shows the ADC0816/17 connected for ratiometric conversion. Because both ends of the 256R resistor ladder are available externally (Fig. 9-1), the ICs are ideally suited for use with ratiometric transducers. A ratiometric transducer is a conversion device in which the output is proportional to some arbitrary full-scale value. The actual value of the transducer output is of no great importance, but the ratio of this output to the full-scale reference is valuable. The prime advantage of a ratiometric transducer is that an accurate reference is not essential. However, the reference should be noise free because the voltage spikes during a conversion could cause inaccurate results. The circuit of Fig. 9-4 uses the existing 5-V supply for reference, thus eliminating the need for a special external reference. Take care to reduce power-supply noise. The supply lines should be bypassed, and separate PC traces should be used to route the 5-V and ground to the reference inputs, and to the supply pins (if practical). NATIONAL SEMICONDUCTOR, APPLICATION NOTE 258, 1994, P. 592.

Analog/digital and digital/analog circuits

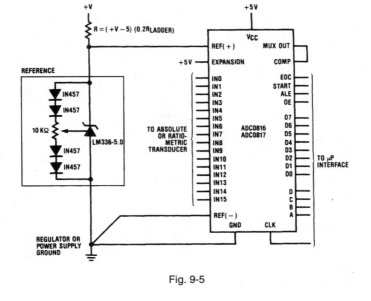

Fig. 9-5

Absolute conversion

Figure 9-5 shows the ADC0816/17 connected for simple absolute conversion. *Absolute conversion* refers to the use of transducers with which the output value is not related to another voltage. The "absolute" value of the output voltage is very important (in contrast to the output voltage of a ratiometric transducer). This implies that the reference must be accurate to determine the value of the absolute output of the transducer. A precise, adjustable reference is provided by an LM336-5.0 and the associated parts. Ratiometric transducers can also be used in this circuit, and in most of the following data-conversion circuits. However, the key point to remember is that accuracy of absolute conversion depends primarily on the accuracy of the reference voltage. With ratiometric systems (Fig. 9-4), accuracy is determined by the transducer characteristics. NATIONAL SEMICONDUCTOR, APPLICATION NOTE 259, 1994, P. 594.

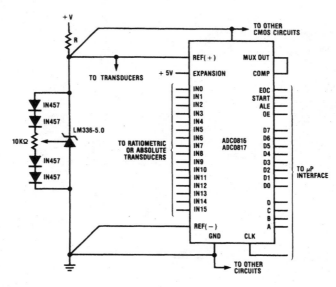

Fig. 9-6

Using the reference as the supply

Figure 9-6 shows the ADC0816/17 connected with the external reference used as the supply. The LM336-5.0 provides regulated 5-V for the ADC power and reference inputs, as well as for the power input of other components in the system. Of course, an unregulated supply greater than 5 V is required for V+. Series resistor R is chosen so that the maximum current needed by the system is supplied, and the LM336-5.0 is kept in regulation. The value of R is found by:

$$R = \frac{(V_S - V_{REF})}{(I_{LAD} + I_{TR} + I_P + I_R)}$$

where V_S = unregulated supply voltage; V_{REF} = reference voltage; $I_{LAD} = V_{REF}/$ 1 kΩ, resistor ladder current; I_{TR} = transducer currents; I_P = system power supply requirements; and I_R = minimum reference current. NATIONAL SEMICONDUCTOR, APPLICATION NOTE 258, 1994, P. 594.

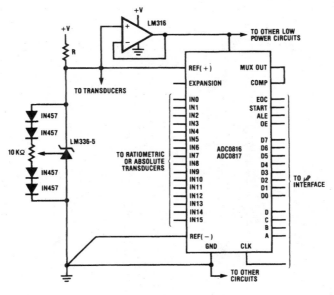

Fig. 9-7

Buffered reference used as a power supply

Figure 9-7 shows the ADC0816/17 used with a buffered reference. This method of buffering the reference provides higher current capabilities than the circuit of Fig. 9-6, and eliminates the IP term in the equation for resistor R. NATIONAL SEMICONDUCTOR, APPLICATION NOTE 258, 1994, P. 594.

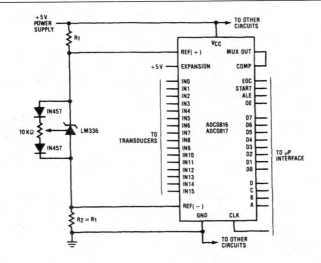

Fig. 9-8

Analog/digital and digital/analog circuit titles and descriptions

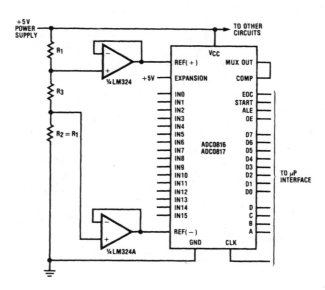

Fig. 9-9

Eliminating input gain adjustments

Figures 9-8 and 9-9 show the ADC0816/17 connected to eliminate gain adjustments on the analog input signals. This is done by varying the ADC REF+ and REF− voltages to get various full-scale ranges. Typically, the reference voltages can be varied from 5 V to about 0.5 V to accommodate various input voltages. However, there is a restriction: the center of the reference voltage must be within ±0.1 V of mid-supply. The reason for this restriction is that the reference ladder is tapped by an N-channel or P-channel MOSFET switch tree (Fig. 9-1). Offsetting the voltage at the center of the switch tree from V_{CC2} causes the transistors to turn off at the wrong point, resulting in inaccurate and erratic conversions. However, if properly applied, this method can reduce parts count and eliminate extra power supplies for the input buffers. In the supply-centered reference circuit of Fig. 9-8, R1 and R2 offset REF+ and REF− from V_{CC} and ground. An LM336-2.5 is shown, but any reference between 0.5 V and 5 V can be used. For odd reference values, use the op-amp circuit of Fig. 9-9. Single-supply op amps, such as the LM324 or LM10, can be used. R1, R2, and R3 form a resistor divider in which R1 and R3 center the reference at V_{CC2}, and R2 can be varied to get the proper reference magnitude. NATIONAL SEMICONDUCTOR, APPLICATION NOTE 258, 1994, P. 595.

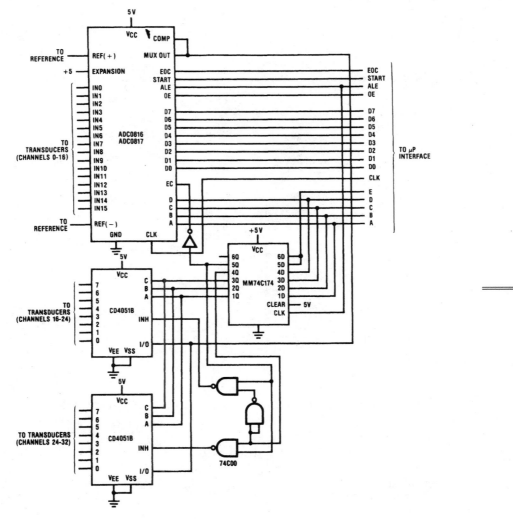

Fig. 9-10

Simple 32-channel ADC

Figure 9-10 shows the ADC0816/17 connected to provide for 32-channel conversion. Such a configuration is possible because of the EC pin, which is actually a multiplexer enable. When the EC signal is low, all switches are inhibited so that another signal can be applied to the comparator input. Additional channels can be implemented as necessary. A total of five address lines are required to address the 32 channels. The lower four bits are applied directly to the A, B, C, and D inputs. All four bits are also applied to an MM74C174 flip-flop which is used as an address latch for the two CD4051s. The 1Q, 2Q, and 3Q outputs of the flip-flop feed the CD4051 address inputs. The 4Q and 5Q outputs are gated to form enable signals for each CD4051. Output 5Q is also applied at the EC input (after inversion) to enable the ADC multiplexer. NATIONAL SEMICONDUCTOR, APPLICATION NOTE 258, 1994, P. 596.

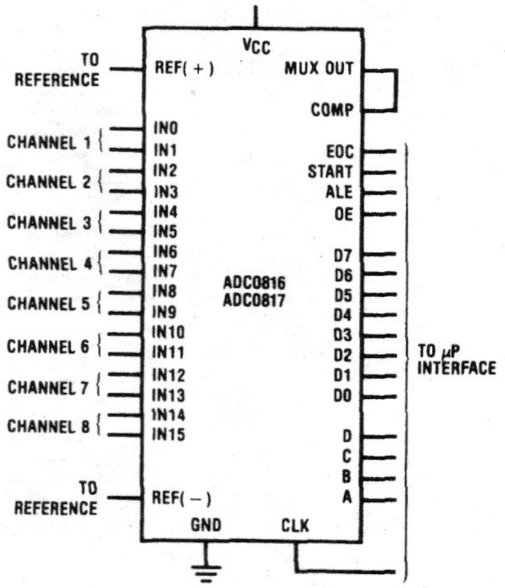

Fig. 9-11

Simple 8-differential-channel ADC
Figure 9-11 shows the ADC0816/17 connected to provide for an ADC with eight differential inputs. The differential inputs are implemented in software. All 16 channels are paired into positive and negative inputs. Then the control logic or microprocessor converts each channel of a differential pair, loads each result, then subtracts the two results. This method requires two single-ended conversions to do one differential conversion. As a result, the effective differential-conversion time is twice that of a single channel, or a little more than 200 μs (assuming a clock of 640 kHz). The differential inputs should be stable throughout both conversions to produce accurate results. NATIONAL SEMICONDUCTOR, APPLICATION NOTE 258, 1994, P. 597.

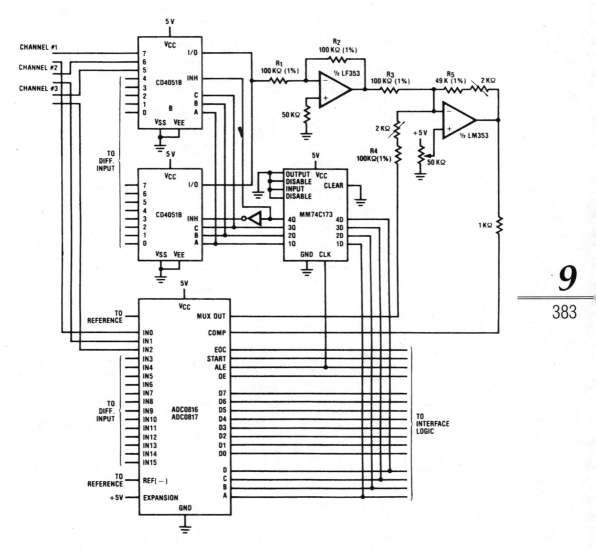

Fig. 9-12

Differential 16-channel converter

Figure 9-12 shows the ADC0816/17 connected to provide for an ADC with 16 differential inputs. This circuit is a modification of the circuit in Fig. 9-10. The CD4051 addressing is changed, and a differential amplifier is added between the multiplexer outputs and the comparator input. The select logic for the CD4051 is modified to enable the switches so that they can be selected in parallel with the ADC. The output of the three multiplexers are connected to a differential amplifier, composed of two inverting amplifiers with gain and offset trimmers. A dual op-amp configuration of inverting amplifiers can be trimmed easily, and has less-stringent feedback-resistor matching requirements (than a single op amp). The transfer equation for the dual op amp shown is:

$$V_{OUTPUT} = \frac{(R_2 R_5)}{(R_1 R_3)} \left[V_1 - \left(\frac{R_5}{R_4} \right) V_2 \right]$$

The propagation delay through the op amps is an important consideration. There must be sufficient time between the analog switch-selection and start-conversion to allow the analog signal at the comparator input to settle. Using the LF353 op amp shown, the delay is about 5 µs. The op-amp gain and offset controls are adjusted to provide the zero and full-scale digital-output readings for the analog-input range or span. NATIONAL SEMICONDUCTOR, APPLICATION NOTE 258, P. 598.

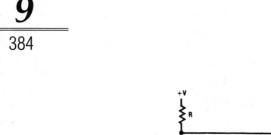

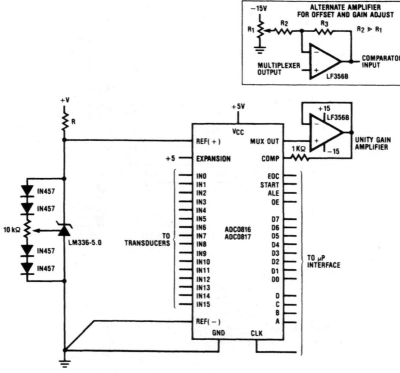

Fig. 9-13

Analog/digital and digital/analog circuits

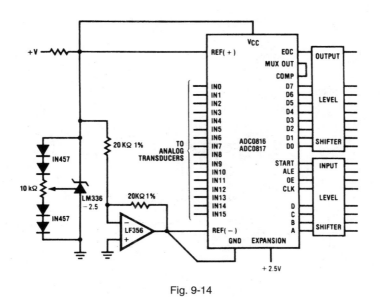

Fig. 9-14

Buffering circuits

Figures 9-13 and 9-14 show two typical buffering circuits for the ADCs. Three basic ranges of input signal levels can occur when ADCs are interfaces to the real world. These are as follows: (1) signals that exceed V_{CC} or go below ground; (2) signals with input ranges less than V_{CC} and ground, but are different from the reference range; and (3) signals that have an input range equal to the reference range. Each of these situations requires different buffering. In the last case (in which the signals are equal to the reference), no buffering is usually required, unless the source impedance of the input signal is very high. In this case, a buffer can be added between the multiplexer output and comparator input (Fig. 9-13). An op amp with high input impedance and low output impedance reduces input leakage (when one views the configuration from the multiplexer). If the input signal is within the supply range, but different from the reference range (or when the reference cannot be manipulated to conform to the full input range), the unity-gain buffer of Fig. 9-13 can be replaced with another op amp (as shown in the inset of Fig. 9-13). This type of amplifier provides gain or offset control to produce a full-scale range equal to the reference. When the input range exceeds V_{CC} or goes below ground, the input signals must be level-shifted before the input can go to the multiplexer. There is a limit to such level shifting when the input voltage range is with 5 V, but outside the 0.5-V supply range. In this case, the supply for the entire chip can be shifted to the input range, and the digital-output signals can be level-shifted to the system 5-V supply. A typical example of level-shifting and buffering is the situation in which the bipolar inputs range from -2.5 V to $+2.5$ V. If the ADCs have the supply and reference provided (as shown in Fig. 9-14), then the ±2.5-V logic outputs can be shifted to 0-V and 5-V logic levels. NATIONAL SEMICONDUCTOR, APPLICATION NOTE 258, 1994, P. 599.

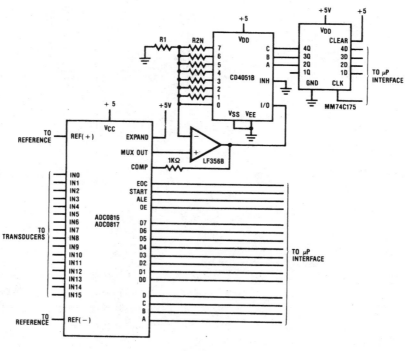

Fig. 9-15

ADC with microprocessor-controlled gain
Figure 9-15 shows the ADC0816/17 connected for an external gain-control under supervision of a microprocessor. The CD4051 analog multiplexer is placed in the feedback loop of a simple noninverting op amp. The op-amp gain is controlled by selecting one of the CD4051 analog switches. This cuts a resistor in and out of the feedback loop. If the resistors (R2N) are of different value, different gains are realized. The gains are given by: *Gain* $(A_v) = 1 + (R_{2N}/R_1)$. A microprocessor (or some control logic) selects a gain by latching the channel address into a MM74C173. The LF356B output must not exceed the power supply. As a result, the op-amp gain must be reduced to a new level before a new channel is selected. The 1-kΩ resistor at the LF356B output helps protect the comparator inputs from accidental overvoltage (or undervoltage). The two back-biased diodes at the input to V_{CC} and ground (1N914 or Schottky) offer further protection. NATIONAL SEMICONDUCTOR, APPLICATION NOTES 258, 1994, P. 602.

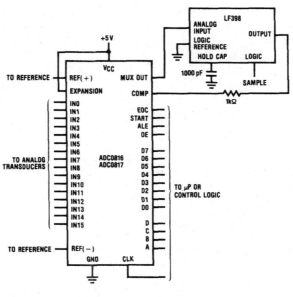

Fig. 9-16

Fig. 9-17

ADC with sample-and-hold (S/H)

Figures 9-16 and 9-17 show the ADC0816/17 connected for S/H operation. (The S/H function is the only major data-acquisition element not included in these ADCs.) If the input signals are fast moving, then an S/H should be used to quickly acquire the signal, then hold the signal while the ADCs convert it to a digital readout. This can be implemented by inserting an S/H function between the multiplexer output and the comparator input. In the simplest form, the multiplexer output is connected to the comparator input, with a capacitor connected to ground (similar to that shown in Fig. 9-16). The expansion-control pin is used as a sample-control input. When EXPAND is high, one switch is on and the capacitor voltage follows the input. When EXPAND is low, all switches are turned off and the capacitor holds the last value. However, this simple solution is not practical. The input bias to the comparator is about 2 μA (worst case, with a clock of 640 kHz). The droop (discharge rate) for a 1000-pF capacitor is about 2000 V/s (about 0.2-V per conversion). This is not practical. If a 0.01-μF capacitor is used instead, the rate is about 20 μV, which might work. However, the acquisition time would be about 100 μs, about the length of a conversion. The circuit of Fig. 9-16 eliminates the problem produced by the high comparator-input leakage. With the LF356 buffer connected between the multiplexer-output and comparator input pins, the leakage is reduced from 2 μA to about 100 nA. The droop-per-conversion is typically less than 1.0 μV per conversion (with the 1000-pF capacitor shown) and the acquisition time is about 20 μs (instead of the 100 μs). The circuit of Fig. 9-16 isolates the capacitor from both the multiplexer and comparator pins using an LF398 IC sample-and-hold. Acquisition time for the LF398 is a typical 4 μs to (0.1%), and droop rate is about 20 μV/conversion. Because the LF398 has its own S/H input, the expansion control of the ADC is free to be used in the normal manner. Use a hold capacitor with minimum dielectric absorption (polypropylene and polystyrene) for best results. NATIONAL SEMICONDUCTOR, APPLICATION NOTE 258, 1994, P. 603, 604.

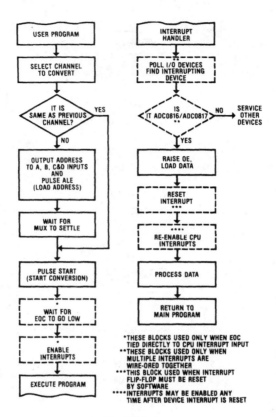

USER PROGRAM

↓

SELECT CHANNEL TO CONVERT

↓

IT IS SAME AS PREVIOUS CHANNEL? — YES →

↓ NO

OUTPUT ADDRESS TO A, B, C&D INPUTS AND PULSE ALE (LOAD ADDRESS)

↓

WAIT FOR MUX TO SETTLE

↓

PULSE START (START CONVERSION)

↓

WAIT FOR EOC TO GO LOW

↓

ENABLE INTERRUPTS

↓

EXECUTE PROGRAM

INTERRUPT HANDLER

↓

POLL I/O DEVICES FIND INTERRUPTING DEVICE

↓

IS IT ADC0816/ADC0817 ** ** — NO → SERVICE OTHER DEVICES

↓ YES

RAISE OE, LOAD DATA

↓

RESET INTERRUPT *

↓

RE-ENABLE CPU INTERRUPTS **

↓

PROCESS DATA

↓

RETURN TO MAIN PROGRAM

*THESE BLOCKS USED ONLY WHEN EOC TIED DIRECTLY TO CPU INTERRUPT INPUT
**THESE BLOCKS USED ONLY WHEN MULTIPLE INTERRUPTS ARE WIRE-ORED TOGETHER
***THIS BLOCK USED WHEN INTERRUPT FLIP-FLOP MUST BE RESET BY SOFTWARE
****INTERRUPTS MAY BE ENABLED ANY TIME AFTER DEVICE INTERRUPT IS RESET

Fig. 9-18

Analog/digital and digital/analog circuit titles and descriptions

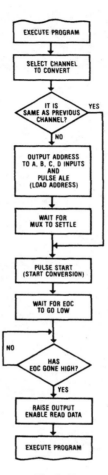

Fig. 9-19

Microprocessor interface considerations

Figures 9-18 and 9-19 show flow charts for the interrupt-control and polled-I/O modes of ADC/microprocessor interface, respectively. Either interface can be used with the ADC0816/17, but the polled-I/O method usually requires fewer external components. With polled-I/O, the microprocessor (or CPU) periodically interrogates the ADC, which looks like an I/O port to the CPU. With interrupt-control, the ADC appears as a memory and interrupts the microprocessor. From a simplified-design standpoint, the major concern is whether the EOC (end of conversion, Fig. 9-3) should be polled by the microprocessor. Even though the actual timing of CPU read and write cycles varies, most microprocessors output the address and data (during write) onto the system buses. A certain time later, the read or write strobes go active for a specified time. The interface logic must detect the state of the address and data buses and initiate the action. For the ADC0816/17, these actions are: (1) load channel address, (2) start conversion, (3) detect EOC, and (4) read the resultant data. These functions are performed by decoding the read-write strobes, address, and data to form ALE and START pulses, then to detect EOC, and finally to read the data. NATIONAL SEMICONDUCTOR, APPLICATION NOTE 258, 1994, P. 605.

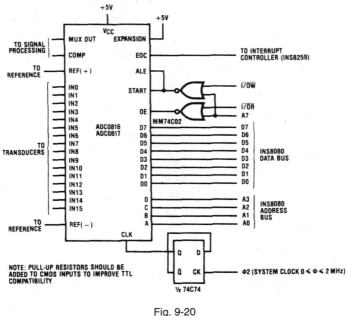

Fig. 9-20

Simple 8080 interface
Figure 9-20 shows connections for interface between the ADC0816/17 and classic 8080-type microprocessors (INS8080/8224/8228). This interfacing is quite simple because the INS8080 CPU has separate I/O read (I/OR) and I/O write (I/OW) strobes (or separate I/O addressing). As a result, in this simple interface systems, little or no address decoding is required. Two NOR gates are used to gate the I/O strobes with the most-significant address bit A7. (The INS8080 has 8 bits of port address, yielding a maximum of four I/O ports if inputs A, B, C, and D are connected to the address bus.) An MM74C74 flip-flop is used as a divide-by-2 to generate a converter clock of 1 MHz. If the system clock is equal to or less than 1 MHz, the flip-flop can be omitted. Typical software for the Fig. 9-20 circuit first writes the channel address to the converter as a start signal. The two start pulses are sent to the ADCs to allow the comparator input to settle. After the second start

pulse, the CPU can execute other program segments until the CPU is interrupted by EOC going high. Depending on interrupt structure, program control is then given to the interrupt handler, which reads the converter data. NATIONAL SEMICONDUCTOR, APPLICATION NOTE 258, 1994, P. 606.

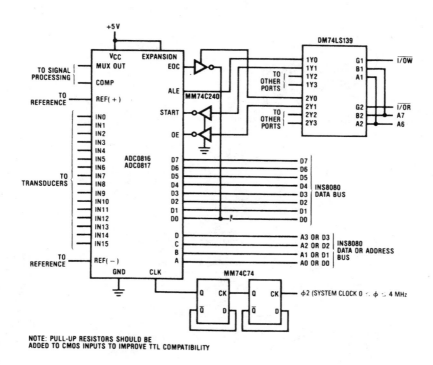

Fig. 9-21

8080 interface with partial decoding
Figure 9-21 shows the ADC0816/17 connected with a DM74LS139 dual 2-4 decoder to form an 8080 interface with partial decoding. One half of the DM74SL139 is used to create read pulses, and the other half to create write pulses. The START and OE inputs are inverted to provide the correct pulse polarity. This interface partially decodes A6 and A7 to provide more I/O capabilities than the Fig. 9-20 circuit. The circuit in Fig. 9-21 also implements a simple polled-I/O structure. The EOC output is placed on the data bus by a tristate inverter when the inverter is enabled by a read pulse from the INS8080.

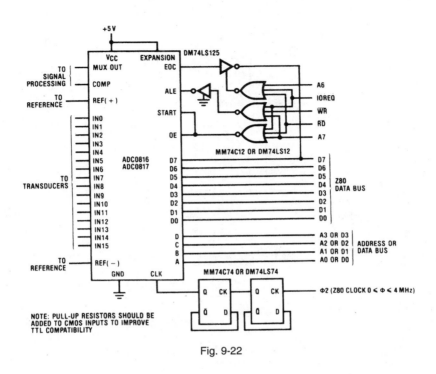

Fig. 9-22

Simple Z80 interface

Figure 9-22 shows the ADC0816/17 connected to form a simple interface with classic Z80-type microprocessors. The Z80, even though architecturally similar to the INS8080, uses slightly different control lines to perform I/O reads and writes. NOR gates are used to strobe the I/O functions. However, the Z80 has RD (read) and WR (write) strobes, which are gated with I_{OREQ} (I/O request). START is connected to OE. This causes a new conversion to be started whenever the data bits are read. (Such a configuration can be useful if the converter is to be continually restarted upon completion of the previous conversion.) Address bit A6 is used to drive a strobe that placed EOC on the data bus to be read by the CPU. NATIONAL SEMICONDUCTOR, APPLICATION NOTE 258, 1994, P. 607.

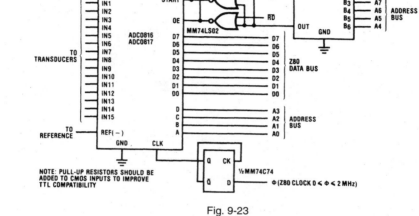

Fig. 9-23

Decoded Z80 interface

Figure 9-23 shows the ADC0816/17 connected to form a decoded Z80 interface. A 6-bit comparator is used to decode A4 through A7 and I_{OREQ}. Two NOR gates are used to gate the ALE/START and OE pulses. This configuration functions the same as that of Fig. 9-22, except that the DM8131 provides much more decoding. NATIONAL SEMICONDUCTOR, APPLICATION NOTE 258, 1994, P. 608.

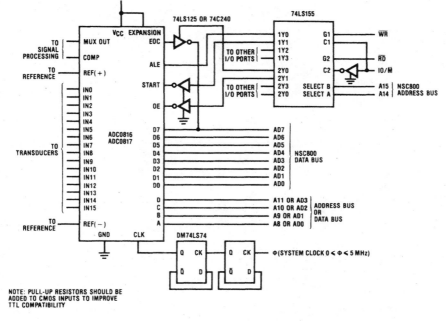

Fig. 9-24

Partially decoded NSC800 interface

Figure 9-24 shows the ADC0816/17 connected to form a partially coded NSC800 interface. This interface is quite similar to that for the 8080—even though the timing is very different. The NSC800 multiplexes the lower 8 address bits on the data bus at the beginning of each cycle. When accessing memory, A0 through A7 must be latched out at the beginning of a read or write cycle. For I/O accessing, the NSC800 duplicates the 8-bit I/O addresses on A8 through A15 address lines. Latches are not necessary because these lines are not multiplexed. The I/O read and write strobes are taken from RD (read) and WR (write) lines and the IO/M signal. A dual 2–4 line decoder decodes A15. A14 is enabled by the read-write strobes. Tristate inverters are used to implement a decoding similar to that of Fig. 9-21. Double pulsing is not required because START and ALE are accessed separately. NATIONAL SEMICONDUCTOR, APPLICATION NOTE 258, 1994, P. 608.

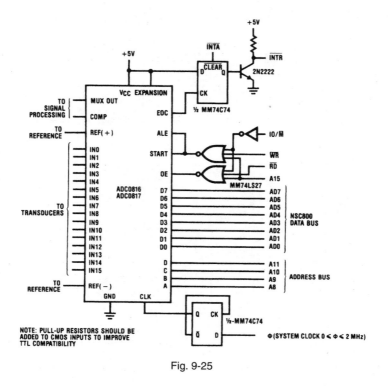

Fig. 9-25

Minimum NAC800 interface

Figure 9-25 shows the ADC0816/17 connected to form a simple or minimum NAC800 interface. This circuit uses NOR gates (similar to that of Fig. 9-20), but with different control signals. When EOC goes high, the flip-flop is set and INTR goes low. When the NSC800 acknowledges the interrupt by lowering INTA, the flip-flop resets. If more than one interrupt can occur simultaneously, either INTA should be gated with EOC, or a signal other than INTA must be used. This is required because the NSC800 can detect another interrupt and clear the ADC interrupt before the ADC signal is detected. NATIONAL SEMICONDUCTOR, APPLICATION NOTE 258, 1994, P. 609.

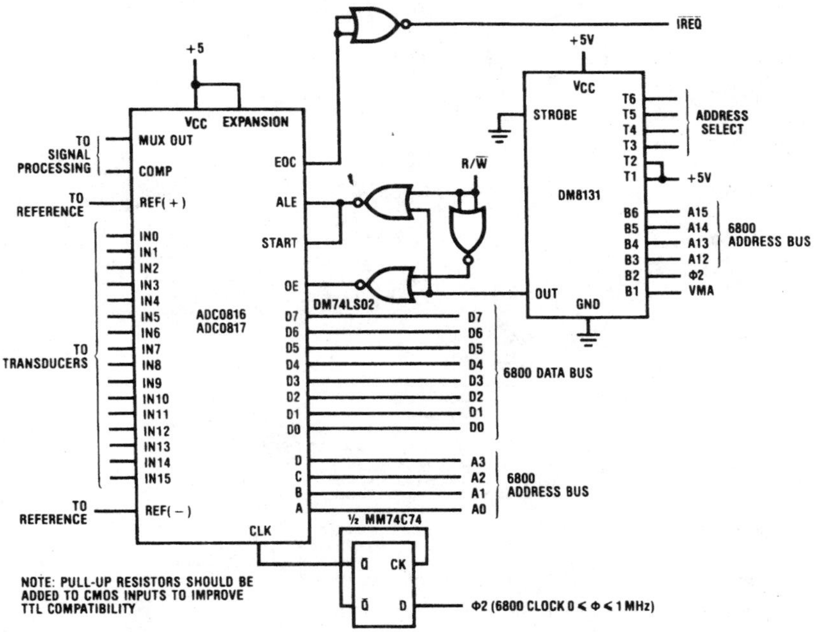

Fig. 9-26

Simple 6800 interface

Figure 9-26 shows the ADC0816/17 connected for a simple or minimum 6800 interface. This circuit uses a DM8131 comparator to particlly decode the A12, A13, A14, and A15 address lines with the phase-2 clock and VMA (valid memory address). This provides an address-decode pulse for the two NOR gates, which in turn generate the START/ALE pulse and the output-enable OE signal. The design locates the ADC in one 4-kb or block. EOC is tied to IREQ interrupt through an inverter, and is usable only in single-interrupt systems because the 6800 has no way of resetting the interrupt (except by starting a new conversion). Because EOC is directly ties to the interrupt input, the controlling software must not re-enable interrupts until eight converter clock periods after the start pulse, when EOC is low. NATIONAL SEMICONDUCTOR, APPLICATION NOTE 258, 1994, P. 610.

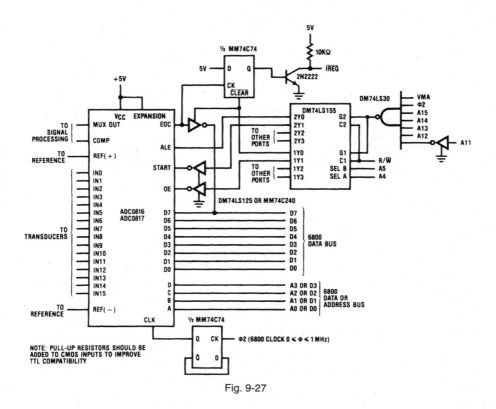

Fig. 9-27

Partially decoded 6800 interface

Figure 9-27 shows the ADC0816/17 connected to form a partially decoded 6800 interface. This interface has more I/O-port strobes than the circuit of Fig. 9-26. A NAND gate and inverter are used to decode the addresses, VMA, and phase-2 clock. The I/O addresses are located at 11110XXXXXAABBBB (binary); where X = don't care; A = 00 (binary) for ALE write or IREQ reset/EOC read and A = 01 for START write or data read; and B = channel-select address, if A, B, C, and D are connected to the address bus and ALE is accessed. A dual 2–4 line decoder is used to generate these strobes. Inverters are used to create the correct logic levels. The 6800 supports only a wired-OR interrupt structure. In a multi-interrupt environment, only one interrupt is received and the interrupt-handler routine must determine which device has cause the interrupt and must service that device. To do this, the EOC is brought out to the data bus so that EOC can be checked by the CPU. NATIONAL SEMICONDUCTOR, APPLICATION NOTE 258, 1994, P. 610.

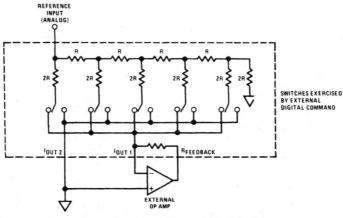

REFERENCE
INPUT
(ANALOG)

SWITCHES EXERCISED
BY EXTERNAL
DIGITAL COMMAND

$I_{OUT\ 2}$ $I_{OUT\ 1}$ $R_{FEEDBACK}$

EXTERNAL
OP AMP

Details (Simplified) of CMOS DAC1020—Last 5 Bits Shown

Other CMOS DACs are similar in the nature of operation but also include internal logic for ease of interface to microprocessor based systems. Typical is the DAC1000 shown below.

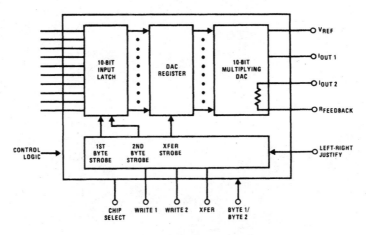

Fig. 9-28

Multiplying DACs

Figure 9-28 shows the internal functions of a multiplying DAC. Because such four-quadrant DACs allow a digital word to operate on an analog input, or vice versa, the output can represent a sophisticated function. CMOS multiplying DACs allow true bipolar analog signals to be applied to the reference input. This feature makes such DACs useful in many applications that are not generally considered data converters. NATIONAL SEMICONDUCTOR, APPLICATION NOTE 269, 1994, P. 659.

Analog/digital and digital/analog circuit titles and descriptions

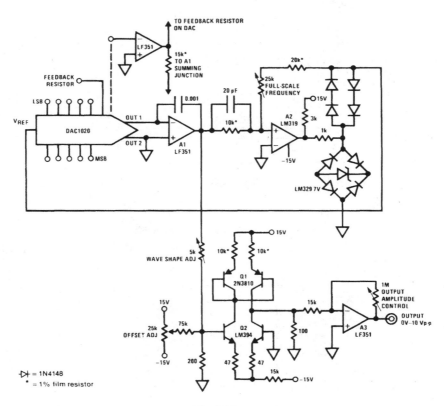

Fig. 9-29

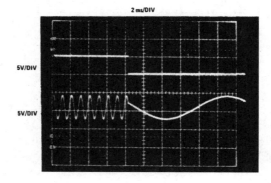

Fig. 9-30

Sine-wave generator with digital control
Figure 9-29 shows a DAC1020 connected to provide a variable-frequency, sine-wave generator. This circuit is capable of producing signals at frequencies up to 30 kHz under digital control. The linearity of the output frequency to the digital-code input is within 0.1% for each of the 1024 discrete output frequencies. To adjust the circuit, set all DAC digital inputs high and trim the 25-kΩ pot for a 30-kHz output (using a frequency counter). Then connect a distortion analyzer to the circuit output and adjust the 5-kΩ and 75-kΩ pots for minimum distortion. Finally, set the 1-MΩ output control for the desired output. The circuit provides rapid switching of the output frequency, as shown in Fig. 9-30. Notice that the output frequency shifts immediately (actually with no undesired delay) by more than an order of magnitude in response to digital commands (top line of Fig. 9-30). If operation over temperature is required, the absolute change in resistance in the DAC internal ladder might cause unacceptable errors. This can be corrected by reversing the A2 inputs and inserting an amplifier (dashed lines in Fig. 9-29) between the DAC and A1. Because this amplifier uses the DAC internal feedback resistor (Fig. 9-28), the temperature error in the ladder is cancelled. This results in more stable operation. NATIONAL SEMICONDUCTOR, APPLICATION NOTE 269, 1994, P. 660.

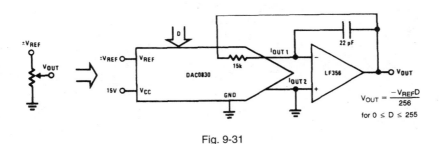

$$V_{OUT} = \frac{-V_{REF}D}{256}$$

for $0 \le D \le 255$

Fig. 9-31

Digital potentiometer
Figure 9-31 shows a DAC0830 connected as a digital potentiometer (pot). The applied digital-input word multiplies the applied reference voltage. The resultant output voltage is the product of this multiplication, normalized to the resolution of the DAC. The op amp converts the DAC output current to a voltage through the 15-kΩ feedback resistor within the DAC. To preserve output linearity, the two current-output pins must be as close to 0 V as possible. Thus, the input-offset voltage of the op amp must be nulled. The amount of linearity-error degradation is about $V_{OS} + V_{REF}$. When the digital pot is used to attenuate ac signals (in audio applications, for example), the DAC linearity over the full range of the applied reference voltage (even if it passes through zero) is good enough to distort a 10-V sine wave by only 0.004%. NATIONAL SEMICONDUCTOR, APPLICATION NOTE 271, 1994, P. 665.

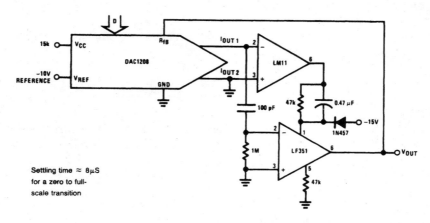

Fig. 9-32

Digital potentiometer with composite amplifier
Figure 9-32 shows a DAC1208 connected as a digital pot with improved characteristics (over those of the Fig. 9-31 circuit). The Fig. 9-32 circuit combines the excellent dc input characteristics of the classic LM11 with the fast response of a LF351 (a combination bipolar device). NATIONAL SEMICONDUCTOR, APPLICATION NOTE 271, 1994, P. 665.

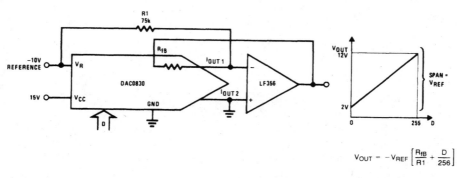

$$V_{OUT} = -V_{REF}\left[\frac{R_{fB}}{R1} + \frac{D}{256}\right]$$

Fig. 9-33

DAC with level-shifted output
Figure 9-33 shows a DAC0830 operated with the output level shifted. The shift is made by summing a fixed current to the DAC current-output terminal, offsetting the output voltage to the op amp. The applied reference voltage then serves as the output-span controller and is added (in fractions) to the output as a function of the applied digital code. NATIONAL SEMICONDUCTOR, APPLICATION NOTE 271, 1994, P. 665.

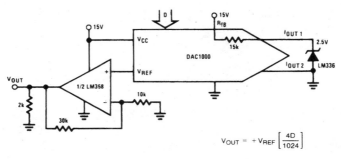

$$V_{OUT} = +V_{REF}\left[\frac{4D}{1024}\right]$$

Fig. 9-34

DAC connected for single-supply operation
Figure 9-34 shows a DAC1000 connected for single-supply operation. The R-2R ladder can be operated as a voltage-switching network to prevent the output-voltage inversion that is so common in the current-switching mode. In this circuit, the reference voltage is applied to the IOUT 1 terminal and is attenuated by the R-2R ladder in proportion to the applied code. The voltage is then output to the V_{REF} terminal with no phase inversion. To ensure linear operation in single-supply modes, the applied voltage must be kept less than 3 V for 10-bit DACs, or less than 5 V for 8-bit DACs. The supply voltage to the DAC must be at least 10 V more positive than the reference voltage to ensure that the CMOS ladder switches have enough voltage overdrive to fully turn on. An external op amp can be added to provide gain to the DAC output voltage for a wide overall output span. This circuit provides generally good linearity for 8-bit and 8-bit DACs, but can have a problem with 12-bit DACs (because of the very low reference required). If 12-bit operation is desired, use a DAC specifically designed for single-supply operation. NATIONAL SEMICONDUCTOR, APPLICATION NOTE 271, 1994, P. 666.

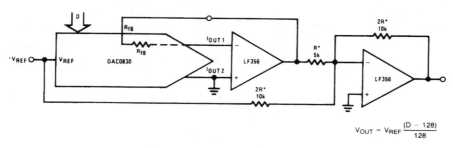

$$V_{OUT} = V_{REF} \frac{(D - 128)}{128}$$

$$1 \text{ LSB} = \frac{|V_{REF}|}{128}$$

Input Code	Ideal V_{OUT}	
MSB . . . LSB	$+ V_{REF}$	$- V_{REF}$
1 1 1 1 1 1 1 1	$V_{REF} - 1 \text{ LSB}$	$-|V_{REF}| + 1 \text{ LSB}$
1 1 0 0 0 0 0 0	$V_{REF}/2$	$-|V_{REF}|/2$
1 0 0 0 0 0 0 0	0	0
0 1 1 1 1 1 1 1	-1 LSB	$+1 \text{ LSB}$
0 0 1 1 1 1 1 1	$-\dfrac{|V_{REF}|}{2} - 1 \text{ LSB}$	$\dfrac{|V_{REF}|}{2} + 1 \text{ LSB}$
0 0 0 0 0 0 0 0	$-|V_{REF}|$	$+|V_{REF}|$

Fig. 9-35

DAC with bipolar output from a fixed reference

Figure 9-35 shows a DAC0830 connected to provide a bipolar output from a fixed reference voltage. This connection is made with a second op amp in the analog-output circuit. In effect, the circuit gives sign significance to the MSB of the digital-input word, allowing four-quadrant multiplication of the reference voltage. The polarity of the reference can still be reversed (or can be an ac signal) to realize full four-quadrant multiplication. NATIONAL SEMICONDUCTOR, APPLICATION NOTE 271, 1994, P. 666.

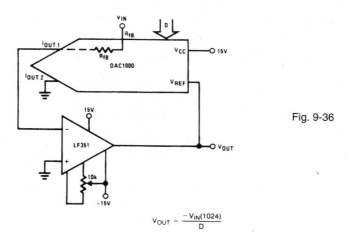

Fig. 9-36

$$V_{OUT} = \frac{-V_{IN}(1024)}{D}$$

Analog/digital and digital/analog circuits

DAC-controlled amplifier
Figure 9-36 shows a DAC1000 connected to control the output of an LF351 amplifier. In this circuit, the DAC is used as the feedback element for an inverting amplifier. The R-2R ladder digitally adjusts the amount of output signal fed back to the amplifier summing junction. The feedback resistance can be thought of as varying from about 15 kΩ to infinity when the input code changes from full-scale to zero. The internal feedback resistor is used as the amplifier input resistor. When the input code is all 0s, the feedback loop is opened and the op-amp output saturates. NATIONAL SEMICONDUCTOR, APPLICATION NOTE 271, 1994, P. 667.

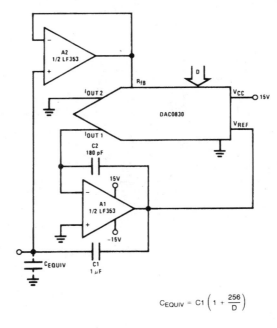

$$C_{EQUIV} = C1 \left(1 + \frac{256}{D} \right)$$

Fig. 9-37

Capacitance multiplier
Figure 9-37 shows a DAC0830 connected as a capacitance multiplier. Actually, the circuit is a DAC-controlled amplifier (used for capacitance multiplication) to give a microprocessor control of system time-domain or frequency-domain response. The microprocessor controls the digital input to the DAC, which controls the amplifier to produce a variable capacitance. The capacitance can be used to vary the time constant of RC circuits, varying either time or frequency. In this circuit, the DAC adjusts the gain of a stage with fixed capacitive feedback. This produces a Miller-effect equivalent input capacitance equal to the fixed capacitance, multiplied by 1 plus the amplifier gain. The voltage across the equivalent input

capacitance to ground is limited to the maximum output voltage of op amp A1, divided by 1 plus $2^n/D$, where: n = the DAC bits of resolution, and D = decimal equivalent of the binary input. NATIONAL SEMICONDUCTOR, APPLICATION NOTE 271, P. 667.

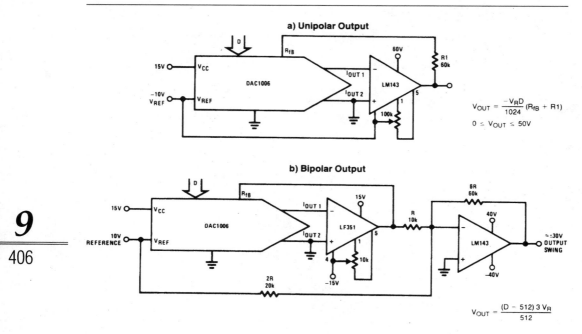

a) Unipolar Output

$$V_{OUT} = \frac{-V_R D}{1024}(R_{fB} + R1)$$

$$0 \le V_{OUT} \le 50V$$

b) Bipolar Output

$$V_{OUT} = \frac{(D - 512)\,3\,V_R}{512}$$

Fig. 9-38

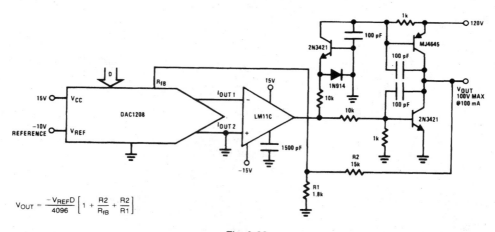

$$V_{OUT} = \frac{-V_{REF} D}{4096}\left[1 + \frac{R2}{R_{fB}} + \frac{R2}{R1}\right]$$

Fig. 9-39

Analog/digital and digital/analog circuits

DACs with high-voltage, and high-current outputs

Figure 9-38 shows DAC1006s connected for high-voltage outputs, both unipolar and bipolar. The output current of these circuits depends on the current limit of the LM143 op amp (typically 20 mA). Figure 9-39 shows how a discrete power stage can be added to further increase output current capability (to 100 mA at 100 V). NATIONAL SEMICONDUCTOR, APPLICATION NOTE 271, 1994, P. 668.

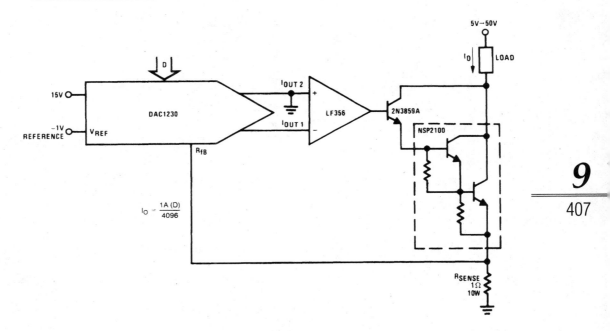

Fig. 9-40

DAC high-current controller

Figure 9-40 shows a DAC1230 connected to provide digital control of a 1-A current sink. Such a circuit can be used for heater control, stepper-motor torque compensation, and automatic test equipment. The largest source of nonlinearity in this circuit is the stability of the current-sensing resistance (with changes in power dissipation). The sensing resistance should be kept as low as possible to minimize this effect. The reference voltage must be reduced (to −1 V, as shown) to maintain the output-current range. The triple Darlington is used to minimize the base-current flowing through the sensing resistance, while simultaneously maintaining the collector current flow to the load. NATIONAL SEMICONDUCTOR, APPLICATION NOTE 271, 1994, P. 669.

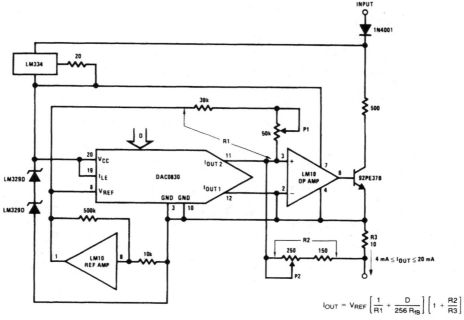

Fig. 9-41

DAC current-loop controller

Figure 9-41 shows a DAC0830 connected to provide digital control of the standard 3-mA to 20-mA industrial-process current loop. The circuit is two-terminal, and all circuit components (including the DAC) are powered directly from the loop. The output transistor conducts whatever current is necessary to keep the voltage across R3 equal to the voltage across R2. This voltage, and therefore the total loop current, is directly proportional to the output current from the DAC. The net resistance of R1 is used to set the zero-code loop current to 4 mA, and R2 is adjusted to provide the 16-mA output span for a full-scale DAC code. The entire circuit floats by operating whatever ground-reference potential is required for the total loop resistance and loop current. The voltage differential between the input and output terminals must be kept in the range of 16 V to 55 V, and the digital inputs to the DAC must be electronically isolated from the ground potential of the controlling microprocessor. This isolation can best be achieved with opto-isolators. In a non-microprocessor system in which the loop-controlling information comes from thumbwheel switches (or a similar mechanical device), the digital input for the DAC can be taken from BCD-to-binary CMOS logic elements (which are ground referenced to the ground potential of the DAC). The total supply current requirements of all circuits used must (of course) be less than 4 mA, and R1 can be adjusted accordingly. NATIONAL SEMICONDUCTOR, APPLICATION NOTE 271, 1994, P. 670.

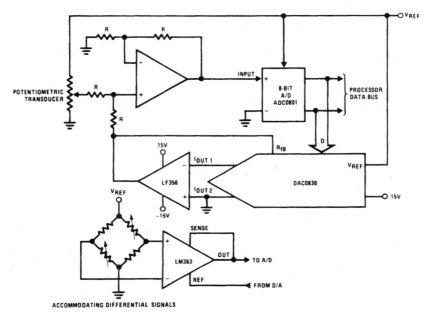

Fig. 9-42

Digital tare compensation
Figure 9-42 shows a DAC0830 and an ADC0801 connected to provide digital tare compensation. Such a function is used in a weighing system in which the weight of the scale platform, and possibly a container, is subtracted automatically from the total weight being measured. This expands the range of weight that can be measured by preventing a premature full-scale reading and allows an automatic indication of the actual unknown quantity. The DAC is initially given a zero code, and the system input is set to a reference quantity. A conversion of the input is performed, then the corresponding code is applied to the DAC. The DAC output is then equal to and of opposite polarity to the input voltage. This forces the amplifier output, and the ADC input, to zero. (In this case, an 8-bit ADC is used.) The DAC output is held constant so that any subsequent ADC conversion will yield a value relative in magnitude to the initial reference quantity. To ensure that the output code from the ADC generates the correct DAC output voltage, the two devices should be driven from the same reference voltage. For differential input signals, an instrumentation amplifier (such as an LM363) can be used. The output reference pin of the LM363 can be driven directly by the DAC, as shown. This will offset the ADC input. NATIONAL SEMICONDUCTOR, APPLICATION NOTE 271, 1994, P. 670.

Index

Wien-bridge oscillators
continued
  sine-wave, I-309
  varactor-tuned, I-291-292
window comparator, op amp,
  programmable, III-438
window detector, II-680,
  III-370, IV-246-247

low-power, III-446-447
write protection, IV-2
write signals, III-337, III-339,
  IV-7-8

Z

Zener diode, II-493

zero-crossing detector, II-661,
  II-668, II-675, III-452-453
  low-input current, II-665
  magnetic transducer, III-431,
    IV-223
  MOS, II-663,
  squaring a sine wave, II-668,
    III-433, IV-225

About the Author

John D. Lenk has been a technical author specializing in practical electronic design/
service guides for over 40 years. He is the best-selling author of more than 90 books
on circuit and consumer electronics, which together have sold over 2 million copies
in nine languages and 33 countries. His most recent titles include *Lenk's Video Handbook, Lenk's Digital Handbook, Lenk's Audio Handbook, Lenk's Laser Handbook, Lenk's RF
Handbook, Lenk's Television Handbook,* the *McGraw-Hill Circuit Encyclopedia, Volumes
1–3,* the *McGraw-Hill Electronic Testing Handbook,* and the *McGraw-Hill Electronic
Troubleshooting Handbook.*